HYBRID, ELECTRIC AND FUEL-CELL VEHICLES

JACK ERJAVEC & JEFF ARIAS

DELMAR
CENGAGE Learning

Australia • Brazil • Japan • Korea • Mexico • Singapore • Spain • United Kingdom • United States

Hybrid, Electric and Fuel-Cell Vehicles
Jack Erjavec, Jeff Arias

Vice President, Technology and Trades ABU:
  David Garza

Director of Learning Solutions: Sandy Clark

Senior Acquisitions Editor: David Boelio

Product Manager: Matthew Thouin

Marketing Director: Deborah Yarnell

Channel Manager: Erin Coffin

Marketing Coordinator: Patti Garrison

Content Project Manager: Cheri Plasse

Technology Project Manager: Kevin Smith

Editorial Assistant: Andrea Domkowski

Cover Image: Courtesy of Toyota Motor Sales,
  U.S.A., Inc.

> For product information and technology assistance, contact us at
> **Cengage Learning Customer & Sales Support, 1-800-354-9706**
> For permission to use material from this text or product,
> submit all requests online at **www.cengage.com/permissions**
> Further permissions questions can be emailed to
> **permissionrequest@cengage.com**

ISBN-13: 978-1-4018-8108-5

ISBN-10: 1-4018-8108-4

Delmar
Executive Woods
5 Maxwell Drive
Clifton Park, NY 12065
USA

Cengage Learning is a leading provider of customized learning solutions with office locations around the globe, including Singapore, the United Kingdom, Australia, Mexico, Brazil, and Japan. Locate your local office at **international.cengage.com/region**

Cengage Learning products are represented in Canada by Nelson Education, Ltd.

For your lifelong learning solutions, visit **delmar.cengage.com**

Visit our corporate website at **www.cengage.com**

Notice to the Reader

Printed in Canada
2 3 4 5 6 7 11 10 09 08

CONTENTS

PREFACE

Auto manufacturers have invested much time and money trying to find practical alternatives to vehicles powered by the internal combustion engine. This has led to the development of battery-operated electric vehicles, hybrid electric vehicles, and fuel cell electric vehicles. These are the main subjects of this book.

There are many reasons the manufacturers, and others, have invested so much on electric vehicles. The most highly advertised reason is fuel savings. However, there are many other reasons that seem to be more important to consumers. One of these is what I call the "green attitude." Vehicles powered by gasoline and diesel fuel emit things that are not good—not good for us or for the generations to come. Driving a clean or green machine is way to say, "I care about the environment and our present health and the health of future generations." There is also the compelling idea that we ought to do something about our dependence on fossil fuel and foreign oil.

The internal combustion engine will never be developed to the point where it emits zero emissions; therefore, not using an internal combustion engine in our vehicles has a strong attraction to those with the green attitude.

Many different alternative fuels have been tested and used in conventional engines to reduce our dependency on fossil fuels and to reduce emission levels. All of these show promise and are briefly discussed in this book. However, the only technology that promises to drastically reduce emissions and provide excellent fuel economy, are electric drive vehicles.

Although few manufacturers are currently developing new battery-operated electric vehicles, this book looks at what they developed in the recent past. Some may say studies on those are irrelevant, but the knowledge gained by those experiments has led to vehicles that are on the road today and will be in the future. Those being hybrid and fuel cell vehicles.

In order to make electric drive vehicles practical, they must be powered by high-voltage systems, in fact the higher the voltage, the more practical they become. With the high-voltages also comes serious safety issues. The voltages of electric drive vehicles are high enough to kill anyone who does not respect them and does not carefully adhere to the precautions given by the manufacturers of these vehicles. If this book has one dominant theme, it is "respect the voltage!" Throughout this book, regardless of the topic, **CAUTIONS**, **NOTES**, and **WARNINGS** are given to remind everyone who reads this book to be very careful while doing any thing on an electric drive vehicle.

Many assume that because some of the vehicle's systems are just like what has been used for years in conventional vehicles that they can just maintain and service vehicles unimpeded. This is not true. To prevent great personal injury and/or damage to the vehicle, you must do what you can to work safely on these vehicles.

Too often, technicians and others take some risks to complete a job quickly. On electric drive vehicles, moving too quickly or proceeding without checking a few things can end a career or life quickly. These messages are not meant to scare anyone from working on electric vehicles; rather they are intended to make one aware of the dangers. Knowing the dangers, I hope that everyone will enjoy the technology and the thrill of working with it.

Electric drive technologies are advancing very quickly. So much has changed between the time I started writing this to the time I thought I was finished. In fact, when I thought it was completed and reviewed what I had written, I saw some vehicles I did not write about that were running on the roads. Unfortunately, this will be the case for quite some time, so I decided to stop. If I waited to stop until the technology cooled down a bit, this book would not have been available for another ten years or so. But I did try to cover the basics to allow you to understand those systems that cannot be covered in this book.

The topics are presented in a progression, from yesterday's technology to tomorrow's. The first chapter focuses on the basics. The various types of electric drive vehicles are defined and described. There is also a discussion of various alternative fuels that can be used in an internal combustion engine. This discussion may seem out of sorts for a book about electric vehicles, but these fuels can be used in hybrid vehicles and as sources of hydrogen for fuel cell vehicles. There is also a quick look at the history of electric vehicles.

The next three chapters provide the basics for the rest of the book. Basic electricity, as it applies to these vehicles, is covered from a theoretical and practical standpoint. The basics of electric motors and batteries are also covered, in separate chapters. Regardless of the type of electric drive vehicle being considered, the two most important items are the motor and battery. Many different designs of both are covered in these chapters because many designs have been and can be used in electric vehicles.

The next chapter covers pure electric vehicles. These battery-operated vehicles were available from many different manufacturers. Although there are currently no new electric cars available today, the technologies developed for them have been modified to work in today's vehicles.

Since hybrid vehicles are quite popular today, there are four chapters dedicated to the subject. All hybrid vehicles available at the time of this writing are described and discussed. These are grouped by system and operational commonalities. Plus there is a chapter on general service to these vehicles. That chapter does not go into extreme detail because the manufacturers do not want technicians going deeply into their systems. However, because of the high-voltages found in these vehicles, many common non-hybrid service procedures need to be modified to work safely. Many of these new procedures are presented in the chapter.

The last chapter is a look into the future. It contains a look at fuel cell vehicles and other potential technologies that may affect the operation of an automobile in the future. Manufacturers have built and tested many fuel cell vehicles and this chapter looks at what worked and what did not in many of these vehicles.

I sincerely hope the information in this book opens doors of thought and rewards for you. The electric drive technology is different, rewarding, and exciting.

Jack Erjavec

I would like to thank the following companies for their help in the preparation of this book:

Aisin AW Co., LTD.
American Honda Motor Co., Inc.
Ballard Power Systems
The Battery Council International
Beta Research & Development Ltd.
The BMW Group
Continental Automotive Systems
DaimlerChrysler Corporation
Dana Corporation
Fluke Corporation
Grainger, Inc.
Johnson Control Battery Group, Inc.
Moteur Development International
The Southern Company
Toyota Motors Corporation
United Technologies Company
The U.S. Department of Energy
Visteon Corporation

I would also like to thank the following individuals who took the time to review the manuscript and to make sure this book had a minimum of errors and that it met its goals:

Elias Alba
Eastfield College
Mesquite, TX

John Ball
Whittier, CA

Dave Hostert
Morton College
Cicero, IL

Scott A. Martin
University of Northwestern Ohio
Lima, Ohio

Terry Nicoletti
Central Missouri State University
Warrensburg, MO

AN INTRODUCTION TO ELECTRIC VEHICLES

After reading and studying this chapter, you should be able to:

❏ Describe the various types of vehicles used for personal transportation.

❏ Describe the differences between vehicles that are powered by electricity and those powered by an internal combustion engine.

❏ Explain the basic advantages for having electric drive vehicles available to the public.

❏ Explain the advantages and disadvantages of using the commonly available alternative fuels in an internal combustion engine.

❏ Describe the basic components of all electric drive vehicles.

❏ Explain what regenerative braking does.

❏ Describe what a battery electric vehicle is.

❏ Describe what a hybrid electric vehicle is.

❏ Explain the basic operation of a fuel cell.

❏ Describe what a fuel cell electric vehicle is.

❏ Discuss the evolution of electric drive vehicles.

Introduction

Imagine a world without motorized transportation! Nearly everything we do depends on some sort of powered vehicle. This thought is obvious when we think of going somewhere that is too far to walk to or when there is inclement weather. However, let us think about the things we use in our daily lives. All of these products were delivered somewhere so we could purchase them. Even if we go to the source to purchase them, we need a means to get to the source. Most of motorized transportation today depends on burning fossil fuel. This book explores the actual costs of this mode of transportation as well as some alternative modes and fuel sources. Because this book is about electric vehicles (EVs), it will stress that electric vehicles offer a legitimate alternative to the internal combustion engine (ICE). In an attempt to categorize vehicles that depend on electricity for mobility, we will initially group the various designs as having electric drive (Figure 1-1).

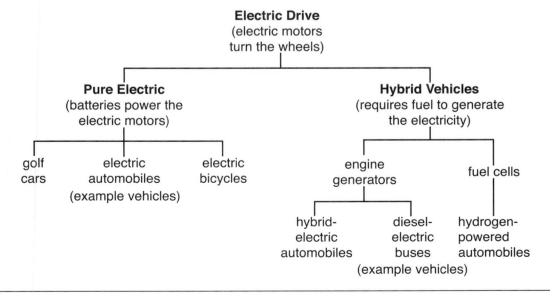

Figure 1-1 The common categories of electric drive vehicles.

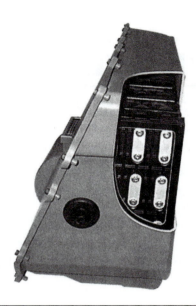

Figure 1-2 A battery pack for an electric drive vehicle, which is comprised of many battery cells.

Electric drive means that electricity is used to move the wheels of a vehicle. Electric drive is used on many different types of vehicles, including golf cars, bicycles, trains, forklifts, and automobiles. Automobiles with pure electric drive have electric motors that are powered only by batteries (Figure 1-2). These batteries are recharged by an external source of electricity, such as a wall plug.

Hybrid vehicles are automobiles with an electric motor and an ICE. The high-voltage batteries used in a hybrid vehicle are recharged by an engine-driven generator and the energy captured during braking. Another type of electric vehicle is the fuel cell electric vehicle. Although only experimental at this time, there is much promise for fuel cell electric vehicles. These vehicles are powered solely by electric motors, but the energy for the motors is produced by fuel cells, which use hydrogen to produce the electricity.

Why Electric Drive?

There is an automatic mental association of automobiles and internal combustion engines. For more than 100 years, drivers only knew gas-powered vehicles. When the cost of fuel is high, consumers want vehicles with better gasoline mileage. Other times, those same consumers think little about the cost and continue to pump in gasoline and drive. Some however, look at the true costs of gas-powered vehicles and know there is a better way.

The cost of using gasoline in our automobiles is not limited to the price per gallon or liter. There are other factors, or costs, that need to be considered: our environment, our dependence on foreign oil supplies, and the depletion of future oil supplies. Any reduction in the use of fossil fuels will have benefits for our generation and generations to come. Electric drive vehicles can have an impact on our fossil fuel dependence, which is why they are again being developed and produced. Before looking at the advantages of electric drive, let us first look at some simple facts:

❑ The number of household vehicles in the United States is growing and nearly tripled from 1969 to 2001. Last year, nearly 17 million new cars and light trucks were sold in the United States. In North America (including the United States, Canada, and Mexico), nearly 20 million new cars and light trucks were sold. These numbers do not include the automobiles on the road that were not bought that year. There are well over 225 million vehicles on the road.

❑ It is estimated that the total miles covered by those automobiles, in one year, is well over 2,000 billion. To put this in perspective, let us assume the average fuel mileage of all those vehicles is 20 miles per gallon (which is well higher than actual). This means over 100 billion gallons of oil are burned by our automobiles each year.

❑ By 2020, oil consumption is expected to grow by nearly 40% and our dependence on foreign oil sources is projected to rise to more than 60%.

❑ Transportation accounted for 66% of all oil consumed in the United States in 2000.

❑ A 10% reduction in fossil fuel consumption by cars and light trucks, achieved by the use of alternative fuels, electric drive, or improving fuel mileage, would result in using 24 million less gallons of oil each day.

❑ Americans spend close to $100,000 per minute to buy foreign oil, and oil purchases are a major contributor to the national trade deficit.

❑ Cars and light trucks are the largest source of urban air pollution (Figure 1-3).

❑ Automobiles and gasoline are major contributors to environmental damage. Not only do automobiles emit pollutants (Figure 1-4), but the extraction, production, and marketing of gasoline also leads to air pollution, water pollution, and oil spills.

❑ Because of the heavy reliance on fossil fuels, the transportation industry is a major source of carbon dioxide and other heat-trapping gases that cause global warming.

Vehicles powered by electric motors have low emissions, consume much less fuel or energy, and lessen our dependence on fossil fuels. The degree to which these are true depends on how the electricity for the vehicle is generated.

Figure 1-3 Cars and light trucks are the largest source of urban air pollution.

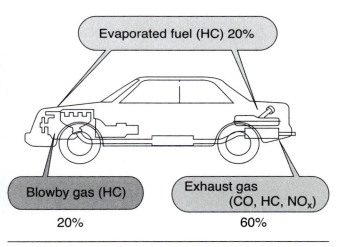

Figure 1-4 Sources of air pollution from an automotive.

Alternative Fuels

It is important to know there are ways to reduce our dependence on foreign oil, other than using electric drive. Much research has and is being conducted on the use of alternative fuels in ICEs. Many of these fuels are also being considered as the fuel of choice for fuel cell electric vehicles. Different kinds of alternative fuels have been tested and some are currently being used in vehicles. By using alternative fuels, we not only reduce our reliance on oil but we reduce emissions and the effects an automobile's exhaust has on global warming.

Propane / LPG vehicles

Propane, also referred to as liquefied petroleum gas (LPG), is the third most commonly used fuel for ICEs. The most common are, obviously, gasoline and diesel fuel. Propane is used by many fleets around the world in taxis, police cars, school buses, and trucks.

Propane is a clean-burning fuel that offers a driving range closer to gasoline than other alternative fuels. Propane is a by-product of natural gas production and the petroleum refining process. In its natural state, propane is a gas. LPG vehicles have special tanks or cylinders to store the gas (Figure 1-5). However, the gas must be stored at about 200 pounds per square inch. Under this pressure, the gas turns into a liquid and is stored as a liquid. When the liquid propane is drawn from the tank, it warms and changes back to a gas before it is burned in the engine.

Ethanol / Methanol Vehicles

Alcohol fuel was used to power Ford's Model T and has been used in a variety of applications since. Two types of alcohol are used in ICEs: methyl alcohol (methanol) and ethyl alcohol (ethanol), the alcohol used in the Model T (Figure 1-6). These fuels are similar but have different chemical compositions.

Figure 1-5 LPG must be stored in special tanks or cylinders.

Figure 1-6 Ford's Model T was designed to use ethyl alcohol (ethanol). *Courtesy of Ford Motor Company*

Ethanol (CH_3CH_2OH), commonly called grain alcohol, is a fuel made from corn, other grains, or biomass waste. Ethanol is a renewable fuel that can be made from nearly anything that contains carbon (Figure 1-7). Ethanol can be used as a high-octane fuel in vehicles and is often mixed with gasoline to boost its octane rating.

Methanol (CH_3OH) is a clean-burning fuel that is most often made from natural gas, but can also be produced from coal and biomass. Because North America has an abundance of these materials, the use of methanol can decrease the dependence on foreign oils. Methanol use as a fuel has declined through the years but may soon be used for fuel cell vehicles. It has been the fuel of choice for Indianapolis-type race cars since 1965. However, beginning in 2007, Indy Racing League (IRL) cars will switch to pure ethanol for all races. Today, these alcohols are mixed with 15 percent gasoline, creating M85 and E85. The small amount of gasoline improves the cold-starting ability of the alcohols.

Flexible-fuel vehicles (FFVs) can use ethanol and/or gasoline, or methanol and/or gasoline. The alcohol fuel and gasoline are stored in the same tank, which enables the use of alcohol when it is available, or regular gasoline when it is not, or a combination of the two.

Natural Gas Vehicles

Natural gas, compressed natural gas (CNG), and liquefied natural gas (LNG), are very clean-burning fuels. There is an abundant supply of natural gas and it is less expensive than gasoline.

Figure 1-7 An ethanol pump. *Courtesy of Missouri Corn Growers Association*

Both of these factors make the use of natural gas an attractive alternative fuel, especially to companies with fleets of vehicles. In fact, most of the natural gas vehicles have been sold to fleets. Typically, CNG (Figure 1-8) is used in light- and medium-duty vehicles whereas LNG is used in transit buses, train locomotives, and long-haul semi-trucks.

Figure 1-8 A compressed natural gas filling station.

CNG must be safely stored in cylinders at pressures of 2,400, 3,000, or 3,600 pounds per square inch, which is the biggest disadvantage of using CNG as a fuel. The space occupied by these cylinders takes away luggage and, sometimes, passenger space. As a result, CNG vehicles have a shorter driving range than comparable gasoline vehicles. Bi-fuel vehicles are equipped to store both CNG and gasoline and will run on either.

Natural gas turns into a liquid when it is cooled to minus 263.2 degrees Fahrenheit (–164 degrees Celsius). Because it is a liquid, a supply of LNG consumes less space in the vehicle than does CNG. Therefore, the driving range of a LNG vehicle is longer than a comparable CNG vehicle. However, the fuel must be dispensed and stored at extremely cold temperatures, which requires refrigeration units that also consume space and makes LNG impractical for personal use.

The Basics of Electric Vehicles

Electric vehicles are commonly used in manufacturing, shipping, and other industrial plants, where the exhaust of an internal combustion engine could cause illness or discomfort to the workers in the area. These vehicles are also used on golf courses, where the quiet operation adds to the relaxing atmosphere. EVs are also commonly used in the downtown areas of large cities and large campuses where peace, quiet, and fresh air are a priority.

EVs are powered by one or more electric motors that are "fueled" by electricity. The source of the electricity may be rechargeable batteries, fuel cells, or photovoltaic (PV) solar cells that convert the sun's energy into electricity (Figure 1-9). The drivetrain of an electric drive vehicle is much more efficient than the drivetrain in an ICE vehicle. They also produce zero or near-zero tailpipe emissions.

When the electricity and fuels used in electric drive vehicles are produced from renewable energy sources, these vehicles provide additional reductions in fossil fuel energy consumption and emissions. An electric drive vehicle's source of power is typically stored in and dispensed from batteries (Figure 1-10).

Regenerative Braking

A law of nature that is critical to an understanding of anything that moves or does work is that energy cannot be created or destroyed. It can only be moved from one point to another. This is a critical law to consider when trying to understand the efficiencies of an electric drive vehicle.

Figure 1-9 This is University of Michigan Solar Car Team's racecar. It is powered by electric motors that receive electrical energy from the sun. This car won the North American Solar Challenge in 2005. To do so, they traveled nearly 2,500 miles with only the sun as an energy source. *Courtesy of University of Michigan Solar Car Team*

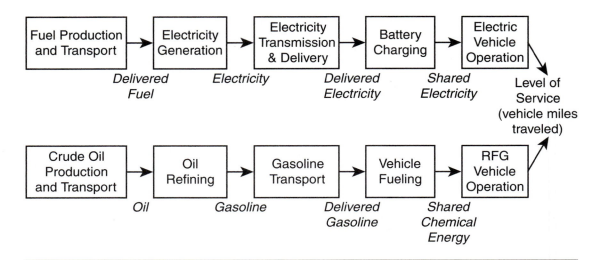

Fuel Production and Transport	→	Electricity Generation	→	Electricity Transmission & Delivery	→	Battery Charging	→	Electric Vehicle Operation

Delivered Fuel   *Electricity*   *Delivered Electricity*   *Shared Electricity*   Level of Service (vehicle miles traveled)

Crude Oil Production and Transport	→	Oil Refining	→	Gasoline Transport	→	Vehicle Fueling	→	RFG Vehicle Operation

Oil   *Gasoline*   *Delivered Gasoline*   *Shared Chemical Energy*

Figure 1-10 Comparison of the energy cycles for an electric drive vehicle and a gasoline vehicle.

The internal combustion engine is a loved, but very inefficient machine (meaning, that much of the energy going into it is wasted). Although there are some energy losses with electric drive, the amount is very low if the vehicle is designed properly.

One of the keys to the overall efficiency of an electric drive vehicle is called regenerative braking. Regenerative braking is the process by which a vehicle's kinetic energy can be captured while it is decelerating and braking. Whenever the driver applies the brakes in a conventional car, friction converts the vehicle's kinetic energy into heat. That heat is useless to the car and becomes lost energy.

The operation of most electric drive vehicles is based on batteries, electric motors, and electric generators (Figure 1-11). Batteries supply the power to operate the motors. The motors take that electrical power and change it to mechanical energy, which rotates the wheels and allows the vehicle to move. A generator takes the kinetic energy, or the energy of something in motion, and changes it to energy that charges the batteries.

When the generator is operating during regenerative braking, it helps slow down the vehicle. The rotation of the wheels turns the generator, which generates a voltage to charge the batteries. Because of the magnetic forces within the generator, the vehicle slows down. A conventional brake system is used in conjunction with the regenerative brake system to bring the vehicle to a safe stop.

Regenerative braking can recover about 30% of the energy normally lost as heat when a vehicle is slowing down or braking. Regenerative braking is unique to electric drive vehicles.

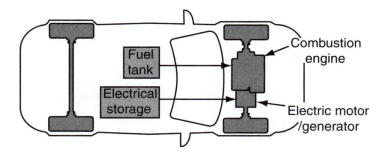

Figure 1-11 The motor/generators drive the wheels and absorb the vehicle's kinetic energy during slow down and braking.

Battery-Operated Electric Vehicles

A battery-operated electric vehicle, sometimes referred to as a battery-electric vehicle (BEV) uses one or more electric motors to turn its drive wheels (Figure 1-12). The electricity for the motors is stored in a battery that must be recharged from an external electrical power source. This technology is used for passenger cars, forklifts, urban buses, airport ground support equipment, and off-the-road industrial equipment. BEVs are zero-emission vehicles because they do not directly pollute the air. The only pollution associated with them is the result of creating the electricity to charge their batteries. Even when those emissions are included, BEVs are more than 99% cleaner than the cleanest ICE vehicle.

Normally, a battery-operated vehicle drives the same as any other, but it is quiet and carries no fossil fuel. However, rather than filling a tank with fuel, you need to recharge the batteries. The batteries are recharged by plugging them into a recharging outlet at home (Figure 1-13) or other locations. The recharging time varies with the type of charger, the size and type of battery, and other factors. Normal recharge time is four to eight hours.

In the recent past, manufacturers offered BEVs to the public. Today there are no BEVs, although some manufacturers are still studying their use. There are many reasons the public did not clamor for them. Perhaps the most prevalent was their high cost. It was hard to justify the purchase or lease costs, in spite of the advantages. Another cost that turned off consumers was the need to replace the vehicles' batteries every few years. These batteries are quite costly. A major factor that led to the demise of the modern electric car was the limited driving range. Most had less than a 120-mile range before the batteries needed to be recharged.

Whether battery-operated electric vehicles return to the market really depends on the development of new batteries. To be practical, electric vehicles need to have much longer driving ranges between recharges and must be able to sustain highway speeds for great distances. Although pure EVs did not become a popular transportation option, the lessons learned from building BEVs have given the manufacturers the technology to move on to hybrid electric and fuel cell electric vehicles.

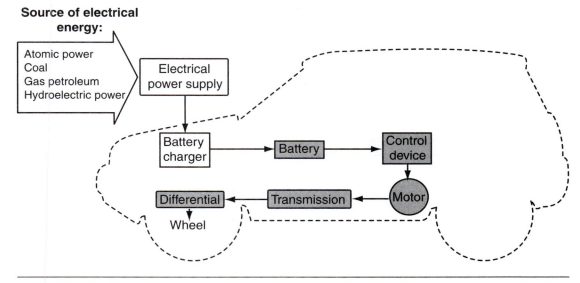

Figure 1-12 A battery-operated electric vehicle uses one or more electric motors to turn its wheels; the energy for the motor is stored in a battery.

Figure 1-13 The batteries of an EV need to be charged by an external source.

Hybrid Electric Vehicles

A hybrid electric vehicle (HEV) uses one or more electric motors and an ICE to propel the vehicle. Depending on the design of the system, the ICE may propel the vehicle by itself, act together with the electric motor to propel the vehicle, or it may drive a generator to charge the vehicle's batteries. The electric motor may propel the vehicle by itself or assist the ICE while it is propelling the vehicle. Some hybrids rely exclusively on the electric motor(s) during slow speed operation, the ICE alone at higher speeds, and both during some driving conditions.

A hybrid's electric motor is powered by batteries, which are continuously recharged by a generator that is driven by the ICE. The battery is also recharged through regenerative braking. Complex electronic controls monitor the operation of the vehicle. Based on the current operating conditions, electronics control the ICE, electric motor, and generator. The system recharges the batteries while driving, therefore plug-in charging is not required.

The engines used in hybrids are specially designed for the vehicle and electric assist. Therefore, they can operate more efficiently, resulting in very good fuel economy and very low tailpipe emissions. Hybrids will never be true zero-emission vehicles, however, because they have an ICE.

HEVs have an extended range, going further than a BEV can on just the charge in its batteries. They also have a longer driving range than a comparable ICE-equipped vehicle. HEVs also provide the same performance, if not better, as the same vehicle equipped with a larger ICE. The delivery of power to the wheels is smooth and very responsive.

There are two major types of hybrids: the parallel and the series designs. A parallel HEV uses either the electric motor or the gas engine to propel the vehicle, or both (Figure 1-14). A true series HEV only uses the ICE to power the generator to keep the batteries charged. The vehicle is powered only by the electric motor(s) (Figure 1-15).

There are current hybrid models based on the series configuration. However, most can be considered as having a series/parallel configuration because they have the features of both designs.

There are several hybrid cars on the market today, with more planned for the near future. The public has accepted these and potential owners are often put on a waiting list because the demand is greater than the supply. Although most current hybrids are focused on fuel economy, the same construction is used to create high-performance vehicles. Hybrid technology will also influence off-the-road performance. By using individual motors at the front and rear drive axles, additional power can be applied to certain drive wheels when needed (Figure 1-16).

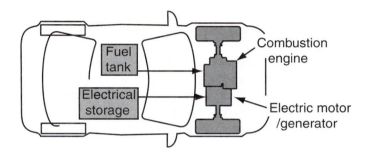

Figure 1-14 The configuration of a typical parallel hybrid vehicle.

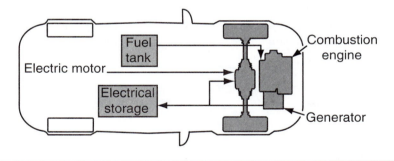

Figure 1-15 The configuration of a typical series hybrid vehicle.

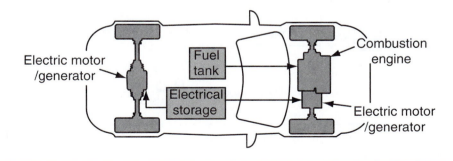

Figure 1-16 This hybrid power train allows for additional 4WD control and power by having a motor/generator attached to the rear axle.

Economics

HEVs are priced slightly higher than comparable ICE models. The difference in price can be offset by fuel savings over time. In addition, there are tax incentives to encourage consumers to purchase an alternative-fuel or advanced-technology vehicle. However, most buyers of hybrid vehicles are motivated by the reduced emission levels and the expected lower annual fuel costs.

Fuel Cell Electric Vehicles

A possible alternative fuel for the future is hydrogen, which is the fuel for fuel cells. Basically, a fuel cell generates electrical power through a chemical reaction. A fuel cell electric vehicle (FCEV) uses the electricity produced by the fuel cell to power motors that drive the vehicle's wheels. FCEVs operate like most EVs, but their batteries do not need to be charged by an external source. FCEVs emit little, if any, pollutants. Fuel cell technology may also be used to provide energy for homes and businesses.

Fuel cells convert chemical energy to electrical energy by combining hydrogen with oxygen from the air (Figure 1-17). The hydrogen can be supplied directly as pure hydrogen gas or through a "fuel reformer" that pulls hydrogen from hydrocarbon fuels such as methanol, natural gas, or gasoline. Simply put, a fuel cell is comprised of two electrodes (the anode and the cathode) located on either side of an electrolyte. As the hydrogen enters the fuel cell, the hydrogen atoms give up electrons at the anode and become hydrogen ions in the electrolyte. The electrons that were released at the anode move through an external circuit to the cathode. As the electrons are moving toward the cathode, they can be diverted and used to power the vehicle. When the electrons and hydrogen ions combine with oxygen molecules at the cathode, water and heat are formed. There are no smog-producing or greenhouse gases generated and only water is emitted from its tail pipe.

A fuel cell power system has many other parts (Figure 1-18), but central to them all is the fuel cell stack. The stack is made of many thin, flat fuel cells layered together. Each cell produces electricity and the total output of all the cells is used to power the vehicle. The entire stack of fuel cells is often referred to as a fuel cell, although that is not technically correct. A fuel cell is one cell whereas the stack is many cells.

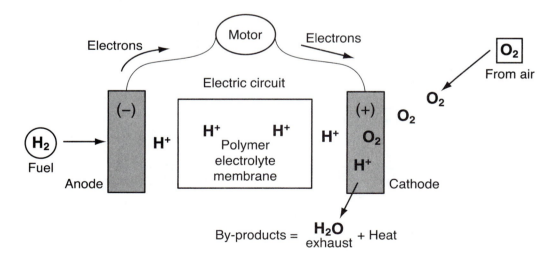

Figure 1-17 In a fuel cell, as hydrogen flows through the anode and oxygen flows through the cathode, charges are produced to power electrical loads.

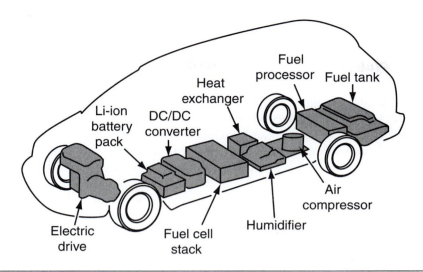

Figure 1-18 Components of a sodium borohydride fuel cell.

Vehicles that run on pure hydrogen are true zero-emission vehicles. FCEVs that have reformers will emit some pollutants, but far less than an ICE vehicle. Without a reformer, FCEVs would not consume any fossil fuels or contribute to global warming. However, currently there are some emissions related to the production of hydrogen. This is an area that must be addressed before FCEVs can move into the mainstream.

Many obstacles need to be overcome before FCEVs become a truly viable option for personal transportation. These include:

❏ Storage—Because hydrogen is a gas, a larger volume of it is needed to travel the same distance as with a tank of gasoline.
❏ Weight and size—Current fuel cells are quite large; both need to be reduced to improve overall fuel efficiency.
❏ Cost—The cost of a fuel cell is high and this must also be reduced.
❏ Start-up time—Fuel operates best at a fixed moderate temperature. Vehicles must be equipped with systems that allow for quick reactions when cold and quick responses to changing operating conditions.
❏ Hydrogen sources—Fuel cells depend on hydrogen fuel. This requires an infrastructure that can supply hydrogen needs and/or clean operating reformers.

Most auto manufacturers are actively researching fuel cell transportation technologies and testing prototype passenger vehicles. Many cities are testing fuel cell-powered transit buses. The advances made by hybrid vehicle technology will benefit fuel cell vehicle development.

A Look at History

Electric drive vehicles have been around for a long time. Early automobiles were mostly electric or steam powered. "Steamers" were the most common until the late 1800s. Early electric vehicles faced the same problems that plague pure EVs today, namely battery technology and cost. The internal combustion engine became popular because it allowed a vehicle to travel great distances, achieve a decent high speed, and was much less expensive to buy.

Let us take a quick look at some interesting developments of electric vehicles throughout history. It is said that the first practical electric road vehicle was probably made either by Thomas Davenport in the United States or by Robert Davidson in Edinburgh in 1842. Both of these vehicles had non-rechargeable electric cells (batteries) and, therefore, had a limited travel range and were not accepted by consumers. In 1865, the storage battery was invented and then further improved in 1881. More significantly, between the years 1890 and 1910, battery technology drastically improved with the development of the modern lead-acid battery by Henri Tudor in 1881. Thomas Edison's nickel-iron alkaline battery was used to power the first electric drive vehicles 100 years ago.

Many different individuals and companies developed electric vehicles. In 1897, the London Electric Cab Company began daily operation using cars designed by Walter Bersey. The Bersey Cab, which used a 40-cell battery and 3-horsepower electric motor, could be driven 50 miles between charges. A year later, Dr. Jacob Ferdinand Porsche, an engineer for Lohner & Company in Austria, built his first car, the Lohner-Porsche. This was the world's first front-wheel-drive vehicle. His second car is of more interest; it was a hybrid vehicle. This Porsche used an ICE to run a generator that provided the power to electric motors that powered individual wheels through the hub of the wheels.

At the beginning of the twentieth century, thousands of hybrid and electric vehicles were made. In fact, they were the people's choice. In 1900, 38% of the cars sold were electrically powered, the others ran on steam or gasoline (just for reference, more steam-powered vehicles were sold than gasoline-powered ones).

Electric vehicles did not have the vibration, smell, and noise of gasoline cars. Starting the engine and changing gears were the most difficult things about driving a gasoline-powered vehicle. Electric drive vehicles did not need to be manually cranked to get going and had no need for a transmission or change of gears. These were the primary reasons the public accepted electric drive over the ICE vehicles.

In 1905, H. Piper filed a patent for a petrol-electric hybrid vehicle. His idea was to use an electric motor to assist an ICE, enabling the vehicle to accelerate to 25 mph in 10 seconds. This is thought to be the first parallel hybrid automobile. This claim may be stretching things a bit, because in 1900 a Belgian company, Pieper, fitted a small vehicle with a very small ICE and an electric motor. When the vehicle was operating under very low load, the electric motor served as a generator to charge the batteries. That generator then became a motor to assist the engine when there were heavy loads on the engine.

One of the companies that had many variations of electric drive was the Woods Motor Vehicle Company. The 1902 Wood's Phaeton was totally electric powered and had a range of 18 miles and a top speed of 14 mph. The 1905 Woods Interurban was also an electric vehicle but it also had an ICE. This vehicle was designed to allow the driver and a mechanic to switch the driveline from the electric motor to the ICE in about 15 minutes. Although this vehicle had alternate power sources, it was not a hybrid. However, in 1917, Woods introduced their Woods Dual Power. This vehicle had an ICE and an electric motor. This was probably the first parallel hybrid vehicle, as the ICE and motor were designed to work together or alone.

The popularity of electric drive automobiles reached its height in 1912, but began to drop because of the advances made with ICE-powered vehicles.

In 1904, Henry Ford overcame some of the common objections to gasoline-powered cars (noise, vibration, and smells) and, thanks to assembly line production, offered gasoline-powered vehicles at very low prices ($500 to $1000). The prices of electric vehicles were much higher and were rising each year. In 1912, an average electric car sold for $1,750, while a gasoline-powered one sold for only $650.

The popularity of electric drive vehicles declined further in 1912, when Charles Franklin Kettering invented the self-starter for automobile engines. This electric starter was first offered on Cadillac cars in 1912 and on many other makes in 1913. Because one of the objections the public had about ICE vehicles was getting the engine started, this invention answered the problem directly. It made it easier for all drivers to start an ICE engine.

Other factors that contributed to the decline in electric drive vehicles include:

❏ Gasoline engines were becoming more refined and efficient.
❏ The United States had a weak infrastructure for supplying electricity for businesses and homes; much less for charging batteries. Additionally, electricity was quite expensive.
❏ The country had developed a system of roads that connected cities, and vehicles needed to provide a longer driving range than that offered by electric vehicles.
❏ The availability of gasoline had increased and it was inexpensive.

For the most part, electric drive road vehicles were a thing of the past from 1920 to 1965. Although there were concerns in the 1960s about the environment and our dependence on foreign oil, there was not a noticeable rebirth of the electric car. However, the use of electric drive in other vehicles made a resurgence. One significant development did take place during this period of time. In 1959, Dr. Harry Karl Ihring created the modern day fuel cell. At the time, it was not certain how this machine would fit in the scheme of things. We know today that it has much promise for powering electric vehicles in the future.

In 1966, the United States Congress introduced the first bills recommending the use of electric vehicles as a way to reduce air pollution. Subsequent laws put mandates on auto manufacturers to clean up emissions. The initial result of these laws was altering the basic ICE with a variety of emission controls.

In 1973, the price of gasoline increased drastically as the result of the Arab oil embargo. The rising cost led to increased attention to the development of electric drive vehicles. Manufacturers worked overtime to reduce fuel consumption to meet the desires of the consumer. However, they did not develop a practical electric vehicle in response. Two small companies did. Sebring-Vanguard produced over 2,000 "CitiCars." The CitiCars were designed for commuters who drove short distances on city streets. The cars had a top speed of 44 miles per hour (mph) and a range of 50 to 60 miles. The Elcar Corporation produced another commuter car called the Elcar, which had a top speed of 45 mph and a range of 60 miles.

In 1975, AM General, a division of American Motor Company, began delivery of 350 electric jeeps to the U.S. Postal Service for testing. These jeeps had a top speed of 50 mph and a range of 40 miles at a speed of 40 mph.

The U.S. Congress passed into public law the Electric and Hybrid Vehicle Research, Development, and Demonstration Act of 1976. One objective of the law was to work with industry to improve batteries, motors, controllers, and other hybrid-electric vehicle components.

Other legislation has been passed through the years that provides incentive to manufacturers to produce cleaner emission vehicles, such as electric drive vehicles. In addition, there have been several legislative and regulatory actions that have renewed development of electric vehicles. The 1990 Clean Air Act Amendment, the 1992 Energy Policy Act, and regulations issued by the California Air Resources Board (CARB) have had the most impact. CARB emissions certification places all passenger vehicles into these major groups. Each group is defined by a number of factors, primarily the measurable emissions of particular substances.

- ❏ LEV—Low Emission Vehicle
- ❏ ULEV—Ultra Low Emission Vehicle
- ❏ SULEV—Super Ultra Low Emission Vehicle
- ❏ PZEV—Partial Zero Emission Vehicle
- ❏ AT PZEV—Advance Technology Partial Zero Emission Vehicle
- ❏ ZEV—Zero Emission Vehicle

In 1990, CARB adopted a requirement that 10% of the new cars offered for sale in California in 2003 and beyond must be zero-emission vehicles (ZEVs). In 1998, the Air Resources Board modified the requirements for 2004 (Figure 1-19). The change allows automobile manufacturers to satisfy up to 6% of their ZEV requirement with automobiles that qualify as partial ZEVs. A ZEV is one that has no tailpipe emissions, no evaporative emissions, no emissions from gasoline refining or sales, and no onboard emission-control systems that can deteriorate over time. Today, only battery-powered electric vehicles qualify as ZEVs. Fuel cell vehicles powered by hydrogen will also qualify as true ZEVs.

In 1991, the United States Advanced Battery Consortium (USABC), a Department of Energy program, began a project that would lead to the production of a battery that would make electric vehicles a viable option for consumers. The initial result was the development of the nickel hydride (NiMH) battery. This battery can accept three times as many charge cycles as lead acid, and can work better in cold weather.

	CV	TLEV	LEV	ULEV	ZEV
NMOG	0.25 *	0.125	0.075	0.040	0.0
CO	3.4	3.4	3.4	1.7	0.0
NOx	0.4	0.4	0.2	0.2	0.0

(*) Emission standards of NMHC

Figure 1-19 CARB's emissions standards in grams per mile at 50,000 miles.

In 1997, Toyota Motor Corporation made the first modern hybrid automobile available to the public in Japan. Also during that time, a few models of all-electric cars were made available in the United States, including Honda's EV Plus, GM's EV1 (Figure 1-20) and S-10 electric pickup, a Ford Ranger pickup, and Toyota's RAV4 EV (Figure 1-21). All of these are no longer available because of poor acceptance by the market.

Testing the market in the United States, Honda released the two-door, two-seat Insight (Figure 1-22). This was the first hybrid car to hit the mass market and received much positive press coverage. It won many awards and was rated at 61 miles per gallon (mpg) city and 70 mpg highway by the U.S. Environmental Protection Agency (EPA). Although it was a different-looking and very small car, it did fairly well in the marketplace.

In 2000, Toyota introduced the Prius to the United States. Again, the public responded positively, in spite of the fact it was another hybrid that looked different. Comfortable with the public's acceptance of hybrid technology, Honda introduced the Honda Civic Hybrid in 2002. The appearance and drivability of the Civic Hybrid was identical to the conventional Civic.

After some success with the first model, Toyota introduced its second-generation Prius in 2004 (Figure 1-23). This model looked normal and was more practical than the previous model. It won numerous awards including *Motor Trend*'s Car of the Year. The demand for this model was and still is great, so great, in fact, that Toyota had to increase production of it and buyers had to wait up to six months to get one.

Since then, Ford has released a hybrid version of its great-selling small SUV, the Escape. This was the first American hybrid and the first SUV hybrid. The demand for the hybrid model has also been great. Chevrolet has released a mild hybrid Silverado. Toyota has released two

Figure 1-20 GM's EV1. *Courtesy of Georgia Power Company*

Figure 1-21 Toyota's RAV4 EV. *Courtesy of Dewhurst Photography and Toyota Motor Sales, U.S.A., Inc.*

Figure 1-22 The Honda Insight. *Courtesy of American Honda Motor Co., Inc.*

Figure 1-23 The Toyota Prius was the world's first mass-produced modern hybrid vehicle.

Figure 1-24 A Honda fuel cell vehicle. *Courtesy of American Honda Motor Co., Inc.*

hybrid SUVs and a Camry hybrid. In addition, several manufacturers are working on fuel cell vehicles. In fact Honda, as well as other manufacturers, has released some to the general public for testing (Figure 1-24). The list of available hybrids and fuel cell vehicles is growing so quickly that it is impossible to include them in this book before it is printed.

Precautions For Working On Electric Drive Vehicles

Electric drive vehicles (battery-operated electric vehicles, hybrid electric vehicles, and fuel cell electric vehicles) have high-voltage electrical systems (from 42 volts to 600 volts). These high voltages can kill you! Fortunately, most high-voltage circuits are identifiable by size and color. The cables have thicker insulation and are typically colored orange. The connectors are also colored orange. On some vehicles, the high-voltage cables are enclosed in an orange shielding or casing; again the orange indicates high voltage. Be careful not to touch these wires. In addition, the high-voltage battery pack and most high-voltage components have "High Voltage" caution labels. Be careful when working around these parts. There are other safety precautions that should always be adhered to when working on an electric drive vehicle:

❏ Always adhere to the safety guidelines given by the vehicle's manufacturer.
❏ If repair operation is incorrectly performed on an EV, an electrical shock, leakage, or explosion could be caused. Be sure to perform each repair operation correctly.
❏ Disable or disconnect the high-voltage system. Do this according to the procedures given by the manufacturer.
❏ Systems may have a high-voltage capacitor that must be discharged after the high-voltage system has been isolated. Make sure to wait the prescribed amount of time (normally about 10 minutes) before working on or around the high-voltage system.
❏ After removing a high-voltage cable, cover the terminal with vinyl electrical tape.
❏ Always use insulated tools.
❏ Always follow the test procedures defined by the equipment manufacturer when using this type of equipment.
❏ Alert other technicians that you are working on the high-voltage systems with a warning sign such as "high-voltage work: do not touch."
❏ Always follow the manufacturer's instructions for removing high-voltage battery packs.
❏ When disconnecting electrical connectors, do not pull on the wires. When reconnecting the connectors, make sure they are securely connected.
❏ Do not wear metallic objects such as rings and necklaces.

- ❏ Do not carry any metal objects such as a mechanical pencil or a measuring tape that could fall and cause a short circuit.
- ❏ Do not leave tools or parts anywhere in the vehicle.
- ❏ Wear insulating gloves, commonly called "lineman's gloves," when working on or around the high-voltage system. Make sure the gloves have no tears, holes, or cracks and that they are dry. The integrity of the gloves should be checked before using them.
- ❏ Always install the correct type of circuit protection device into a high-voltage circuit.
- ❏ Use only the tools and test equipment specified by the manufacturer.
- ❏ Many electric motors have a strong permanent magnet in them; individuals with a pacemaker should not handle these parts.
- ❏ Before doing any service to an electric drive vehicle, make sure the power to the electric motor is disconnected or disabled.
- ❏ Any time the engine is running in a hybrid vehicle, the generator is producing high-voltage and care must be taken to prevent being shocked.
- ❏ When an electric drive vehicle needs to be towed into the shop for repairs, make sure it is not towed on its drive wheels. Doing this will drive the generator(s), which can overcharge the batteries and cause them to explode. Always tow these vehicles with the drive wheels off the ground or move them on a flat bed.
- ❏ In the case of a fire, use a Class ABC powder type extinguisher or very large quantities of water.

Battery Precautions

Because the electrical power for an electric drive vehicle is stored in a battery or battery pack, special handling precautions must be followed when working with or near batteries.

- ❏ Make sure you are wearing safety glasses (preferably a face shield) and protective clothing when working around and with batteries.
- ❏ Keep all flames, sparks, and excessive heat away from the battery at all times, especially when it is being charged.
- ❏ Never smoke near the top of a battery and never use a lighter or match as a flashlight.
- ❏ Remove wristwatches and rings before servicing any part of the electrical system. This helps prevent the possibility of electrical arcing and burns.
- ❏ Never lay metal tools or other objects on the battery because a short circuit across the terminals can result.
- ❏ All batteries have an electrolyte, which is very corrosive. It can cause severe injuries if it comes in contact with your skin or eye. If electrolyte gets on you, immediately wash with baking soda and water. If the acid gets in your eyes, immediately flush them with cool water for a minimum of 15 minutes and get immediate medical attention.
- ❏ Lead-acid batteries use sulfuric acid as the electrolyte. Sulfuric acid is poisonous, highly corrosive, and produces gases that can explode in high heat.
- ❏ Acid from the battery damages a vehicle's paint and metal surfaces and harms shop equipment. Neutralize any electrolyte spills during servicing.
- ❏ When removing a battery from a vehicle, always disconnect the battery ground cable first. When installing a battery, connect the ground cable last.
- ❏ Always use a battery carrier or lifting strap to make moving and handling batteries easier and safer.
- ❏ Always disconnect the battery's ground cable when working on the electrical system or engine. This prevents sparks from short circuits and prevents accidental starting of the engine.
- ❏ Always charge a battery in well-ventilated areas.
- ❏ Never connect or disconnect charger leads when the charger is turned on. This generates a dangerous spark.
- ❏ Never recharge the battery when the system is on.

- ❏ Turn off all accessories before charging the battery and correct any parasitic drain problems.
- ❏ Make sure the charger's power switch is off when you are connecting or disconnecting the charger cables to the battery.
- ❏ Always disconnect the battery ground cable before fast charging the battery on the vehicle.
- ❏ Never attempt to use a charger as a boost to start the engine.

Review Questions

1. Define the term "electric drive."
2. True or False. Electric vehicles were very much a part of the development of the modern automobile; however by 1900, nearly all automobiles sold were gasoline powered.
3. List three alternative fuels for an ICE and briefly explain where each comes from.
4. What is the basic difference between a series hybrid vehicle and a parallel one?
5. Which is the basic fuel for a fuel cell?
   A. hydrogen
   B. methanol
   C. electricity
   D. gasoline
6. What is regenerative braking?
7. List three factors that have elicited a renewed interest in developing electric drive vehicles.
8. What is a Zero Emission Vehicle (ZEV)?
9. Name three possible energy sources for future electric vehicles.
10. Explain why hybrid vehicles will never be considered ZEVs.

ELECTRICAL BASICS

After reading and studying this chapter, you should be able to:

❏ Explain the basic principles of electricity.

❏ Define the terms normally used to describe electricity.

❏ Explain what is defined by Ohm's law.

❏ Explain the differences between AC and DC.

❏ Describe the differences between a series and a parallel circuit.

❏ Name the various electrical components and their uses in electrical circuits.

❏ Explain the principles of magnetism and electromagnetism.

Introduction

To understand how electric drive vehicles work and how to maintain them, you must have an understanding of electricity. This chapter will not cover the topic in depth; rather, it covers the fundamentals.

All things are made up of atoms, which are extremely small particles. In the center of every atom is a nucleus. The nucleus contains positively charged particles called protons and particles called neutrons that have no charge. Negatively charged particles called electrons orbit around every nucleus. The electrons stay in orbit around the nucleus because they are naturally attracted to the protons.

Electricity is the flow of electrons from one atom to another (Figure 2-1). The release of energy as one electron leaves the orbit of one atom and jumps into the orbit of another is electrical energy. The key to creating electricity is to provide a reason for the electrons to move to another atom.

Electrons have a negative charge and are attracted to something with a positive charge. When an electron leaves the orbit of an atom, the atom then has a positive charge. An electron moves from one atom to another because the atom next to it appears to be more positive than the one it is orbiting around. To have a continuous flow of electricity, three things must be present: an excess of electrons in one place, a lack of electrons in another place, and a path between the two places.

Electricity is stored in batteries (Figure 2-2). Batteries have two terminals, a positive and a negative terminal. The chemical reaction in the battery causes a lack of electrons at the positive terminal and an excess at the negative terminal. This creates an electrical imbalance, causing the electrons to flow through the path provided by the vehicle's wiring.

The chemical process in the battery continues to provide electrons until the chemicals become weak. At that time, either the battery has run out of electrons or all of the protons are matched with an electron. When this happens, there is no longer a reason for the electrons to want to move to the positive side of the battery. Charging the battery restores the supply of electrons.

Key Terms

Alternating current (AC)
Capacitors
Chassis
Closed circuit
Conductors
Continuity
Current
Dielectric
Direct current (DC)
Electromotive force (EMF)
Fixed valve resistors
Flux density
Flux field
Ground wire
Induction
Insulators
Ohm
Open circuit
Parallel circuit
Potentiometers
Relay
Reluctance
Resistance
Rheostats
Series circuit
Sine wave
Solenoids
Stepped resistors
Tapped
Thermistor
Variable resistors
Voltage

Conductor

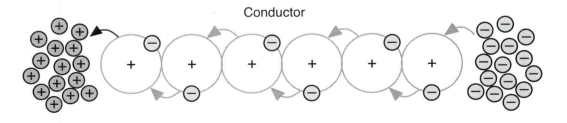

Figure 2-1 Electricity results from the flow of electrons from one atom to another.

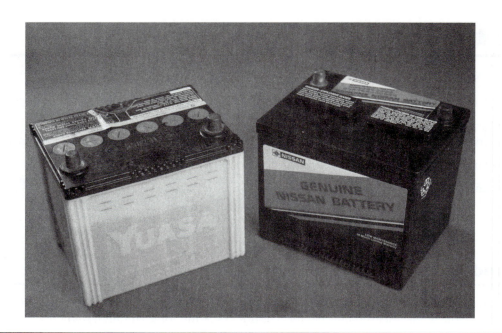

Figure 2-2 The electrical energy needed to drive all automobiles is stored in a battery.

Electrical Terms

Electrical **current** describes the movement or flow of electricity. The greater the number of electrons flowing past a given point in a given amount of time, the more current the circuit has. The unit for measuring electrical current is the ampere, usually called an amp. The instrument used to measure electrical current flow in a circuit is called an ammeter (Figure 2-3).

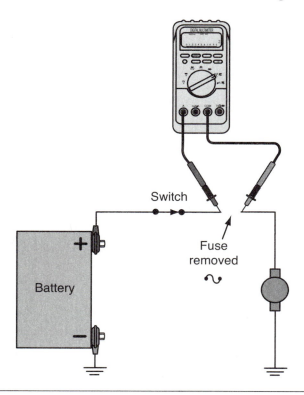

Figure 2-3 To measure the current or amperage in a circuit, the ammeter should be connected in series with the circuit.

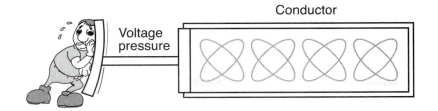

Figure 2-4 Voltage is electrical pressure.

There are two types of current: **direct current (DC)** and **alternating current (AC)**. In direct current, the electrons flow in one direction only. In alternating current, the electrons change direction at a fixed rate. Typically, an automobile uses DC, whereas the current in homes and buildings is AC. Some components of the automobile generate or use AC. Most motors used in electrical vehicles are powered by AC. The storage batteries in automobiles are DC devices; therefore, to use the stored electricity to run the AC devices, the DC must be converted to AC. Likewise, to use the electrical energy generated by an AC device to charge a battery, AC must be changed to DC.

Voltage is electrical pressure (Figure 2-4). It is the force developed by the attraction of the electrons to protons. The more positive one side of the circuit is, the more voltage is present in the circuit. Voltage does not flow; it is the pressure that causes current flow. This force is the pressure that exists between a positive and negative point within an electrical circuit. This force or pressure, also called **electromotive force (EMF)**, is measured in units called volts. One volt is the amount of pressure required to move 1 ampere of current through a resistance of 1 ohm. Voltage is measured by an instrument called a voltmeter.

When any substance flows, it meets resistance. The **resistance** to electrical flow produces heat and can be measured. A unit of measured resistance is called an **ohm**. Resistance can be measured by an instrument called an ohmmeter.

Ohm's Law

In 1827, a German mathematics professor, Georg Ohm, published a book that included his explanation of the behavior of electricity. His thoughts have become the basis for a true understanding of electricity. He found that it takes one volt of electrical pressure to push one ampere of electrical current through one ohm of resistance. This statement is the basic law of electricity and is known as **Ohm's law**.

In any electrical circuit, current (I), resistance (R), and voltage (E) work together in a mathematical relationship. This relationship is expressed in a mathematical statement of Ohm's law (Figure 2-5). Ohm's law can be applied to the entire circuit or to any part of a circuit. When any two factors are known, the third factor can be found by using Ohm's law.

Voltage (E) = Current (I) times Resistance (R), therefore

$$E = I \times R.$$

Current (I) = Voltage (E) divided by Resistance (R), therefore

$$I = E / R.$$

Resistance (R) = Voltage (E) divided by Current (I), therefore

$$R = E / I.$$

Figure 2-5 The mathematical expression for Ohm's Law.

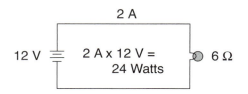

Figure 2-6 The power of the lamp in this simple circuit is 24 watts.

Power

Electrical power, or the rate of work, is found by multiplying the amount of voltage by the amount of current flow (Power = Voltage × Amperage). Power is measured in watts (Figure 2-6). Although power measurements are rarely, if ever, needed in automotive service, knowing the power requirements of light bulbs, electric motors, and other components is sometimes useful when troubleshooting electrical systems.

It is also important to understand why system voltage increases with various electrical and electronic applications. As voltage increases, the current for a load decreases. This means cables and wires can be deceased in size to accommodate high-voltage components, such as the motors to move a vehicle.

Circuit Terminology

An electrical circuit is considered complete when there is a path that connects the positive and negative terminals of the electrical power source. A completed circuit is called a **closed circuit**, whereas an incomplete circuit is called an **open circuit**. When a circuit is complete, it is said to have **continuity**.

In many wiring diagrams or electrical schematics, the return wire from the load or resistor is shown as being connected directly to the negative terminal of the battery. If this were the case in an actual vehicle, there would be literally hundreds of wires connected to the negative battery terminal. To avoid this, auto manufacturers use a wiring style that involves using the vehicle's metal frame components as part of the return circuit. Using the chassis as the negative wire is often referred to as **grounding** and the connection is called a **chassis ground**. The wire or metal mounting that serves as the contact to the chassis is commonly called the **ground wire** or lead (Figure 2-7).

An electrical component may be mounted directly to the engine block, transmission case, or frame. This direct mounting effectively grounds the component without the use of a separate ground wire. In other cases, however, a separate ground wire must be run from the component to the frame or another metal part to ensure a good connection for the return path. The increased use of plastics and other nonmetallic materials in body panels and engine parts has made electrical grounding more difficult. To ensure good grounding back to the battery, some manufacturers now use a network of common grounding terminals and wires.

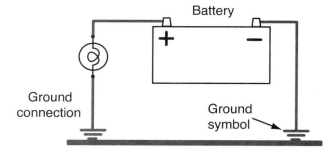

Figure 2-7 A simple light circuit that uses the vehicle as the negative conductor through ground connections.

In a complete circuit, the flow of electricity can be controlled and applied to do useful work, such as light a headlamp or turn on a motor. Components that use electrical power put a load on the circuit and consume electrical energy. These components are often referred to as electrical loads.

The amount of current that flows in a circuit is determined by the resistance in that circuit. As resistance goes up, the current goes down. The total resistance in a circuit determines how much current will flow through the circuit. The energy used by a load is measured in volts. Amperage stays constant in a circuit but the voltage is dropped as it powers a load. Measuring voltage drop tells you how much electrical energy is being changed to heat by the load (Figure 2-8).

Alternating Current

Alternating current is current that constantly changes in voltage and direction. DC, on the other hand, always moves in the same direction and the voltage is constant until it reaches a resistance. Direct current always moves from a point of higher potential (voltage) to a point of lower potential (voltage).

If a graph is used to represent the amount of DC voltage available from a battery during a fixed period time, the line on the graph will be flat, which represents a constant voltage. If AC voltage is shown on a graph, it will appear as a **sine wave**. The sine wave shows AC changing in amplitude (strength) and direction (Figure 2-9). The highest positive voltage equals the highest negative voltage. The movement of the AC from its peak at the positive side of the graph to the negative side and then back to the positive peak is commonly referred to as "peak-to-peak" value. This value represents the amount of voltage available at a point. During each complete cycle of AC, there are always two maximum or peak values, one for the positive half-cycle and the other for the negative half-cycle. The difference between the peak positive value and the peak negative value is used to measure AC voltages.

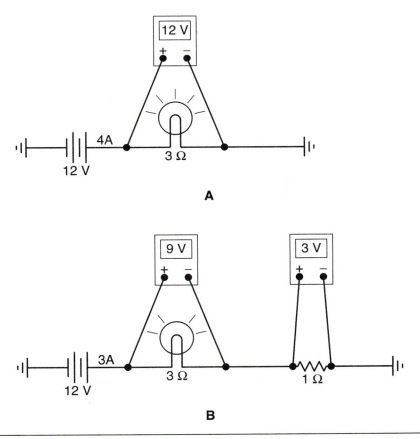

Figure 2-8 Connecting a voltmeter across a load measures the voltage drop across that load.

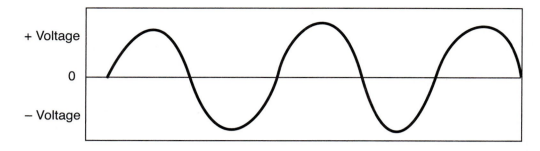

Figure 2-9 This sine wave results from alternating current.

AC does not have a constant value; therefore, as it passes through a resistance, nearly 29% less heat is produced when compared to DC. This is one reason AC is preferred over DC when powering motors and other electrical devices.

The lack of heat also causes us to look at AC values differently than the same values in a DC circuit. An alternating current has an effective value of one ampere when it produces heat in a given resistance at the same rate, as does one ampere of direct current. The effective value of an alternating current is equal to 0.707 times its maximum or peak current value. Because alternating current is caused by an alternating voltage, the ratio of the effective voltage value to the maximum voltage value is the same as the ratio of the effective current to the maximum current or 0.707 times the maximum value.

According to Ohm's law, current is directly proportional to the voltage applied to the circuit. This remains true for AC circuits as well. However, AC voltage and current change constantly and its values must be viewed as average or effective. When AC is applied to a resistance, as the actual voltage changes in value and direction, so does the current. In fact, the change of current is in phase with the change in voltage. An "in-phase" condition exists when the sine waves of voltage and current are precisely in step with one another. The two sine waves go through their maximum and minimum points at the same time and in the same direction. In some circuits, several sine waves can be in phase with one another.

If a circuit has two or more voltage pulses but each has its own sine wave that begins and ends its cycle at a different time, the waves are "out-of-phase." If two sine waves are 180 degrees out-of-phase, they will cancel each other out if they are of the same voltage and current. If two or more sine waves are out-of-phase but do not cancel each other, the effective voltage and current is determined by the position and direction of the sine wave at a given point within the circuit.

Conductors and Insulators

Controlling and routing the flow of electricity requires the use of materials known as conductors and insulators. **Conductors** are materials with a low resistance to the flow of current. Most metals, such as copper, silver, and aluminum, are excellent conductors.

Insulators resist the flow of current. Thermal plastics are the most common electrical insulators used today. They can resist heat, moisture, and corrosion without breaking down. The insulation of the vehicle's various wires is colored or marked to allow for circuit identification (Figure 2-10).

Copper wire is by far the most popular conductor used in automotive electrical systems. Where flexibility is required, the copper wire will be made of a large number of very small strands of wire woven together.

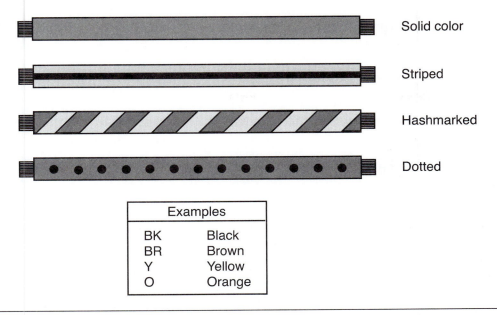

Examples	
BK	Black
BR	Brown
Y	Yellow
O	Orange

Figure 2-10 The insulation of the vehicle's various wires is colored or marked to allow for circuit identification.

The resistance of a uniform, circular copper wire depends on the length of the wire, the diameter of the wire, and the temperature of the wire. If the length is doubled, the resistance between the wire ends is doubled. The longer the wire, the greater the resistance. If the diameter of a wire is doubled, the resistance for any given length is cut in half. The larger the wire's diameter, the lower the resistance.

Heat is developed in any wire carrying current because of the resistance in the wire. If the heat becomes excessive, the insulation will be damaged. Resistance occurs when electrons collide as current flows through the conductor. These collisions cause friction that in turn generates heat.

Circuits

Most automotive circuits contain four basic parts.

1. Power sources, such as a battery, that provides the energy needed to cause electron flow.
2. Conductors, such as copper wires, that provide a path for current flow.
3. Loads, which are devices that use electricity to perform work, such as light bulbs, electric motors, or resistors.
4. Controllers, such as switches or relays, that control or direct the flow of electrons.

There are also two basic types of circuits used in automotive electrical systems: series and parallel circuits. Each circuit type has its own characteristics regarding amperage, voltage, and resistance.

A **series circuit** consists of one or more resistors (loads) connected to a voltage source with only one path for electron flow (Figure 2-11). The word "series" is given to a circuit in which the same amount of current is present throughout the circuit. The current that flows through one resistor also flows through other resistors in the circuit. As that amount of current leaves the battery, it flows through the conductor to the first resistor. At the resistor, some electrical energy or voltage is consumed as the current flows through it. The decreased amount of voltage is then applied to the next resistor as current flows to it. By the time the current is flowing in the conductor leading back to the battery, all voltage has been consumed. All of the source voltage available to the circuit is dropped by the resistors in the circuit.

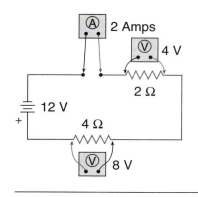

Figure 2-11 The characteristics of a series circuit.

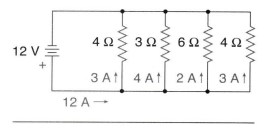

Figure 2-12 The characteristics of a parallel circuit.

In a series circuit, the total amount of resistance in the circuit is equal to the sum of all the individual resistors, and the sum of all voltage drops in a series circuit must equal the source voltage.

A **parallel circuit** provides two or more different paths for the current to flow through. Each path has separate resistors (loads) and can operate independently of the other paths (Figure 2-12). The different paths for current flow are commonly called the legs of a parallel circuit.

A parallel circuit is characterized by the following facts.

❏ Total circuit resistance is always lower than the resistance of the leg with the lowest total resistance.
❏ The current through each leg will be different if the resistance values are different.
❏ The sum of the current on each leg equals the total circuit current.
❏ The voltage applied to each leg of the circuit will be dropped across the legs if there are no loads in series with the parallel circuit.

Circuit Components

Automotive electrical circuits contain a number of different types of electrical devices. The more common components are outlined here.

Resistors

Resistors are used to limit current flow (and thereby voltage) in circuits where full current flow and voltage are not needed or desired. Resistors are devices specially constructed to put a specific amount of resistance into a circuit. In addition, some other components use resistance to produce heat and even light. An electric window defroster is a specialized type of resistor that produces heat. Electric lights are resistors that get so hot they produce light.

Fixed value resistors are designed to have only one rating, which should not change. These resistors are used to decrease the amount of voltage applied to a component. **Tapped** or **stepped resistors** are designed to have two or more fixed values, available by connecting wires to the several taps of the resistor (Figure 2-13). Heater motor resistor packs, which provide for different fan speeds, are an example of this type of resistor.

Variable resistors are designed to have a range of resistances available through two or more taps and a control. Two examples of this type of resistor are **rheostats** and **potentiometers**. Rheostats have two connections, one to the fixed end of a resistor and one to a sliding contact with the resistor (Figure 2-14). Moving the control moves the sliding contact away from or toward the fixed end tap, increasing or decreasing the resistance. Potentiometers have three connections, one

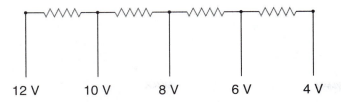

12 V   10 V   8 V   6 V   4 V

Figure 2-13 A row of resistors with taps. Notice how the available voltage is different at each tap.

Figure 2-14
A rheostat.

Figure 2-15
A potentiometer.

at each end of the resistance and one connected to a sliding contact with the resistor (Figure 2-15). Moving the control moves the sliding contact away from one end of the resistance, but toward the other end. These are called potentiometers because different amounts of potential or voltage can be sent to another circuit. As the sliding contact moves, it picks up a voltage that is equal to the source voltage minus the amount dropped by the resistor, so far.

Another type of variable resistor is the **thermistor**. This type of resistor is designed to change its resistance value as its temperature changes (Figure 2-16). Although most resistors are

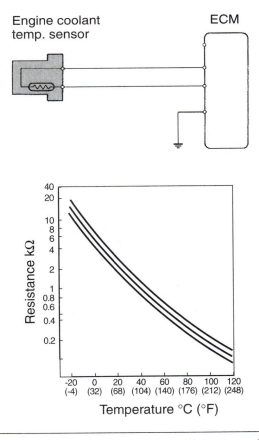

Figure 2-16 This engine coolant temperature sensor is an example of a thermistor. Notice how the resistance changes with temperature.

Figure 2-17 Various circuit protection devices.

■ **CAUTION:**

For 42-volt and higher systems, such as those used in electric and hybrid vehicles, there is a unique problem with circuit protection; most circuit protection devices used in 12-volt systems are actually rated at 32 volts. If these protection devices were used in a 42+-volt system, problems such as severe damage to the vehicle's wiring and electrical components could result. The burning of the components and wiring could also cause a fire. Higher voltage systems must be protected with devices that have a higher voltage rating than the normal system voltage.

carefully constructed to maintain their rating within a few ohms through a range of temperatures, the thermistor is designed to change its rating. Thermistors are used to provide compensating voltage in components or to determine temperature.

Circuit Protective Devices

When overloads or shorts in a circuit cause too much current to flow, the wiring in the circuit heats up, the insulation melts, and a fire can result, unless the circuit has some kind of protective device. Fuses, fuse links, maxi-fuses, and circuit breakers are designed to provide protection from high current (Figure 2-17). They may be used singularly or in combination.

Switches

Electrical circuits are usually controlled by some type of switch. Switches do two things. They turn the circuit on or off (Figure 2-18), or they direct the flow of current in a circuit. Switches can be under the control of the driver or can be self-operating through a condition of the circuit, the

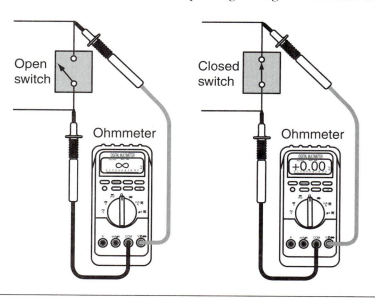

Figure 2-18 The action of a simple switch.

vehicle, or the environment. A temperature-sensitive switch usually contains a bimetallic element heated either electrically or by some component where the switch is used as a **sensor**. When engine coolant is below or at normal operating temperature, the engine coolant temperature sensor is in its normally open condition. If the coolant exceeds the temperature limit, the bimetallic element bends the two contacts together and the switch is closed to the indicator or the instrument panel. Other applications for heat-sensitive switches are time-delay switches and flashers.

Relays

A **relay** is an electric switch that allows a small amount of current to control a high-current circuit (Figure 2-19). When the control circuit switch is open, no current flows to the coil of the relay, so the windings are de-energized. When the switch is closed, the coil is energized, turning the soft iron core into an electromagnet and drawing the armature down. This closes the power circuit contacts, connecting power to the load circuit. When the control switch is opened, current stops flowing in the coil and the electromagnet disappears. This releases the armature, which breaks the power circuit contacts.

Solenoids

Solenoids are also electromagnets with movable cores used to change electrical current flow into mechanical movement. They can also close contacts, acting as a relay at the same time.

Capacitors

Capacitors (condensers) are constructed from two or more sheets of electrically conducting material with a non-conducting or **dielectric** (anti-electric) material placed between them. Conductors are connected to the two sheets. Capacitors are devices that oppose a change of voltage.

If a battery is connected to a capacitor, the capacitor will be charged when current flows from the battery to the plates (Figure 2-20). This current flow will continue until the plates have the same voltage as the battery. At this time, the capacitor is charged and remains charged until a circuit is completed between the two plates.

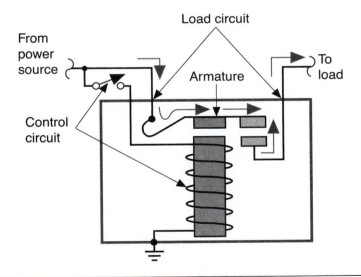

Figure 2-19 A relay.

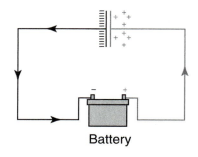

Battery

Figure 2-20 The action of a capacitor.

Electromagnetism Basics

Electricity and magnetism are related. One can be used to create the other. Current flowing through a wire creates a magnetic field around the wire. Moving a wire through a magnetic field creates current flow in the wire. Many automotive components, such as generators, operate using these principles of electromagnetism.

Fundamentals of Magnetism

A substance is said to be a magnet if it has the property of magnetism—the ability to attract substances such as iron, steel, nickel, or cobalt. These are called magnetic materials. A magnet has two points of maximum attraction, one at each end of the magnet. These points are called poles; with one designated as the north pole and the other the south pole. When two magnets are brought together, opposite poles attract, whereas similar poles repel each other (Figure 2-21).

A magnetic field, called a **flux field**, exists around every magnet. The field consists of imaginary lines along which the magnetic force acts. These lines emerge from the north pole and enter the south pole, returning to the north pole through the magnet itself. All lines of force leave the magnet at right angles to the magnet. None of the lines crosses each other and all lines are complete.

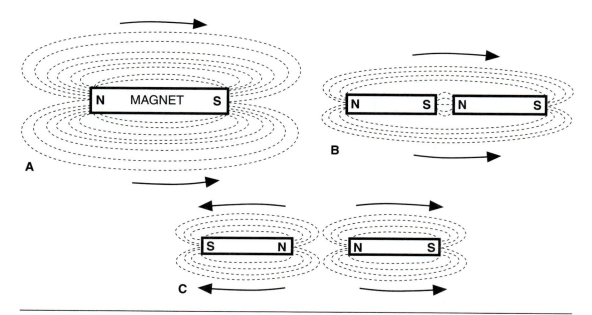

Figure 2-21 (A) The field of flux around a magnet. (B) Unlike poles attract each other. (C) Like poles repel.

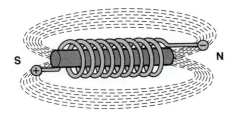

Figure 2-22 An electromagnet.

Magnets can occur naturally in the form of a mineral called magnetite. Artificial magnets can also be made by inserting a bar of magnetic material inside a coil of insulated wire and passing direct current through the coil (Figure 2-22). This principle is important in understanding certain automotive electrical components. Another way of creating a magnet is by stroking the magnetic material with a bar magnet. Both methods force the randomly arranged molecules of the magnetic material to align themselves along north and south poles.

Artificial magnets can be either temporary or permanent. Temporary magnets are usually made of soft iron. They are easy to magnetize but quickly lose their magnetism when the magnetizing force is removed. Permanent magnets are difficult to magnetize. However, once magnetized, they retain this property for long periods.

The more flux lines, the stronger the magnetic field at that point. Increasing current will increase **flux density**. Also, two conducting wires lying side by side carrying equal currents in the same direction create a magnetic field equal in strength to one conductor carrying twice the current. Adding more wires also increases the magnetic field (Figure 2-23).

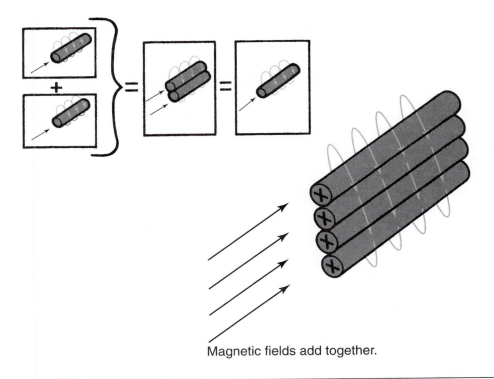

Magnetic fields add together.

Figure 2-23 When conducting wires carrying equal currents in the same direction are placed next to each other, the strength of the magnetic field increases.

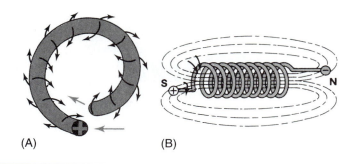

(A)   (B)

Figure 2-24 Looping a wire into a coil concentrates the lines of force around the wire.

Looping a wire into a coil concentrates the lines of force around the wire. The resulting magnetic field is the sum of all the single-loop magnetic fields (Figure 2-24). The overall effect is the same as placing many wires side by side, each carrying current in the same direction.

Magnetic Circuits and Reluctance

Just as current can only flow through a complete circuit, the lines of flux created by a magnet can only occupy a closed magnetic circuit. The resistance that a magnetic circuit offers to a line of flux is called **reluctance**. Magnetic reluctance can be compared to electrical resistance.

When a coil is wound around an iron core, it becomes a usable electromagnet. The strength of the magnetic poles in an electromagnet is directly proportional to the number of turns of wire and the current flowing through them. The behavior of an electromagnet can be summarized by these points:

❏ Field strength increases if current through the coil increases.
❏ Field strength increases if the number of coil turns increases.
❏ If reluctance increases, field strength decreases.

Induced Voltage

Now that we have explained how current can be used to generate a magnetic field, it is time to examine the opposite effect of how magnetic fields can produce electricity through a process called **induction**.

Figure 2-25 shows a straight piece of wire with the terminals of a voltmeter attached to both ends. If the wire is moved across a magnetic field, the voltmeter registers a small voltage reading. A voltage has been induced in the wire.

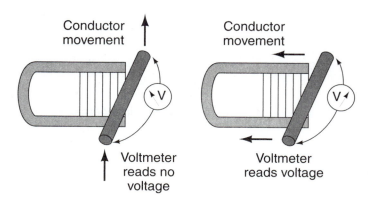

Figure 2-25 Moving a conductor through a magnetic field generates a voltage in the conductor.

It is important to realize that the wire must cut across the flux lines to induce a voltage. Moving the wire parallel to the lines of flux does not induce voltage. Voltage can also be induced by holding the wire still and moving the magnetic field at right angles to the wire. This is the exact setup used in most generators. A magnetic field is made to cut across stationary conductors to produce voltage.

The wire becomes a source of electricity and has a polarity or distinct positive and negative end. However, this polarity can be switched depending on the relative direction of movement between the wire and magnetic field. This is why charging devices produce alternating current.

The amount of voltage that is induced depends on four factors.

❏ The stronger the magnetic field, the stronger the induced voltage.
❏ The faster the field is being cut, the more lines of flux are cut and the stronger the voltage induced.
❏ The greater the number of conductors, the greater the voltage induced.
❏ The closer the conductor(s) and magnetic field are to right angles (perpendicular) to one another, the greater the induced voltage.

Electrical Systems

Normal automotive voltages are approximately 12 volts. In a typical ICE vehicle, this voltage is used to operate everything electrical on the car. Although this is called a 12-volt system, the battery actually stores about 14 volts and the charging system puts out 14 to 15 volts while the engine is running. Because the primary source of electrical power when the engine is running is the charging system, it is fair to say an automobile's electrical system is a 14-volt system.

Today's vehicles, with their electronics, put a good deal of drain on a 12-volt battery. This has led to the development of 42-volt systems (Figure 2-26), which represent three 12-volt batteries (3 times 14 volts equals 42 volts). A 42-volt system is also desirable for safety reasons. Sixty volts can stop a person's heart from beating; therefore, 42-volt systems allow for a margin of safety. The higher voltages also allow for smaller conductors because operating current is lower.

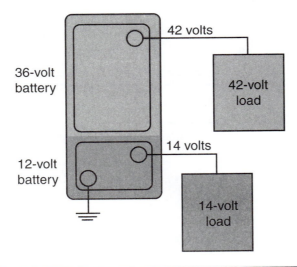

Figure 2-26 A 42-volt battery with a tap for 14-volt loads.

High-Voltage Systems

High-voltage systems are those electrical systems that operate with more than 50 volts, including BEVs, HEVs, and FCEVs. These vehicles need very high voltages to operate the electric motors. Most electric drive vehicles also have a separate 12-volt system to operate the lights and other common accessories (Figure 2-27). The high voltages range from 144 to 300, depending on the system. With these high voltages come special precautions, including the following:

❏ If repair operation is incorrectly performed on an electric vehicle (EV), an electrical shock, leakage, or explosion could be caused. Be sure to perform repair operation correctly.

❏ Always follow the manufacturer's procedures to disconnect the power source before working on the high-voltage system

❏ Systems may have a high-voltage capacitor that must discharge after the high-voltage system has been isolated. Make sure to wait the prescribed amount of time (about 10 minutes) before working on or around the high-voltage system.

❏ Wear insulating gloves when working on or around the high-voltage system. Make sure the gloves have no tears, holes, or cracks and that they are dry.

❏ The wire harnesses and connectors with high-voltage circuits are colored orange. In addition, high-voltage parts have a "high voltage" caution label. Be careful not to touch these wires.

❏ After removing a high-voltage cable, cover the terminal with vinyl tape.

❏ Always use insulated tools.

❏ Alert other technicians that you are working on high-voltage systems with a warning sign such as "High-voltage work: do not touch."

❏ Do not carry any metal objects such as a mechanical pencil or a measuring tape that could fall and cause a short circuit.

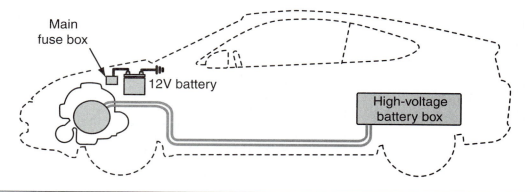

Figure 2-27 A hybrid car with two batteries, one low-voltage and the other high-voltage.

Review Questions

1. What is the name for the formula E = IR?
2. True or False. In a series circuit, circuit current is the same throughout the circuit.
3. What type of wire is most commonly used in automobiles?
4. What are the four factors that determine how much voltage is induced by a magnet?
5. How can you identify high-voltage circuits in an electric drive vehicle?
6. The process in which a conductor cuts across a magnetic field and produces a voltage is _____.
7. What happens in an electrical circuit when the resistance increases?
8. What is the difference between voltage and current?
9. True or False. The strength of the magnetic poles in an electromagnet decreases with an increase in the number of turns of wire and the current flowing through them.
10. Which of the following is a characteristic of all parallel circuits?
    A. Total circuit resistance is always higher than the resistance of the leg with the lowest total resistance.
    B. The current through each leg will be different if the resistance values are different.
    C. The sum of the resistance on each leg equals the total circuit resistance.
    D. The voltage applied to each leg of the circuit will be dropped across the legs if there are loads in series with the parallel circuit.

MOTOR AND GENERATOR BASICS

After reading and studying this chapter, you should be able to:

❏ Describe the basic operation of all electric motors.

❏ Understand the importance of magnetic principles in the operation of a motor and generator.

❏ Summarize the principles of magnetism described by Faraday's and Lenz's laws.

❏ Identify the major parts of a DC motor.

❏ Compare the operation of a brushless DC motor to a brushed DC motor.

❏ Understand the characteristics of three-phase AC voltage and describe the operation of a three-phase AC motor.

❏ Explain the differences between a motor and a generator (AC and DC).

❏ Explain the purposes of a controller in a motor/generator circuit.

❏ Define the purpose of an inverter.

Key Terms

Delta
Duty cycle
Full-wave
   rectification
Half-wave
   rectification
Pulse width modula-
   tion (PWM)
Sensing voltage
Sine wave
Synchronous motor
Three-phase voltage
Torque
Wye configuration

Introduction

The operation of an electric motor (and generator) is based on the basic principles of magnetism. The most important principle is that like magnetic poles repel each other and unlike poles are attracted to each other (Figure 3-1). Motors use the interaction of magnetic fields to change electrical energy into mechanical energy. A generator, which is constructed much like a motor, changes mechanical energy into electrical energy.

In nature, some metals have magnetic qualities. This means they are magnetic in their natural state or they have the ability to become magnets when stimulated. Metals such as iron, nickel, and cobalt have a natural ability to become magnets. When these metals are placed

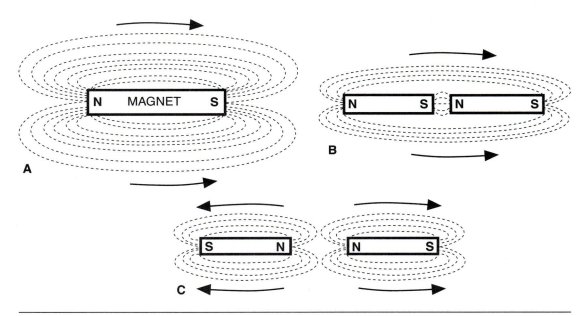

Figure 3-1 (A) The field of flux around a magnet. (B) Unlike poles attract each other. (C) Like poles repel.

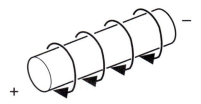

Figure 3-2 When electric current passes through a conductor, a magnetic field is created around that conductor.

into an external magnetic field, their magnetic properties align and they become strong permanent magnets. Other metals do not have the same magnetic properties but can become magnets when electrical current is passed through them. Examples of these metals are aluminum and copper.

When electric current passes through a conductor, a magnetic field is created around that conductor (Figure 3-2). The magnetic field around a single straight wire forms loops around the wire—the current's magnetic field would push a magnetic pole near it around in a circle about the wire. If loops of current-carrying wire are wound around something that is easily magnetized, the metal becomes magnetized. This current-carrying coil develops a north pole at one end of the coil and a south pole at the other. The polarity of the coil depends on the direction of current flow through the loop.

If a conductor is passed through a magnetic field, voltage is induced in that wire. This is the basis for the operation of a generator.

For both generators and motors there is a relationship between the direction of the magnetic flux lines, the direction of motion of the conductor or force acting on the conductor, and the direction of the applied voltage.

Simple Explanation of Basic Motor Types

Electric drive vehicles can use AC voltage or DC voltage motors:

- ❏ If the motor is a DC motor, then it may run on anything from 36 to 192 volts. Many of the DC motors used in electric cars come from the electric forklift industry.
- ❏ If it is an AC motor, then it probably is a three-phase AC motor. The operating voltages vary with manufacturer and application. For example, the Toyota Hybrid Highlander has three motor/generators—all are permanent magnet AC units running at a maximum of 650 volts.

The Distant Past

To get an accurate look at the history of motors and generators, we need to look at the origin of certain terms. The first generators and motors were called "dynamos" or "dynamoelectric" machines. Dynamo is from the Greek word *dynamis*, which means power. Dynamoelectric is defined as "relating to the conversion of mechanical energy into electrical energy or vice versa." The word motor is from the Latin word *motus*, which means one that imparts motion or prime mover. The dynamo was the result of the efforts of several people, in different countries in the mid-nineteenth century, to make electricity work for them. A generator is a device that changes mechanical energy into electrical energy. Although the terms AC and DC generator are alike, a generator is normally considered as a device that provides DC current. An alternator is a device that changes mechanical energy into an alternating current electrical energy, which means it is an AC generator. Here are some important dates regarding the development of motors and generators:

- ❏ 1820 The discovery of electromagnetism by Hans Christian Oersted.
- ❏ 1827 The statement of the law of electric conduction, Ohm's law, by George S. Ohm.
- ❏ 1830 The discovery of electromagnetic induction by Joseph Henry and Michael Faraday.
- ❏ 1867 The development of the first practical dynamo.

These developments took place after another important event in history. During the eighteenth century in England, Thomas Newcomen invented a steam engine that was used to pump water from coalmines. In 1769, James Watt from Scotland improved the steam engine and applied it to other uses. In order to sell the practicality of the engine, he needed to show that his engine could do the work of a horse. Through much testing, he determined a standard that is commonly used today. He found that a horse can work at a rate of 33,000 foot-pounds per minute. This is the standard for measuring horsepower. Today, electric motors are rated in horsepower and Watts (voltage multiplied by current output [1000 W = 1kW, which equals approximately 1-⅓ HP]).

Basic Motor Operation

An electric motor converts electric energy into mechanical energy. Through the years, electric motors have changed substantially in design; however, the basic operational principles have remained the same. That principle is easily observed by taking two bar magnets and placing them end-to-end with each other. If the ends have the same polarity, they will push away from each other. If the ends have the opposite polarity, they will move toward each other and form one magnet.

If we put a pivot through the center of one of the magnets to allow it to spin, and moved the other magnet toward it, the first magnet will either rotate away from the second or move toward it (Figure 3-3). This is basically how a motor works. Although we do not observe a complete rotation, we do see part of one, perhaps a half turn. If we could change the polarity of the second magnet, we would get another half turn. So in order to keep the first magnet spinning, we need to change the polarity immediately after it moves half way. If we continued to do this, we would have a motor.

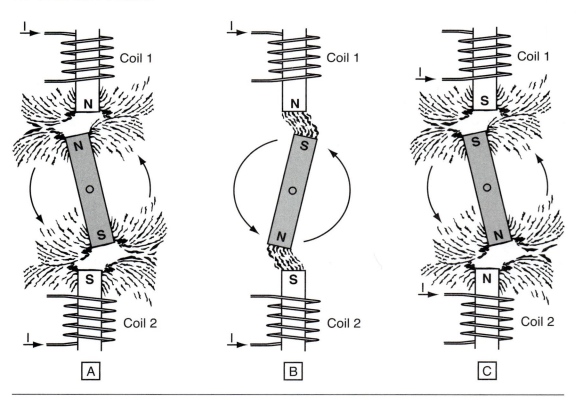

Figure 3-3 (A) Like poles repel, (B) unlike poles are attracted to each other, and if we change the polarity of the coils, (C) the like poles again repel.

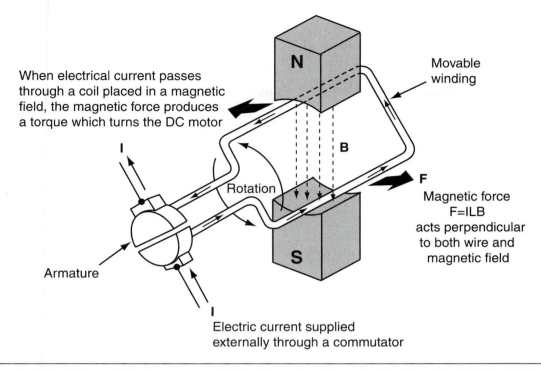

When electrical current passes through a coil placed in a magnetic field, the magnetic force produces a torque which turns the DC motor

Movable winding

Magnetic force $F=ILB$ acts perpendicular to both wire and magnetic field

Armature

Electric current supplied externally through a commutator

Figure 3-4 A simple look at what keeps a motor running.

In a real motor, an electromagnet is fitted on a shaft. The shaft is supported by bearings or bushings to allow it to spin and to keep it in the center of the motor. Surrounding, but not touching, this inner magnet is a stationary permanent magnet or an electromagnet. Actually, there are more than one magnets or magnetic fields in both components. The polarity of these magnetic fields is quickly switched and we have a constant opposition and attraction of magnetic fields. Therefore, we have a constantly rotating inner magnetic field, the shaft of which can do work due to the forces causing it to rotate (Figure 3-4). This force is called **torque**. The torque of a motor varies with rotational speed, motor design, and the amount of current draw the motor has. The speed of the rotation depends on a number of factors, such as the current draw of the motor, the design of the motor, and the load on the motor's rotating shaft.

Other basic principles must be explained before going into detail about the operation of the basic types of motors: DC and AC.

Electromagnets

When electrical current passes through a wire, a magnetic field is formed around that wire. The flux lines of the magnetic field form in concentric circles around the wire. The direction of the magnetic field can be determined by the "Left Hand Rule." This rule states that if you point the thumb of your left hand in the direction of the current flow, your fingers will point in the direction of the magnetic field (Figure 3-5). Remember, the attraction of poles is from North to South.

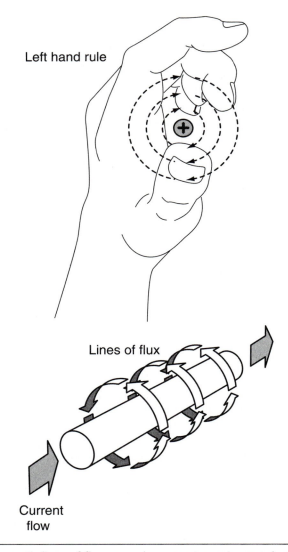

Left hand rule

Lines of flux

Current
flow

Figure 3-5 The magnetic lines of flux around a current carrying conductor leave from the north pole and reenter at the south pole.

When the wire is shaped into a coil or winding, the individual flux lines produced by each section of wire join together to form one large magnetic field around the total coil. The magnetic field around the coil can be strengthened by placing a core of iron or similar metal in the center of the coil (Figure 3-6). The iron core presents less resistance to the lines of flux than air, and the magnetic field's strength increases.

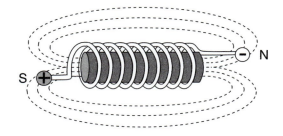

Figure 3-6 The magnetic field around the coil can be strengthened by placing a core of iron or similar metal in the center of the coil.

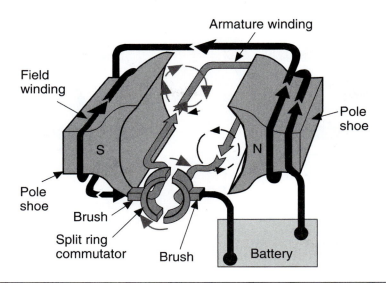

Figure 3-7 A simple DC motor.

The basic components of a motor are the stator or field windings that are the stationary part of the motor and the rotor or armature that is the rotating part (Figure 3-7). The stator is comprised of slotted cores made of thin sections of soft iron wound with insulated copper wire to form one or more pairs of magnetic poles. Some motors have the field windings wound around iron anchors, called pole shoes. The rotor is comprised of loops of current-carrying wire, or it can be a series of permanent magnets. The magnetic fields in the rotor are pushed away by the magnetic field in the stator, causing the rotor to rotate away from the stator field.

The use of an electromagnet in a motor makes it easy to change polarity in a magnetic field and keep a motor spinning. By changing the direction of current flow, the magnetic polarities are changed.

Generators

Just as electricity can be used to create a magnetic field, a magnetic field can be used to produce electricity. If a conductor is passed through the flux lines of a magnet, voltage is induced. Electrical current will result if that voltage has a path to and from a load and if the voltage is strong enough to push through the load.

The polarity of the induced voltage depends on the direction of the conductor through the magnetic field. In one direction a positive voltage is induced; in the other, a negative voltage is induced Figure 3-8). Moving the conductor back and forth through the field produces both, which is AC voltage. The amount of voltage induced increases as the strength of the magnetic field increases, the faster the field is being cut or the more lines of flux are cut, and the closer the conductor(s) and magnetic field are to right angles (perpendicular) to one another.

Faraday's Law

Faraday's law is a summary of the ways a voltage can be generated by a changing magnetic field. Michael Faraday (1791–1867) was a British physicist and chemist and is best known for his laws regarding electromagnetic induction and electrolysis.

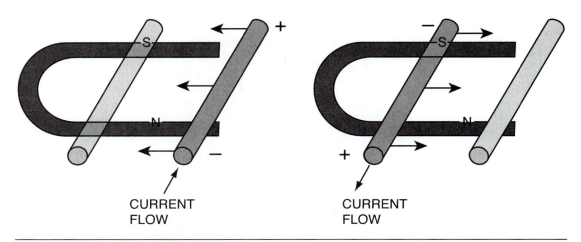

CURRENT FLOW   CURRENT FLOW

Figure 3-8 The polarity of the induced voltage depends on the direction in which the conductor moves as it cuts across the magnetic field.

Faraday's law states that, "the EMF (electromotive force [voltage]) induced between the ends of a loop or coil is proportional to the rate of change of magnetic flux enclosed by the coil; or the EMF induced between the ends of a bar conductor is proportional to the time rate at which magnetic flux is cut by the conductor." Further, he discovered that voltage can be established between the ends of a conductor by doing three things:

❏ By a conductor moving or cutting across a stationary magnetic field. (DC Generator)
❏ By a moving magnetic field cutting across a stationary conductor. (AC Generator)
❏ By a change in the number of magnetic lines enclosed by a stationary loop or coil. (Transformer or ignition coil)

Lenz's Law

In 1834, Heinrich Lenz of Germany further defined the characteristics of electromagnetic induction. Lenz's law states that when a voltage is generated by a change in magnetic flux according to Faraday's law, the polarity of the induced voltage will produce a current with a magnetic field that opposes the change that produces it.

Counter EMF (CEMF) is evidence of Lenz's law. Counter EMF is a force that opposes EMF or voltage. CEMF limits the voltage that is induced and serves as an electrical resistance to current flow in a motor. CEMF limits current in a coil of wire because as a magnetic field is formed around the loops or wire, the magnetic fields form in opposite directions on the sides of the loop. This action induces a small voltage in the wire and that voltage opposes the regular voltage of the circuit (Figure 3-9).

Current through a length of straight wire will be much greater than the current through a coiled wire of the same length. This is the result of CEMF. This occurrence affects the current draw of a motor and the output of a generator.

Both, Faraday's and Lenz's laws explain why the output of a generator is naturally limited and why the torque and output speed of a motor is also limited. During operation, a magnetic field or CEMF is created that limits both. The faster the rotor or armature rotates within the windings or stator, the more CEMF is produced. The higher CEMF limits or reduces the amount of current that can flow through the windings. The same is true with generators.

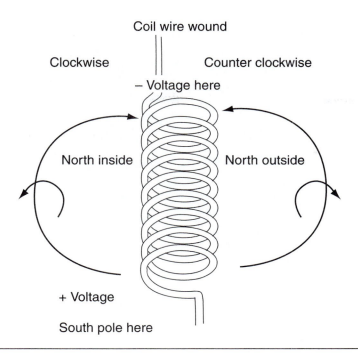

Figure 3-9 The formation of counter EMF.

DC Motors

The design of a DC motor is quite simple. It has a housing, field coils (windings), an armature, a commutator and brushes, bearings, brush supports, and end frames. The motor has a stationary magnetic field and a rotating magnetic field. The stationary field is created by permanent magnets or electromagnetic windings. When current flows through the armature (rotating) windings, a magnetic field is present around the armature windings. Because the armature windings are formed in loops or coils, current flows outward in one direction and returns in the opposite direction. Because of this, the magnetic lines of force are oriented in opposite directions in each of the two sides of the loop. When placed into the field coils, one part of the armature coil is pushed in one direction and the other is pushed in the opposite direction. This causes the coil and its shaft to rotate.

Motor Housing

The motor housing or frame encloses the internal components and protects them from damage, moisture, and foreign materials. The housing also holds the field coils The housing and end frames are usually made of steel, aluminum, or magnesium. The armature is mounted on a steel shaft supported by two bushings or bearings in the end frames. The shaft's bushings and bearings are typically lubricated by grease or oil. An end frame is a metal plate that bolts to the end of the motor housing.

Field Windings

The field windings or coils are normally made of copper wire and are insulated from but wrapped around metal plates, called pole shoes. The field coils and their pole shoes are securely attached to the inside of the housing (Figure 3-10). The field coils are designed to produce strong stationary electromagnetic fields as current is passed through them. These magnetic fields are concentrated at the pole shoes. Fields have a North or South magnetic polarity depending on the direction of current flow. The coils are wound around respective pole shoes in opposite

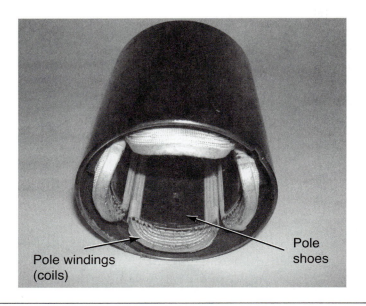

Pole windings
(coils)

Pole
shoes

Figure 3-10 The field (pole) windings are secured to the inside of the motor's housing.

directions to generate opposing magnetic fields (Figure 3-11). The field coils may be either shunt windings (in parallel with the armature winding), series windings (in series with the armature winding), or a combination of both.

Permanent Magnet Windings Often motors use permanent magnets instead of electromagnets for the field windings. Doing this eliminates the need for an electrical circuit to power the windings. Permanent magnets provide consistent magnetic flux lines and can last many years. Permanent magnet motors require special handling because the permanent magnet material is quite brittle and can be destroyed with a sharp blow or if the motor is dropped.

Armatures

The armature is made by coiling wire around two or more poles of the metal core fixed to a shaft. The armature is the only rotating component of a motor. It is located in the center of the motor housing. One factor that determines the power output of a motor is the number of loops or windings in the armature.

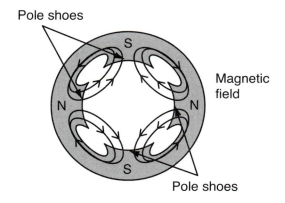

Pole shoes

Magnetic
field

Pole shoes

Figure 3-11 The coils are wound around respective pole shoes in opposite directions to generate opposing magnetic fields.

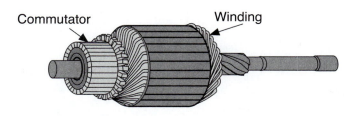

Figure 3-12 A typical armature for a DC motor.

The armature is made up of two main components: the armature windings and the commutator (Figure 3-12). The armature windings are usually made of heavy round or rectangular copper strips to handle high current and are shaped into a single loop. The ends of the loops fit into slots in the armature core or shaft, but they are insulated from it. Between the windings are metal strips that increase the strength of the magnetic field.

The commutator is attached to one end of the shaft and provides a contact for the ends of each loop. The coils connect to each other and to the commutator so that electrical current can flow through all of the armature windings at the same time. The windings of the armature begin and end at the commutator. The commutator has slots to receive the ends, as well as a surface for the brushes to ride on. The number and size of slots depend upon the design and purpose of the motor. Normally, there are as many coils as there are slots. This means that each slot holds two coil ends, one side of each coil being at the top of a slot and the other at the bottom of a slot. Each coil may consist of one or more turns depending on the applied voltage of the motor.

Commutator

In order to keep the armature of a motor rotating, the polarity of its magnetic field must change. To do this, the armature of DC motors is fitted with a commutator. The commutator has plates connected to each of the armature loops. Electrical current enters and leaves the armature through a set of brushes that slide over the commutator's plates or segments. As the brushes pass over one segment of the commutator to another, current flow through a loop of the armature is established. As the armature rotates, the brushes connect with a different segment of the commutator and a magnetic field is set up in another loop or the field in the previously excited loop is reversed.

As the armature rotates due to magnetic influences, the commutator segment attached to each coil end has traveled past one brush and is now in contact with the other. As one segment rotates past the brushes, another segment immediately takes its place. The turning motion is made uniform and the torque needed to rotate the armature is constant rather than fluctuating as it would be if only a few armature coils were used.

The segments of the commutator are made of copper and are insulated from each other and the armature shaft. The segments connect to the ends of the armature windings. The brushes are graphite-copper or carbon contacts that ride on the commutator. Most motors have two to six brushes that carry the current flow from the field windings or a power source to the armature windings.

Brush holders are used to keep the brushes in position around the commutator. These holders may be separate metal or plastic assemblies or are part of the end frame of the motor. Springs are used to keep the brushes against the segments of the commutator with the correct pressure (Figure 3-13). The brush holders for the positive brushes are insulated from the housing and end frame of the motor. The negative brush holders may be grounded directly to the housing or end frame.

Figure 3-13 Brushes are placed in holders and have springs so they can keep good contact with the commutator.

Field Winding Designs

The number of armature coils and brushes vary with the design and purpose of the motor, as does how the armature and fields coils are wired together (Figure 3-14). The armature may be wired in series with the coils (series motor); the coils may be wired parallel or shunted across the armature (shunt motors); or a combination of series and shunt wiring may be used (compound motors). In addition to the field winding designs, permanent magnet fields are used on some motors (Figure 3-15). Also, some motors have separate power sources for the field windings and the armature. The windings of these motors are referred to as "separately excited windings."

Series Windings When the windings are connected in series with the armature, all current that passes through the field windings also passes through the armature windings. This type winding allows the motor to develop maximum torque output at the time of initial start and when it is overloaded. As the speed of the motor increases, however, its torque output decreases. This is due to the counter-EMF (CEMF) created by self-induction. The speed of a series wound motor changes greatly with a change in load. A series motor is often used in applications where there are heavy starting loads.

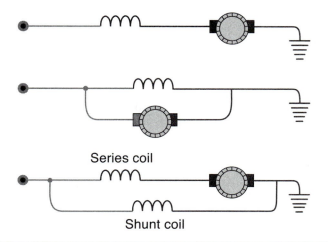

Series coil

Shunt coil

Figure 3-14 The various ways the field windings may be connected to the brushes in DC motors.

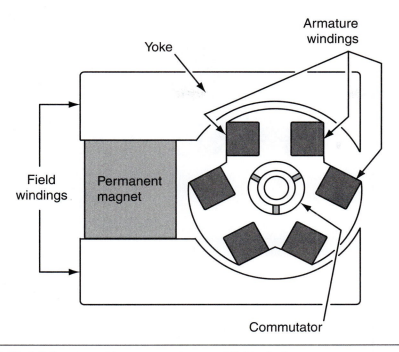

Figure 3-15 A DC motor with permanent magnet field windings.

Shunt Windings Shunt motors have field windings wired in parallel with the armature. A shunt winding usually consists of a large number of turns of thin wire, but fewer numbers of turns than the series winding. A shunt motor does not decrease in its torque output as speeds increase. This is because the CEMF produced in the armature does not decrease the field coil strength. Shunt motors develop considerably less start-up torque but maintain a constant speed at all operating loads.

Compound Windings This type of winding has a shunt winding and a series winding. This type of winding has good starting and constant speed torque. The compound design is used where heavy loads are suddenly applied. In a compound motor, some of the field coils are connected to the armature in series, and the rest are connected in parallel with the battery and the armature.

Work

By controlling the voltage to the armature, the speed of a DC motor is controlled. The higher the voltage to the armature, the faster it will rotate. Likewise, the torque output of a DC motor is controlled by the current to the windings and armature.

Counter EMF also affects the torque output of a motor. It limits the current flow based on the load on the motor. When the load is increased, the rotation of the armature will slow down. This drop in rotational speed causes a decrease in CEMF, which allows for an increase in current flow. As a result, the motor turns with more torque. The reverse is also true; if the load on the motor is decreased, the armature speed increases. More CEMF is then produced and less current flows through the motor. There is less torque from the motor because less is needed.

A motor that has a constant speed, regardless of load is called a **synchronous motor**. Although these are often shunt wound DC motors, they can also be brushless DC or AC motors.

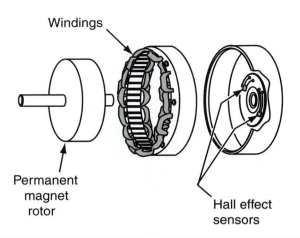

Windings

Permanent
magnet
rotor

Hall effect
sensors

Figure 3-16 The main components of a brushless DC motor.

Brushless DC Motors

A brushless DC motor is like a brushed DC motor, but the purpose of the rotor and field windings (stator) are reversed. The rotor is made up of a set of permanent magnets and the stator has controllable electromagnets (Figure 3-16). Obviously, a brushless motor has no brushes and no commutator. The electrical arcing that takes place between the brushes and commutator is also eliminated with the brushless design. This arcing not only decreased the usable life of the motor but also created electromagnetic interference that is detrimental to advanced electronic systems.

Rather than depend on brushes, an electronic circuit switches current flow to the different stator's windings when needed to keep the rotor turning. The reversing of current flow through the windings is done by power transistors that switch according to the position of the rotor. Many brushless DC motors use Hall effect sensors to monitor the position of the rotor. Other brushless motors monitor the CEMF in unexcited field windings to determine the position of the rotor. The current to the various stator windings is typically controlled by a **pulse width modulation (PWM)** frequency inverter. The voltage to the windings is changed by altering the duty cycle.

The **duty cycle** of something is the length of time the device is turned on compared to the time it is off (Figure 3-17). Duty cycle can be expressed as a ratio or as a percentage. By quickly opening and closing the power circuit to the motor, the speed of the motor is controlled. This is called pulse width modulation. These power pulses vary in duration to change the speed of the motor. The longer the pulses, the faster the motor turns.

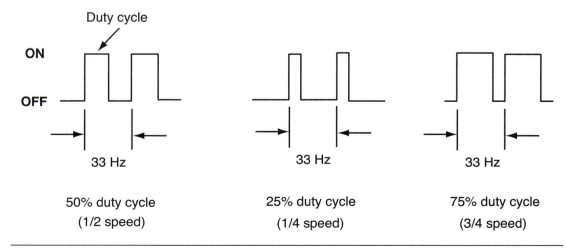

Figure 3-17 The action of various duty cycles.

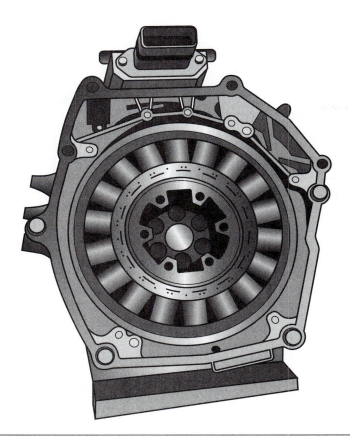

Figure 3-18 Honda's hybrid vehicles use a brushless motor.

Brushless DC motors, when compared to brushed DC motors, are more reliable, more powerful, and more expensive. The expense is largely due to the cost of the electronic controls. High-output brushless DC motors are used in some electric drive vehicles (Figure 3-18).

AC Motors

With AC voltage, the direction of current flow changes but does not change immediately. Rather, as the current is getting ready to change directions, it decreases until it reaches zero and then gradually builds up in the other direction. Therefore, the amount of current in an AC circuit always varies. When AC is given as a value, it is an average rating, not peak current. This average rating is referred to as a "root mean square" value (Figure 3-19).

Basic construction

An AC motor has two basic electrical parts: a stator and a rotor (Figure 3-20). The stator, the stationary field, is comprised of individual electromagnets electrically connected to each other or connected in groups. The rotor is the rotating magnetic field and can be an electromagnet or a permanent magnet. The rotor is located within the stator fields. Like in a DC motor, the rotor will rotate as a result of the repulsion and attraction of the magnetic poles. The way this works is quite different from how a DC motor works.

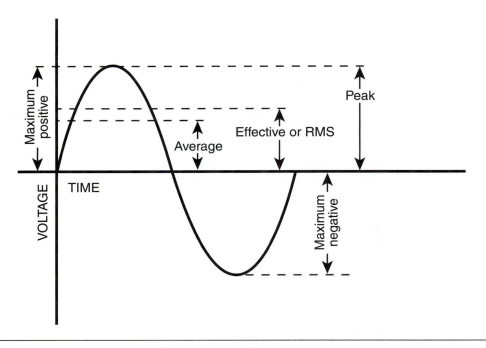

Figure 3-19 An explanation of how AC is rated.

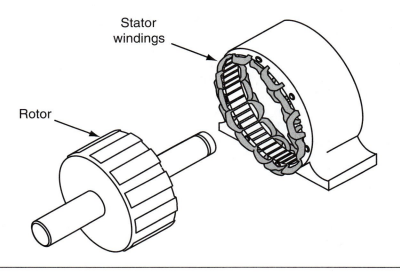

Figure 3-20 The basic construction of an AC motor.

Basic operation

A current is passed through the stator and rotor, causing the rotor to spin. Because the current is alternating, the polarity in the windings constantly changes. A synchronous AC motor will run at the frequency of the AC voltage. Many AC motors are induction types. In these motors, electrical current is induced in the rotor as it rotates, rather than having current delivered to it from an external source. Obviously, this type of motor needs to begin spinning before the rotor induces current, so they are equipped with a variety of starting aids.

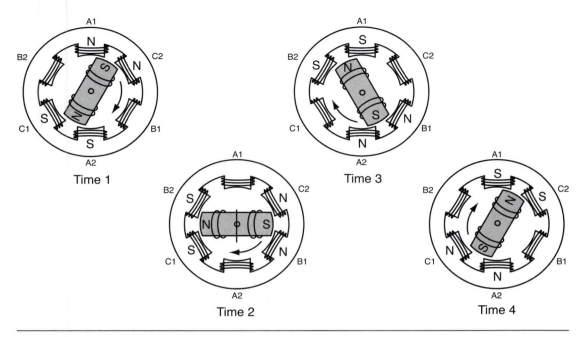

Figure 3-21 Notice how the polarity of the stator and rotor changes over time.

The rotor in an AC motor rotates because it is pulled along by a rotating magnetic field in the stator. The stator does not physically move. The magnetic field does. If the windings of the stator are wired in series, current passes through them one at a time and because it is AC, the polarity and strength of the field around them is constantly changing (Figure 3-21). The magnetic field of the rotor reacts and moves along with the "rotating magnetic field" of the stator.

To better understand this concept, let us look at a three-phase motor. Three-phase AC voltage is commonly used in motors because it provides a smoother and more constant supply of power. Three-phase AC voltage is much like having three independent AC power sources, which have the same amplitude and frequency but are 120 degrees out of phase with each other (Figure 3-22).

To produce a rotating magnetic field in the stator of a three-phase AC motor, each phase of the three-phase power source is connected to separate stator windings. Because each phase reaches its peak at successively later times, the magnetic field is at its strongest point in each winding in succession, as well. This creates the effect of the magnetic field continually moving around the stator. This rotating magnetic field will rotate around the stator once for every cycle of the voltage in each phase (Figure 3-23). This means the field is rotating at the frequency of the source voltage. Remember that as the magnetic field moves, new magnetic polarities are present. As each polarity change is made, the poles of the rotor are attracted by the opposite poles on the stator. Therefore, as the magnetic field of the stator rotates, the rotor rotates with it.

In most cases, the speed of an AC motor depends on:

1. The number of windings and poles built into the motor.
2. The frequency of the AC supply voltage. Controllers are used to change this frequency and allow for a change in motor speed.
3. The load on the rotor's shaft.

Synchronous Motor

A synchronous motor operates at a constant speed regardless of load. Rotor speed is equal to the speed of the stator's rotating magnetic field. A synchronous motor is used when the exact speed of a motor must be maintained. Often, synchronous motors have magnets built into the rotor assembly. These magnets allow the rotor to easily align itself with the rotating magnetic field of the stator.

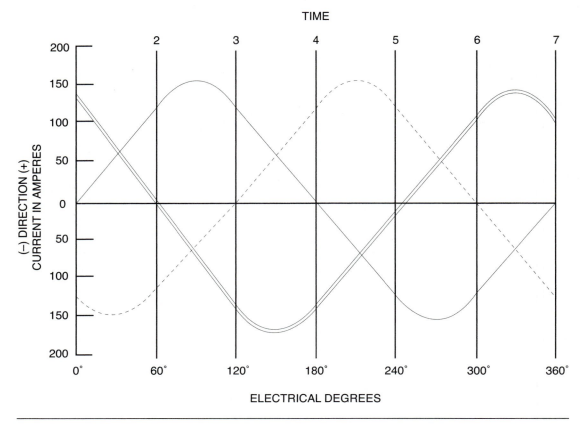

Figure 3-22 A look at the three separate phases of the three-phase AC.

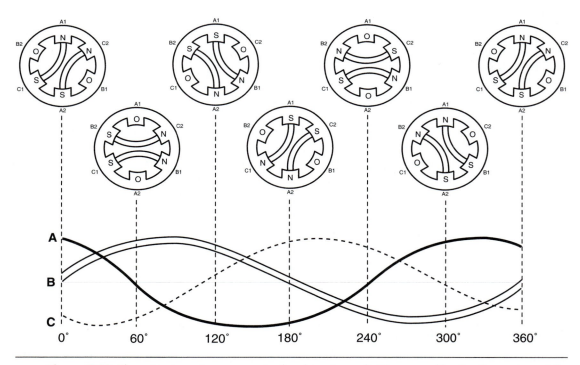

Figure 3-23 The rotor turns in response to the changing polarities caused by the three-phase AC.

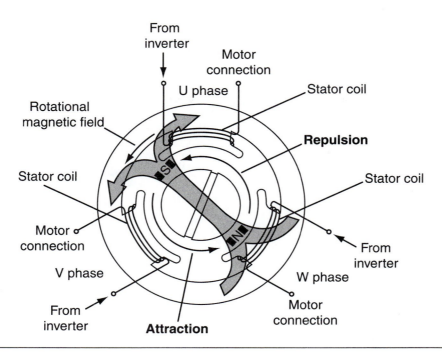

Figure 3-24 The synchronous motor used in Toyota's hybrid vehicles is controlled by the vehicle's inverter, which is part of the hybrid control system.

When three-phase AC is fed to the three sets of windings in the stator coil, a rotating magnetic field is present around the stator. The rotor simply rotates with that rotating magnetic field. The torque output of the rotor, therefore, is dependent on the strength of the magnetic field around the stator. The speed of the rotor is determined by the frequency of the AC input to the stator. Synchronous motors are available with outputs up to thousands of horsepower.

One of the disadvantages of most synchronous motors is they cannot be started by merely applying three-phase AC power to the stator. When AC is applied to the stator, a high-speed rotating magnetic field is present immediately. This field rushes past the rotor so quickly that the rotor cannot get started. The rotor is first repelled in one direction and then, very quickly, in another. There many ways of addressing this issue, but for hybrid and electric vehicles the problem is solved by complex electronics that begin rotating magnetic field in such a way and at such speed that the rotor simply follows the field (Figure 3-24). Once the rotor is spinning, normal synchronous operation begins.

On some industrial motors, a squirrel-type winding (Figure 3-25) is added to the rotor to get the motor going. The windings are heavy copper bars connected by copper rings. When a

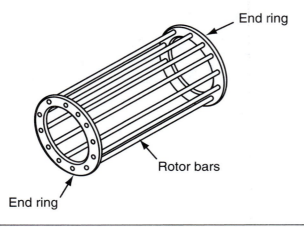

Figure 3-25 A squirrel cage type rotor.

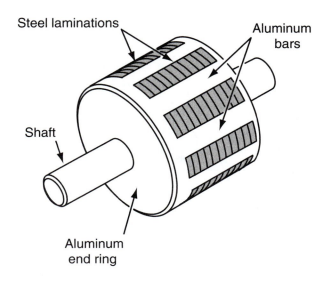

Steel laminations

Aluminum bars

Shaft

Aluminum end ring

Figure 3-26 The construction of a rotor for an AC induction motor.

low voltage is induced in these shorted windings by the rotating stator field, a relatively large current flows in the squirrel cage. This forms a magnetic field that gets the rotor spinning. Because it follows the rotating magnetic field of the stator, the motor starts. Once the rotor spins, the current to the squirrel cage is stopped.

Induction motor

The most common industrial motor is the three-phase AC induction motor. This motor has a low cost and a simple design. The stator is connected to the power source and the rotor (Figure 3-26). The three-phase AC sets up a rotating magnetic field around the stator. The rotor is not connected to an external source of voltage.

An induction motor generates its own rotor current. The current is induced in the rotor windings as the rotor cuts through the magnetic flux lines of the rotating stator field (Figure 3-27). The induced current causes each rotor conductor to act like the permanent magnet. As the magnetic field of the stator rotates, the magnetic field of the rotor follows the rotating magnetic field of the stator. It should be obvious that this type of motor needs some rotation of the rotor before it can rotate on its own. Various methods are used to start these motors, including capacitors and separate starting windings.

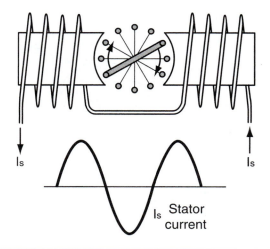

I_s

I_s

I_s Stator current

Figure 3-27 This is a look at a simple induction motor during rotation. The direction of the current and the voltage induced in the rotor are shown.

It is impossible for the rotor of an induction motor to rotate at synchronous speed. If the rotor were to turn at the same speed as the rotating field, there would be no relative motion between the stator and rotor fields. As a result, no lines of force would be cut by the rotor's conductors, and there would be no induced voltage in the rotor. In an induction motor, the rotor must rotate at a speed slower than that of the rotating magnetic field.

The difference between the synchronous speed and actual rotor speed is called **slip**. Slip is directly proportional to the load on the motor. When loads are on the rotor's shaft, the rotor tends to slow and slip increases. The slip then induces more current in the rotor and the rotor turns with more torque, but at a slower speed and therefore produces less CEMF.

Switched Reluctance Motors

A variable switched reluctance motor can be powered by AC or DC. Like other motors, it has a rotor and a coil winding in the stator. The toothed rotor has no coil windings or permanent magnets. The stator typically has slots containing a series of coil windings. The energizing of the stator is done by an electronic controller. The controller establishes a rotating magnetic field around the stator as it activates one coil set in the stator at one time. The timing of this activation is based on rotor angle; therefore, sensors monitoring the position of the rotor are used.

When one coil winding is energized, a magnetic field is formed around it. The metal rotor tooth that is closest to the magnetic field moves toward that field. When the tooth is close, current is switched to another winding in the stator and the tooth moves to it. As the current is sent to the consecutively placed windings, the rotor rotates. By controlling the current and timing through the stator windings, the rotor can be forced to rotate at any desired speed and torque.

Generators

A generator is similar to a motor. However, a motor changes electrical energy to mechanical energy, whereas a generator changes mechanical energy to electrical energy. To generate voltage, a conductor is moved through a magnetic field or a magnetic field is moved over a conductor. The conductor or field is moved by mechanical energy.

DC Generators

A DC generator provides direct current (DC). The biggest difference between a generator and a motor is the wiring to the armature. In a motor, the armature receives current from a power source. This creates the magnetic field that opposes the magnetic fields in the motor's coils, which causes the armature to rotate. The armature in a DC generator is driven by the engine or other device. It is not magnetized and the windings simply rotate through the stationary magnetic field of the field windings. This induces an AC voltage in the armature.

To provide DC voltage, it is necessary to reverse the polarity of the generator's output wires at the same time the voltage in the armature is reversed. This is accomplished by the commutator. The commutator has segments and brushes ride over those segments, just like a brushed DC motor. There is always at least one positive brush and at least one negative brush. Wires deliver the DC output from the armature to an external electrical system (Figure 3-28).

The voltage output from a DC generator is pulsed (Figure 3-29). This is caused by the segmentation of the commutator.

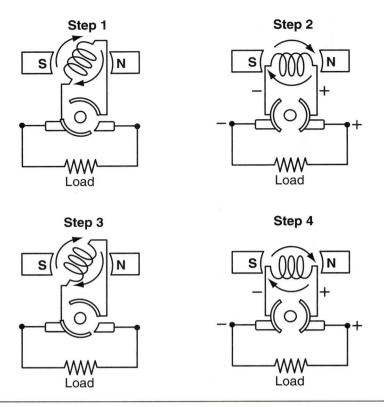

Figure 3-28 The basic operation of a simple DC generator.

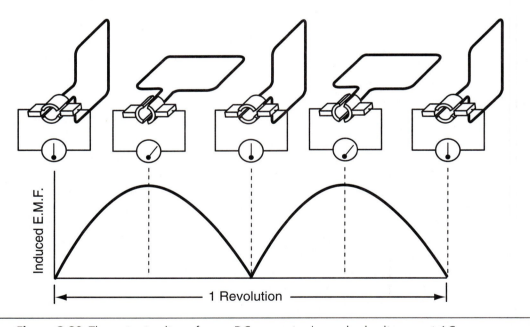

Figure 3-29 The output voltage from a DC generator is a pulsed voltage, not AC.

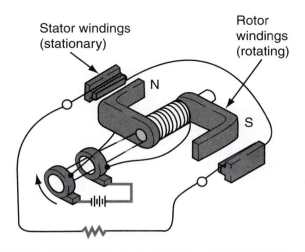

Figure 3-30 A simplified AC generator.

AC Generators

AC generators (Figure 3-30) use a design that is basically the reverse of a DC generator. In an AC generator, a spinning magnetic field rotates inside the stator. As the spinning north and south poles of the magnetic field pass the conductors in the stator, they induce a voltage that first flows in one direction and then in the opposite direction (AC voltage).

The rotor assembly may have electromagnets or permanent magnets. A small air gap separates the rotor and stator. The rotor's magnetic field induces a voltage in all of the stator windings at the same time. Therefore, the generation of AC can be quite high, if needed. The output can be controlled by controlling the current flow through the rotor. Of course, this cannot be done if the rotor is a permanent magnet. If the rotor is an electromagnet, it has slip rings. The slip rings are not segmented and provide an uninterrupted surface for the brushes. It is through the slip rings and brushes that the rotor is energized (Figure 3-31).

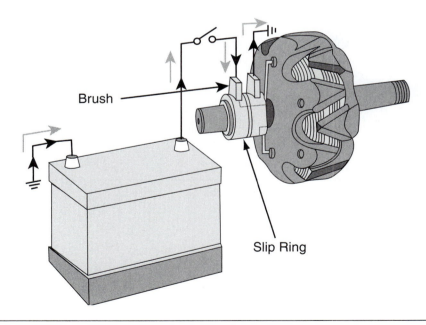

Figure 3-31 Current is carried to the rotor's windings through brushes in contact with the slip rings on the rotor shaft.

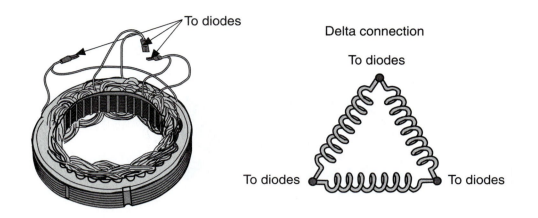

Figure 3-32 A delta-connected stator winding.

The **stator** is the stationary member of the AC generator. It is made up of a number of conductors, or wires, into which voltage is induced by the rotating magnetic field. Most AC generators use three windings to generate the required output. They are placed in slightly different positions so their electrical pulses are staggered in either a **delta** configuration (Figure 3-32) or a **wye configuration** (Figure 3-33). The delta winding received its name because its shape resembles the Greek letter delta. The wye winding resembles the letter Y. Usually, a wye winding is used in applications where high charging voltages at low engine speeds is required. AC generators with delta windings are capable of putting out higher amperages.

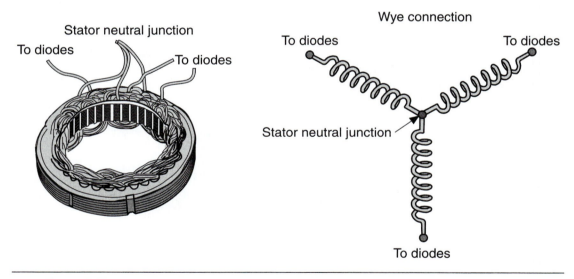

Figure 3-33 A wye-connected stator winding.

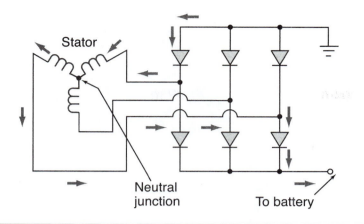

Figure 3-34 Diodes are connected between the output of the windings and the output of the AC generator.

To store the generated AC voltage, some component must convert the AC to DC before it is sent to the batteries. The generated AC can be used to directly power AC equipment and then converted to DC for storage. The AC to DC conversion can be simply done through an arrangement of diodes that are placed between the output of the windings and the output of the AC generator (Figure 3-34).

Alternating current produces a positive pulse and then a negative pulse. The resultant waveform is known as a **sine wave**. The complete waveform starts at zero, goes positive, then drops back to zero before turning negative. Remember there are three overlapped stator windings. This produces overlapping sine waves. This voltage, because it was produced by three windings, is called **three-phase voltage**.

A diode is an electronic device that allows current to flow only in one direction; therefore, if AC runs through a positive diode, the negative pulses are blocked off. If the diode is reversed, it blocks off current during the positive pulse and allows the negative pulse to flow. When only half of the AC current pulses (either the positive or the negative) are able to pass through the diodes, **half-wave rectification** has taken place. When all of the AC is rectified, **full-wave rectification** occurs.

A voltage regulator controls the voltage output of an AC generator. Regulation of voltage occurs by varying the amount of field current flowing through the rotor. The higher the field current, the higher the voltage output. The regulator receives an output signal from the generator, which is called the **sensing voltage**. If the sensing voltage is below the regulator setting, an increase in field current is allowed. Higher sensing voltage will result in a decrease in field current and voltage output.

Pulse width modulation can be used to control the generator's output by varying the amount of time the rotor is energized. When voltage regulation is controlled by a computer or electronic control module, the computer switches or pulses field current at a fixed frequency. By varying on-off times, the correct average field current is produced to provide correct AC generator output.

Motor/Generators

The main difference between a generator and a motor is that a motor has two magnetic fields that oppose each other, whereas a generator has one magnetic field and wires are moved through the field. Using electronics to control the current to and from the battery, engineers have developed a motor that also works as a generator. A motor/generator may be based on two sets of windings and brushes, a brushless design with a permanent magnet, or switched reluctance.

Figure 3-35 A motor/generator (alternator) assembly that is mounted external to the engine.

A motor/generator (Figure 3-35) can be mounted externally to the engine and connected to the crankshaft with a drive belt. In these applications, the unit can function as the engine's starter motor, as well as a generator driven by the engine. Some allow for some regenerative braking as well.

When belt-driven, a belt tensioner is mechanically or electrically controlled to allow the motor/generator to drive or be driven by the belt. One system has an electromagnetic clutch fitted to the crankshaft pulley. With the engine running, the clutch is engaged. This allows the unit to be a generator. When the vehicle is stopped, the crank pulley clutch disengages, and the unit is ready to act as the starting motor when the vehicle is ready to move again.

Motor/generators may also be mounted directly on the crankshaft between the engine and the transmission or integrated into the flywheel. This design works as a starter by spinning the crankshaft during starting and serves as a generator, charging both directly from the engine and during braking (regenerative braking). Many hybrid vehicles use this design. In some cases, the motor/generator may be part of a drive axle assembly. These designs can drive the axles as well as serve as generators during regenerative braking.

Motor/generators are capable of high charging outputs and allow for other features that make the vehicle more efficient:

❑ Stop-start—When the engine is not needed, such as when sitting at a stoplight, it is automatically turned off. The engine restarts as soon as the control module senses a need for engine power.

❑ Electrical assist—The motor/generator can add power to the engine during initial and hard acceleration or when the vehicle is operating under a heavy load.
❑ Regenerative braking—This feature captures some the energy during deceleration and braking and uses it to recharge the vehicle's batteries. Regenerative braking also helps the conventional brakes to slow down and stop the vehicle.

Controllers

A **controller** is used to manage the flow of electricity from the batteries and thereby controls the speed of the electric motor. A sensor located by or connected to the throttle pedal (also called gas pedal or the accelerator) sends the driver's input to the controller. The controller then sends the appropriate amount of voltage to the motor.

A simple controller is a variable resistor or potentiometer connected to the accelerator. When there is no pressure on the accelerator, resistance of the potentiometer is too high to allow voltage to the motor. When the accelerator is fully opened, the resistance is very low and full battery voltage is delivered to the motor. The various positions of the accelerator between closed and wide open allow corresponding amounts of voltage to the motor. This type of system does not provide for smooth control of the motor.

Using electronics, the same principle can be used but with more positive results. A sensor monitors the accelerator and sends information to a control unit. The control unit monitors that signal plus other inputs regarding the operating conditions of the vehicle. The control unit then duty cycles the voltage to the motor. In this way, the voltage is pulsed and more precise motor speed control is possible. Most controllers pulse the voltage more than 15,000 times per second. Pulsing the voltage causes the motor to vibrate at the frequency of the voltage. If the frequency is faster than 15,000 cycles per second, it cannot be heard.

A controller has an additional role when the electric drive motor is an AC-type motor. It is responsible for converting the DC voltage from the battery to a three-phase AC voltage for the motor. This is done by sets of power transistors. The transistors pulse the voltage and reverse it at a fixed frequency. The voltage may also be increased (boosted) before it is sent to the motor (Figure 3-36).

Inverter

An inverter may be part of the controller or be a separate unit. The inverter (Figure 3-37) converts the DC voltage from the batteries into three-phase AC voltage for the motor. The inverter may also convert the AC voltage generated during regenerative braking or by the generator to DC to charge the batteries. Built into the inverter is a DC-to-DC converter that drops some of the high DC voltage to the low voltage required to recharge the 12-volt battery that is used to power accessories such as sound systems, lights, blower fans, and the controller.

The inverter also sends information to the controller. These inputs include generator voltage and current outputs and are used to switch the transistors in the controller.

The inverter may also include an air conditioning inverter. This inverter changes DC battery voltage to AC voltage, but that voltage is not boosted, as is the AC for the motor. This AC voltage is used to operate the air conditioning system's compressor.

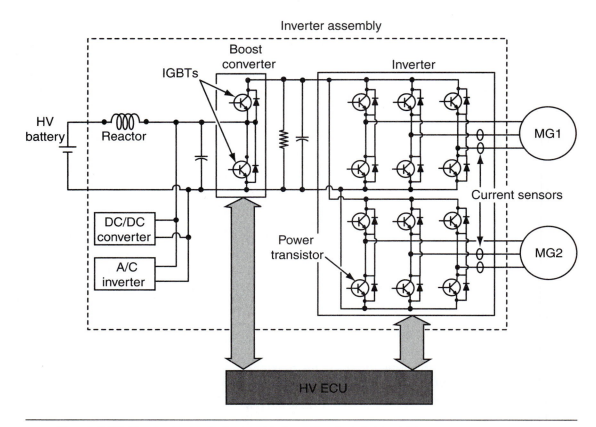

Figure 3-36 The motor control assembly for a Toyota hybrid vehicle.

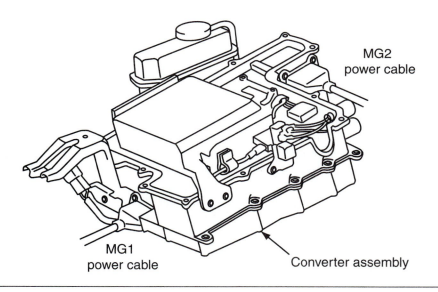

Figure 3-37 The inverter assembly for a Toyota hybrid vehicle.

Review Questions

1. What is a synchronous motor?
2. If a permanent magnet is used in a DC motor, where is it most likely to be and why?
3. Describe the basic difference between a DC motor and a DC generator.
4. All of these statements describe how voltage can be generated, EXCEPT:
   A. By a change in the number of magnetic lines enclosed by a stationary loop or coil.
   B. By a change in current flow through a conductor.
   C. By a conductor moving or cutting across a stationary magnetic field.
   D. By a moving magnetic field cutting across a stationary conductor.
5. An induction motor has all of the following parts, EXCEPT:
   A. Stator windings.
   B. Rotor.
   C. Rotor windings.
   D. Permanent magnet.
6. True or False. The strength of the magnetic poles in an electromagnet decreases with an increase in the number of turns of wire and the current flowing through them.
7. Why is three-phase AC used for many motors and what is it?
8. List five functions of a typical controller/inverter for an electric drive vehicle.
9. All of these factors influence the output speed of an AC motor, EXCEPT:
   A. The number of brushes on the commutator.
   B. The number of windings and poles built into the motor.
   C. The frequency of the AC supply voltage.
   D. The load on the rotor's shaft.
10. Briefly explain the differences between a brushed and a brushless DC motor.

BATTERY BASICS

After reading and studying this chapter, you should be able to:

❏ Explain the purpose of a battery.

❏ Describe how a battery works.

❏ Describe the basic construction of an electrochemical cell.

❏ Explain how electrochemical cells can be connected together to increase voltage and current.

❏ Explain the different methods used to recharge a battery.

❏ List and describe the various ways a battery may be rated.

❏ List and describe the various types of batteries, according to their chemistries, that may be used in automobiles.

❏ Explain why electric drive vehicles operate with high voltages.

❏ List the precautions that must be adhered to when working with or around high-voltage systems.

❏ Describe the construction and operation of a lead-acid battery.

❏ Describe the various types of lead-acid batteries that are available today.

❏ Describe the basic service procedures for maintaining a lead-acid battery.

❏ Explain how a lead-acid battery should be tested.

❏ Describe the safety precautions that should be followed when working with lead-acid batteries.

❏ Describe the construction and operation of a nickel-metal hydride battery.

❏ Describe the construction and operation of a nickel-cadmium battery.

❏ Describe the construction and operation of a lithium-ion battery.

❏ Describe the construction and operation of a lithium-ion polymer battery.

❏ Explain how a capacitor stores electrical energy.

❏ Describe the construction and operation of an ultra-capacitor.

Key Terms

Absorbed glass mat (AGM)
Ampere-hour (AH) rating
Anode
Batteries
Cathode
Capacitance
Capacitor
Cold cranking amps (CCA Rating)
Conductance test
Cylindrical cell
Deep cycling
Electrochemical
Electrolyte
Element
Farad (F)

Introduction

The battery in an electric drive vehicle serves a different purpose than a battery in a conventional vehicle. In an ICE-powered vehicle, the battery's primary purpose is to provide a short powerful burst of power to start the engine. This type of battery is typically called a **starting battery**. In electric drive vehicles, however, the batteries provide continuous current to power electric motors for a long period of time.

A starting battery is also found in hybrids, in addition to a **high voltage (HV) battery** (Figure 4-1). The starting battery is used to start the engine, whereas the HV battery is used to power the electric motors. The HV system may also be used to power some accessories. In these vehicles, the starting battery is also used as the power source for normal accessories, such as lights.

Obviously, in a battery electric vehicle (BEV) and a fuel cell electric vehicle (FCEV), there is no need for a starting battery. These vehicles have drive or traction batteries constructed in the same way as those used in hybrids. The primary downfall of the BEV was short driving range and long recharge times. Actually, the range limitations of BEVs was due to battery technology. Perhaps one day batteries will be developed to make BEVs a practical alternative to ICE-powered vehicles.

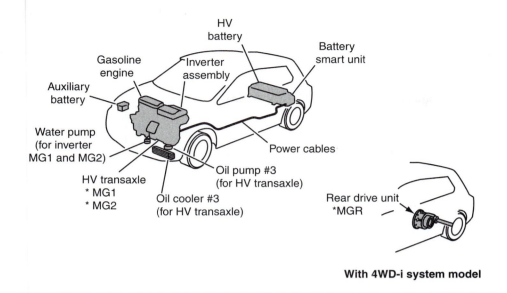

With 4WD-i system model

Figure 4-1 The layout of a hybrid vehicle showing the location of the auxiliary and HV (hybrid) batteries, as well as the other major components.

Electric drive vehicles not only need high-power, high-voltage batteries, they also need batteries that can be totally discharged and recharged often. This is a requirement called **deep cycling**, and a battery designed to do this is called a deep cycle battery. These batteries tend to have less instant power than a starting battery but can deliver electrical energy for longer periods of time, as well as go through many deep cycles.

Electric drive vehicles typically run on 100 to 600 volts. The batteries may be 6- or 12-volt batteries connected in series. If the electric motor requires 240 volts, the vehicle needs forty 6-volt batteries or twenty 12-volt batteries. In many cases, hundreds of individual battery cells, each about the size of a flashlight battery, are connected together to store and provide the needed power. Many different types of batteries are available and under development to exceed the needs of electric drive vehicles. In addition to batteries, some electric drive vehicles are equipped with ultra-capacitors, which also store and provide electrical energy. Each of these energy-storing devices will be discussed in this chapter.

Basic Battery Theory

Electrical current is caused by the movement of electrons from something negative to something positive. The strength of the attraction of the electrons (negative) to the protons (positive) determines the amount of voltage present. When a path is not present for the electrons to travel through, voltage is still present but there is no current flow. When there is a path, the electrons move and there is current. This is the basic operation of batteries.

Construction

A battery stores DC voltage and releases it when it is connected to a circuit. Inside the battery are two different types of electrodes or plates. One of the plates has an abundance of electrons (negative plate) and the other has a lack of electrons (positive plate). The electrons want to move to the positive plate and do so when a circuit connects the two plates. Batteries have two terminals, a positive that is connected to the positive plate and a negative connected to the negative plate.

The plates are surrounded by an **electrolyte**. Electrolytes are chemical solutions that react with the metals used to construct the plates. These chemical reactions cause a lack of electrons on the positive electrode and an excess on the negative electrode. When connected

Key Terms (continued)

Fast charging
Grids
High voltage (HV) battery
Hydrometer
Lithium ion (Li-Ion)
Lithium polymer (Li-Poly or LiPo) battery
Load test
Low-maintenance battery
Maintenance-free battery
Nickel-Cadmium (NiCad)
Nickel Metal Hydride (NMH)
Open circuit voltage
Operating voltage
Parasitic drain
Primary batteries
Prismatic cells
Recombinant battery
Reserve capacity (RC) rating
Secondary batteries
Slow charging
Specific gravity
Starting battery
Trickle charging
Watt-hour rating

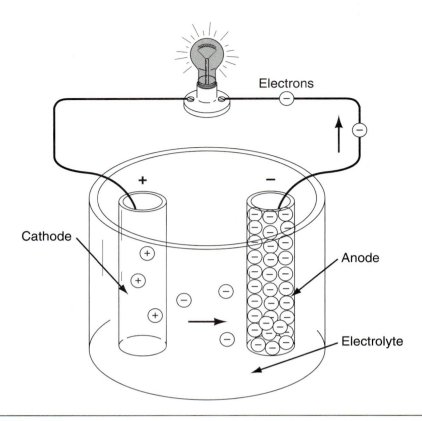

Electrons

Cathode

Anode

Electrolyte

Figure 4-2 The basic components of an electrochemical battery cell. The positive electrode is the cathode and the negative one is the anode. Both are placed in an electrolyte.

into a circuit, the electrons move out of the negative terminal and the chemical reactions begin. The reactions continue to provide electrons for current flow until the circuit is opened or the chemicals inside the battery become weak. At that time, either the battery has run out of electrons or all of the protons are matched with an electron. Recharging simply moves the electrons that moved to the positive electrode back to the negative electrode.

Batteries are devices that convert chemical energy into electrical energy. Chemical reactions that produce electrons are called **electrochemical** reactions. Batteries are normally made up of electrochemical cells connected together. Each electrochemical cell has three major parts: an **anode** (negative electrode), a **cathode** (positive electrode), and electrolyte (Figure 4-2).

Effects of Temperature

Temperature affects the chemical reaction in batteries. For all battery designs, there is an ideal temperature range. Discharging or charging a battery outside its ideal temperature range shortens the life of the battery, reduces its ability to supply power, and can create an unsafe condition.

Battery Hardware

In order to connect the battery to the vehicle's electrical system, battery cables are used. They must safely handle the voltage and current demands of the vehicle. Battery hold-downs are used to prevent damage to the battery and heat shields are sometimes used to keep battery temperatures down. Most high-voltage battery packs are enclosed in a box that serves to secure the pack and to keep it cool.

Battery Cables Battery cables must be able to carry the current required to meet all demands. The normal 12-volt cable size is 4 or 6 gauge. Various forms of clamps and terminals are used to

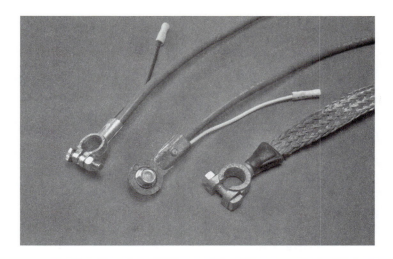

Figure 4-3 A variety of battery cable and connectors.

assure a good electrical connection at each end of the cable (Figure 4-3). Connections must be clean and tight to prevent arcing and corrosion. The positive cable is normally red and the negative cable is black.

High-voltage systems typically have their battery cables marked or colored in such a way that they can be easily identified. The high-voltage cables in nearly all hybrid vehicles are colored orange and have markings on them. Sometimes the cables are enclosed in an orange casing, again to identify them. It is important to remember that some hybrids power other accessories with high-voltage; these cables are orange just like the battery cables (Figure 4-4).

Battery Hold-Downs All batteries must be held securely in the vehicle to prevent damage to the battery and to prevent the terminals from shorting to the vehicle. Battery hold-downs are made of metal or plastic (Figure 4-5).

The performance and durability of a battery, especially high-voltage battery packs, is heavily dependent on maintaining desired temperatures. Batteries may be housed in a box or container with a cooling fan. The box not only secures the batteries, but also serves as a conduit for

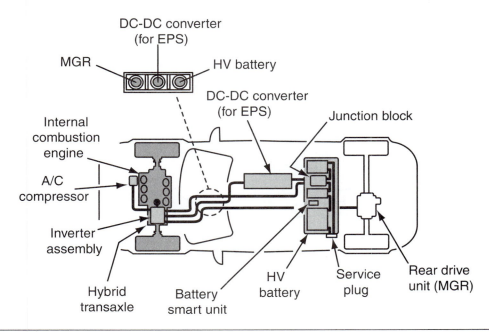

Figure 4-4 The main power cables in a hybrid vehicle.

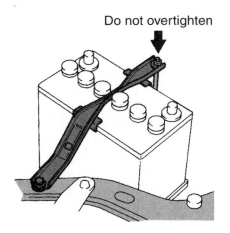

Figure 4-5 Battery hold-downs should be secure, but do not over tighten them as they can damage the battery's case.

the air from the fan (Figure 4-6). Some battery's designs work best when they are warm. For these designs, the battery box also has a heater. Remember, each battery design has its own optimal temperature range.

Heat Shields The efficiency of a battery is directly affected by its temperature. Some batteries may have a heat shield made of plastic or another material to protect the starting battery from high underhood temperatures. Vehicles equipped for cold climates may have a battery blanket or heater to keep the battery warm during extremely cold weather.

Battery Arrangements

The voltage produced by an individual battery cell varies with the chemicals and materials used to construct the cell. Most cells produce between 1.2 to 3.0 volts. To provide higher voltages, cells must be connected. In addition, there is a limited amount of current available from an individual cell so to increase available current, cells are connected together. Cells can be connected in series, parallel, or both.

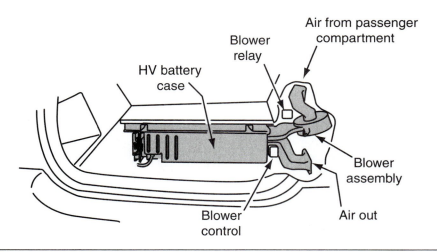

Figure 4-6 This assembly holds the battery pack and provides airflow to keep the batteries cool.

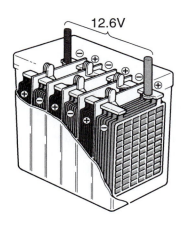

Figure 4-7 The six 2.1-volt cells are connected in series to provide 12.6 volts.

Series Connections Cells are connected in series to have higher voltages. In this arrangement, the total voltage is the sum of the voltages in each cell. For example, a lead-acid cell, commonly used in starting batteries, produces about 2.1 volts. By connecting six cells together in series, the battery has a voltage of 12.6 volts. In an EV or HEV battery pack, hundreds of cells can be connected together to provide the required high voltages (Figure 4-7). Series connections have the positive terminal of one cell connected to the negative terminal of another and the positive terminal of that cell connected to the negative of another, and so on. Individual batteries can also be connected in series.

Parallel Connections Cells are connected in parallel to increase the available current (amperage) of the pack of cells. The positive terminals are connected, and all the negative terminals are connected. The total amperage is the sum of amperages from each cell. The voltage is equal the voltage of an individual cell.

Series-Parallel Connections In this arrangement, groups of cells are wired in parallel and then those groups are connected in series. This arrangement provides for increases in voltage and amperage. Any number of cells can be connected in parallel, as long as each group of parallel cells that are wired in series has the same power output.

Charging

Charging a battery restores the chemical nature of the battery. To do this, a chemical reaction takes place inside the battery. Charging does little more than cause current flow in the electrochemical cells. Discharging allows for current flow outside the cell. To understand the charging process, remember that current flows from a higher potential (voltage) to a lower potential. If the voltage applied to the battery is higher than the voltage of the battery, current will flow into the battery. This means the charging voltage must be higher than the battery's voltage in order to charge it.

Each battery design has its own charging requirements. It is very important to follow the correct procedure for the battery being charged. It is also important to prevent the battery from overheating during charging and to use the correct type of charger; these too vary with battery designs. Using the wrong charger can destroy the batteries or charger.

Chargers Battery chargers are designed to supply a constant voltage, a constant current, or a mixture of the two. Constant voltage chargers provide a specific amount of voltage to the battery. The current varies with voltage of the battery. When the potential difference between the charger's voltage and the battery's voltage is great, the current is high. As the battery charges, its voltage increases and the charging current drops off. A constant current charger varies the voltage applied to the battery in order to maintain a constant current.

Both of these techniques work fine as long as the temperature of the battery is maintained. Some chargers have a thermometer to monitor battery temperature. These chargers reduce the charging voltage and/or current in response to rising temperatures.

Many battery chargers are "smart" or "intelligent." These chargers are designed to charge a battery in three basic steps: bulk, absorption, and float. During bulk charging, current is sent to the battery at a maximum safe rate until the voltage reaches approximately 80% of its capacity. Once the battery reaches that voltage level, the charger begins the absorption step. During this time, charging voltage is held constant while the current changes according to the battery's voltage. Once the battery is fully charged, the charger switches to the float step. During this step, the charger supplies a constant voltage equal to slightly more than the voltage of the battery. The current flow is very low. This step is a maintenance charge and is intended to keep a battery charged while it is not being used.

Chargers can also be designed to supply voltage and current to a battery according to the needs of the vehicle, and often the needs of the customer. **Fast charging** quickly charges a battery. This supplies large amounts of voltage and current. Although this charges the battery quickly, it also can overheat the battery if it is not closely monitored. This technique is best used when a battery is low on charge and will be installed into the vehicle in a short time. **Slow** or **trickle charging** applies low current to the battery and takes quite some time to fully charge it. However, it is unlikely that the battery will overheat and the battery has a good chance to be completely charged. The chemicals used in the construction of the battery should always be considered before fast or slow charging a battery.

Recycling Batteries

■ **CAUTION:**
A battery should never be incinerated. Doing this can cause an explosion.

The materials used to make a battery can be used in the future through recycling. Batteries should not be discarded with regular trash, as they contain metals and chemicals that are hazardous to the environment.

In 1994, the Rechargeable Battery Recycling Corporation (RBRC) was established to promote recycling of rechargeable batteries in North America. RBRC is a non-profit organization that collects batteries from consumers and businesses and sends them to recycling companies. Collected batteries are sorted by their chemical make up. Then they are broken apart and their elements separated. The chemicals or materials are further separated and then collected.

Ninety-eight percent of all lead-acid batteries (Figure 4-8) are recycled. During the recycling process, the lead, plastic, and acid are separated. The electrolyte (sulfuric acid) can be reused or is discarded after it has been neutralized. The plastic casing is cut into small pieces, scrubbed, and melted to make new battery cases and other parts. The lead is also melted and poured into ingots to be used in new batteries.

RECYCLING FOR A BETTER ENVIRONMENT

TRANSPORTATION

The same transportation network used to distribute new batteries, safely trucks spent batteries from the point of exchange to recycling plants.

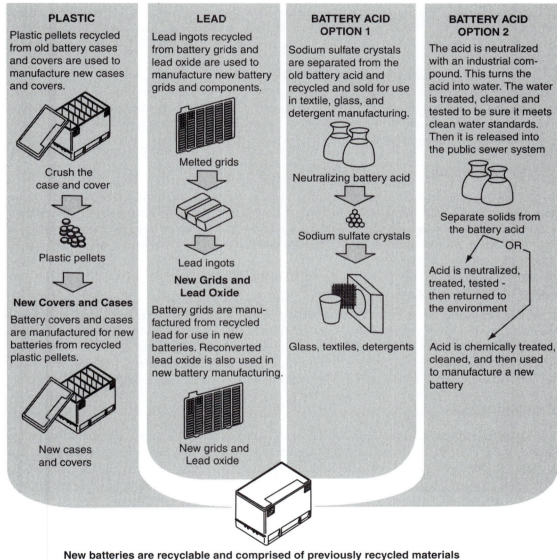

New batteries are recyclable and comprised of previously recycled materials

Figure 4-8 The recycling process for a lead-acid battery.

Battery Ratings

The voltage rating of a battery may be expressed as open circuit or operating voltage. **Open circuit voltage** is the voltage measured across the battery when there is no load on the battery. **Operating voltage** is the voltage measured across the battery when it is under a load.

The available current from a battery is expressed as the battery's capacity to provide a certain amount of current for a certain amount of time and at a certain temperature. Basically, a capacity rating expresses how much electrical energy a battery can store.

Ampere-Hour

A commonly used capacity rating is the ampere-hour rating. In the past, this was the common rating method for lead-acid batteries. However, these batteries are now rated in an different manner. Other battery designs are still rated in ampere-hours or milliamp-hours for cell phone applications.

The **ampere-hour (AH) rating** is the amount of steady current that a fully charged battery can supply for 20 hours at 80°F (26.7°) without the cell's voltage dropping below a predetermined level. For example, if a 12-volt battery can be discharged for 20 hours at a rate of 4.0 amperes before its voltage drops to 10.5 volts, it would be rated at 80 ampere-hours. A 100 AH battery will provide 1 amp for 100 hours, or 10 amps for 10 hours.

Watt-Hour Rating

Some battery manufacturers rate their batteries in watt-hours. The **watt-hour rating** is determined at 0°F (-17.7°C) because the battery's capacity changes with temperature. The rating is calculated by multiplying a battery's amp-hour rating by the battery's voltage. The watt-hour rating of a battery may be listed in units of kilowatts. If a battery can deliver 5 AH at 200-volts, it would be rated at 1 kilowatt-hour.

Cold-Cranking Amps

The **cold cranking amps (CCA) rating** is the common method of rating most automotive starting batteries. It is determined by the load, in amperes, that a battery is able to deliver for 30 seconds at 0°F (-17.7°C) without its voltage dropping below a predetermined level. That voltage level for a 12-volt battery is 7.2 volts. The normal range for passenger car and light truck batteries is between 300 and 600 CCA; some batteries have a rating as high as 1100 CCA.

Cranking Amps

The **cranking amps (CA) rating** is similar to CCA and is a measure of the current a battery can deliver at 32°F (0° C) for 30 seconds and maintain voltage at a predetermined level. Again, this level is 1.2 volts per cell (7.2 volts) for a 12-volt battery. This rating is commonly used in climates that are not subjected to extremely cold weather. Typically, the CCA rating of a battery is about 20% less than its CA rating.

Reserve Capacity

The **reserve capacity (RC) rating** is determined by the length of time, in minutes, that a fully charged starting battery at 80°F (26.7°) can be discharged at 25 amperes before battery voltage drops below 10.5 volts. This rating gives an indication of how long the vehicle can be driven with the headlights on, if the charging system fails. A battery with a reserve capacity of 120 would be able to deliver 25 amps for 120 minutes before its voltage drops below 10.5 volts.

Common Types Of Batteries

In addition to their use in automobiles, batteries are used in many other applications. As a result, there are many different types and designs of batteries available. Most battery manufacturers divide batteries into two separate groups: primary and secondary (Figure 4-9). **Primary batteries** are not rechargeable. These batteries are discarded once they become weak. **Secondary batteries** are rechargeable. The number of times they can be recharged is primarily dependent on the chemicals used in their construction.

Batteries differ in size, from small single cells to large battery packs, comprised of many cells. They also have different ratings (not always dependent on size) and service life. The primary difference between batteries is the chemicals used in the cells.

Battery Chemistry

The following battery types can be or are being used in electric drive vehicles. There are many other types available, but they are not relevant to the topic of this book. Some of the batteries in the list below will be discussed in more detail later in this chapter.

Lead-Acid Lead-acid batteries are the most commonly used starting battery. This type of battery is rechargeable. Several lead-acid cells are connected in series to provide high-voltage in some electric vehicles. There are many variations to the basic design, but all work and are constructed in the same way. The lead-acid cell has electrodes made of lead and lead-oxide with an electrolyte that is a strong acid. The lead-acid battery is one of the oldest battery designs.

Nickel-Cadmium (NiCad) NiCad batteries are most used in portable radios, emergency medical equipment, professional video cameras, and power tools. They provide great power and are normally the battery of choice for power tools. The electrodes in a nickel-cadmium cell are nickel-hydroxide and cadmium. The electrolyte is potassium hydroxide. NiCad batteries are economical and have a long service life. However, cadmium is an environmentally unfriendly metal, which is why NiCad batteries are being replaced by other designs.

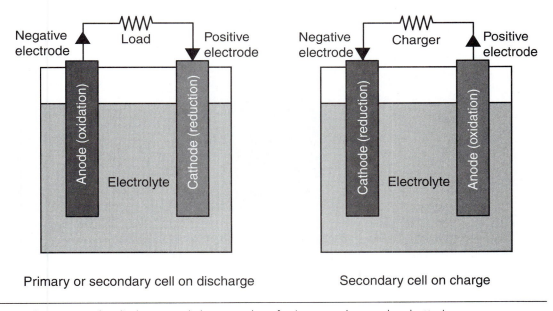

Figure 4-9 The discharge and charge cycles of primary and secondary batteries.

Nickel-Metal Hydride (NiMH) Nickel-metal hydride batteries are very common; they are rapidly replacing nickel-cadmium batteries because they are more environmentally friendly. NiMH batteries also have more capacity than a NiCad but have a lower current capacity under load. These batteries are commonly used in today's hybrid vehicles. The cells have electrodes made of a metal hydride and nickel-hydroxide. The electrolyte is potassium hydroxide.

Sodium-Sulfur (NaS) The electrodes in a sodium-sulfur battery cell are composed of molten sodium (negative electrode) and liquid sulfur (positive electrode). The plates are separated by a solid ceramic electrolyte, made from aluminum. The battery must be kept at about 570°F (300°C) to discharge and recharge. This design of battery is very efficient and is currently being researched for possible use in electric drive vehicles.

Sodium- Nickel-Chloride The electrodes in a sodium-nickel chloride cell are made with nickel and iron powders and sodium chloride (table salt). The electrodes are separated by a ceramic electrolyte. Sodium-Nickel-Chloride batteries are also known as "ZEBRA" batteries (Figure 4-10). These batteries have nearly five times the energy density as a lead-acid battery and are totally recyclable. However, they must operate at high temperatures and the required thermal management system greatly raises their production cost. These batteries were designed to be used in electric drive vehicles, including automobiles and trains.

Lithium-Ion (Li-ion) The electrodes in lithium-ion cells are made of carbon (graphite) and a metal oxide. The electrodes are submersed in lithium salt. Overheating these cells may produce pure lithium in the cells. This metal is very reactive and can explode when hot. To prevent overheating, Li-ion cells have built-in protective electronics and/or fuses to prevent reverse polarity and overcharging. Li-Ion batteries have very good power-to-weight ratios. They are often found in laptop computers, video cameras, and cell phones. Research is also being done on the use of these batteries in electric vehicles.

Lithium-Polymer (Li-Poly) The lithium-polymer battery is nearly identical to a lithium-ion battery. In fact, it evolved from the Li-Ion design. Like the Li-Ion, the electrodes are made of carbon (graphite) and a metal oxide. However, the lithium salt electrolyte is held in a thin solid, plastic-like polymer rather than as a liquid. The solid polymer electrolyte is not flammable, therefore

Figure 4-10 The sodium- nickel-chloride battery, made from nickel and iron powders, sodium chloride, and a ceramic electrolyte, is also known as the "ZEBRA" battery. *Courtesy of Beta Research & Development Ltd., UK*

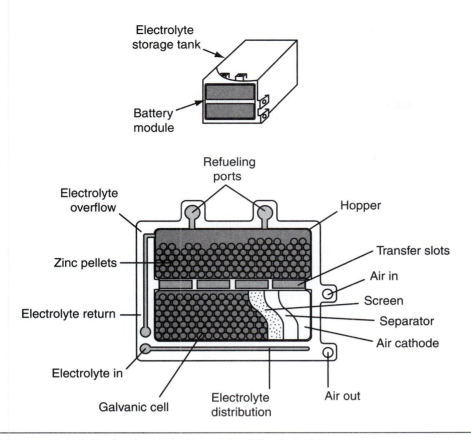

Figure 4-11 Details of a zinc-air battery unit cell (end view)

these batteries are less hazardous if they are mistreated. These batteries can be shaped to fit the application and can store much more energy than a lead-acid battery.

Zinc-Air The zinc-air cell (Figure 4-11) is commonly used in hearing aids, but has been tested and modified for possible use in electric drive vehicles. The interesting characteristic of this battery is that oxygen from the outside air is used as the cathode. The anode is a replaceable cassette made of zinc particles in an electrolyte solution of potassium hydroxide. Within the cell, a chemical reaction produces electrical energy and the cells are not electrically rechargeable. They can, however, be recharged by replacing the anode cassette. This type battery is lightweight and has very high energy density.

Nickel-Zinc Nickel-zinc battery cells are also being researched and tested for possible use in electric drive vehicles. These batteries have a high specific energy and power capability, good deep cycle capability, can operate within a wide range of temperatures, are made of abundant low-cost materials, and are environmentally friendly. The nickel-zinc battery is an alkaline rechargeable system. These cells use a nickel/nickel-oxide electrode as the cathode and the zinc/zinc-oxide electrode as the anode. The electrolyte is normally potassium hydroxide.

Ultra-Capacitors An ultra-capacitor is not a battery; however, it can function much like one. This device stores and releases electrical energy electrostatically, rather than electrochemically. Ultra-capacitors have the ability to quickly discharge high voltages and then be quickly recharged. These characteristics make them ideal for adding boost energy to motors when a vehicle needs extra power for acceleration or to overcome heavy loads. Ultra-capacitors are also very good at absorbing the energy from regenerative braking. Some current hybrid vehicles use ultra-capacitors for both purposes.

High-Voltage Batteries

Electric drive vehicles require high voltages. These high-voltage batteries are assemblies of many cells and are called battery packs. The cells are one of two configurations: cylindrical or prismatic (Figure 4-12). In a **cylindrical cell**, the electrodes are rolled together and fit into a metal cylinder. A separator soaked in the electrolyte is sandwiched between the plates. This design requires much storage space. When several cylinders are assembled together, there is much wasted space between the cylinders. **Prismatic cells** do not have this problem. They have flat electrodes placed into a box with separators between them. Prismatic cells tend to be more expensive to produce than cylindrical cells. Lead-acid batteries have prismatic cells and some NiMH batteries found in hybrids are also constructed this way.

Applications

To understand the need for high-voltage systems, let us take a quick look at some history. In 1954, General Motors equipped its Cadillacs with a 12-volt system. Prior to that, vehicles had 6-volt systems. The electrical demands of accessories, such as power windows and seats, put a severe strain on the 6-volt battery. With the introduction of 12-volt systems, there was half as much strain and drain on the battery than before. This means the charging system had to work less hard to keep the battery charged and there was plenty of electrical power for the electrical accessories.

The increase in voltage also allowed wire sizes to decrease, which lead to less vehicle weight. Wire sizes were able to decrease because the amperage required to power things was reduced. To explain this, consider an accessory that required 20 amps to operate when it was in a 6-volt system. This means it needed 120-watts to operate. When the voltage was increased to 12-volts, the system only drew 10-amperes. Required wire size is dictated by current.

Today we are faced with the same situation. The use of computers and the need to keep their memories fresh has put a drain on the battery. Plus, the number of electrical accessories found on today's vehicles has and will continue to grow. The 12-volt system is becoming over-burdened by these.

Today's vehicles are very sensitive to voltage change. In fact, the overall efficiency of a vehicle depends on a constant voltage. The demands of new technology make it difficult to maintain a constant voltage, and engineers have determined system voltage must be increased to meet those demands and maintain vehicle efficiencies. As our vehicles evolve, emission, fuel economy, comfort, convenience, and safety features will require more electrical power than they do now. This increased demand is the result of converting purely mechanical systems into electromechanical systems, such as steering, suspension, and braking systems, as well as new

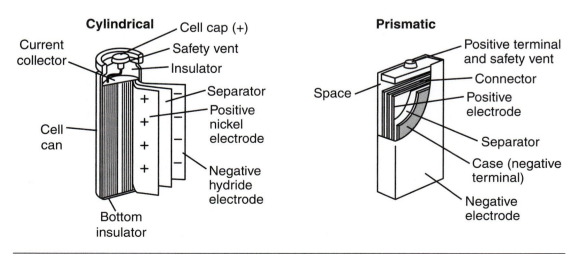

Figure 4-12 The basic construction of cylindrical and prismatic cells.

safety and communication systems. It is has been estimated that the continuous electrical power demand in a few years will be 3000 W to 7000 W. The current 12-volt systems are rated at 800 W to 1500 W.

These demands can be met by increasing the amperage capacity of the battery and charging system or by increasing system voltage. Larger capacity batteries and generators are only a band-aid to the problem. Because the generator is driven by engine power, more power from the engine will be required to keep the higher capacity battery charged. Therefore, overall efficiency will decrease. By moving to a higher system voltage, the battery will need to be larger and heavier; however, because system amperage will be lower, wire size will be smaller, and perhaps the weight gain at the battery will be offset by the decreased weight of the wiring.

All of the advantages of moving from 6- to 12-volt systems apply to the move from 12 volts to 42 volts. But why 42 volts? The starting battery in today's vehicles is rated at 12 volts but stores about 14 volts. Also, the charging system puts out 14 to 15 volts while the engine is running. Because the primary source of electrical power when the engine is running is the charging system, it is fair to say an automobile's electrical system is a 14-volt system. This is the logic used by engineers in planning for 42-volt automotive systems. Forty-two volts represent three 12-volt batteries. Because a 12-volt battery actually holds a 14-volt charge, engineers decided to take advantage of this (3 times 14 volts equals 42 volts).

Forty-two-volt systems are based on a single 36-volt battery but have dual voltage systems (Figure 4-13). Part of the vehicle is powered by 12 to 14 volts and the rest by 36 to 42 volts. The battery has two positive connectors, one for each voltage or the voltage is divided by a converter. The split voltage system provides 42 volts for high-voltage applications such as the starter/generator, power steering, air conditioning, traction control, brake, and engine cooling systems. The 14-volt system powers low load systems, such as lights, power door locks, radios, navigation systems, and cell phones.

Again, the same logic is followed when designing electric drive vehicles. High voltage is needed to prevent the need for extremely large cables and wiring. Also, keeping the required amperage low is easier on the batteries. The voltage used by electric vehicles varies, as does the chemistry of the battery.

The first generation of General Motors' EV1, a BEV, used twenty-six 12-volt lead-acid batteries. The individual batteries were connected in series. The battery pack provided 312 volts and weighed 1310 pounds (595 kg). The driving range between battery recharges was 55 to 95 miles (88 to 153 km). The next, and last, generation EV1 used NiMH batteries and had a slightly longer range.

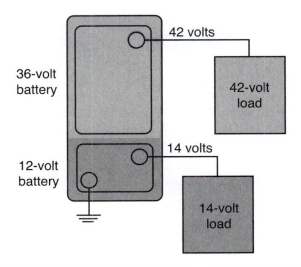

Figure 4-13 A split voltage arrangement providing 14 and 42 volts from a 36-volt battery.

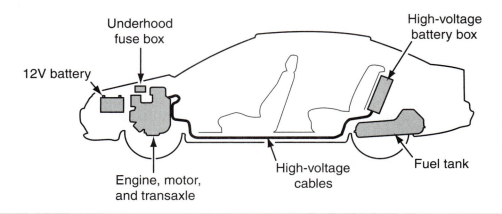

Figure 4-14 Location of the batteries in a Honda Civic Hybrid.

For the most part, production of BEVs has stopped, but the basic technology paved the way for the development of hybrid vehicles. A vehicle equipped with a starter/generator can be considered a mild hybrid and they rely on a 42-volt system. Functions such as, stop-start, regenerative braking, and electrical assist are common to full and mild hybrid vehicles.

Only full hybrids have the ability to move in an electric-only mode. A full hybrid vehicle has a much higher voltage system than a mild hybrid; therefore, the motor is capable of providing much more power, more frequently, and for longer periods of time. Hybrid vehicles may also have a separate starting battery for the ICE (Figure 4-14). The starting battery also is the power source for the lights and other accessories.

The battery pack in a hybrid vehicle is typically made up of several cylindrical cells (Figure 4-15) or prismatic cells (Figure 4-16). These battery packs are often called HV batteries.

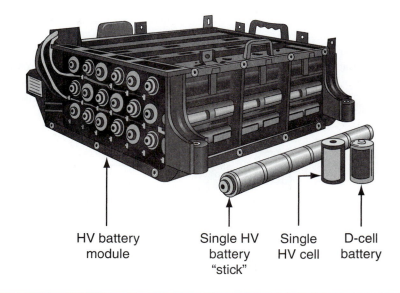

Figure 4-15 A high-voltage battery module made of several cylindrical battery cells.

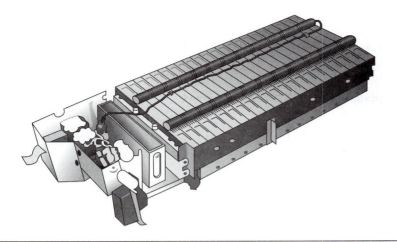

Figure 4-16 The battery pack of a Toyota Prius.

Some do not regard most Honda hybrid vehicles as full hybrids. The electric motor never powers the vehicle on its own. Honda's integrated motor assist (IMA) is a thin brushless electric motor (Figure 4-17) placed between the engine and the transaxle. It assists the engine during acceleration, functions as a generator to recharge the battery pack during deceleration, and serves as the engine's starter motor. Some recent Honda hybrids do have the ability to move by electrical power alone.

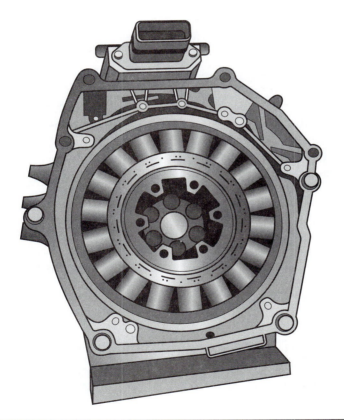

Figure 4-17 Honda's IMA unit functions as a starter or generator and adds power, when needed, to the engine's output.

Safety Issues

High-voltage circuits are identifiable by size and color. The cables have thicker insulation and are colored orange. The connectors are also colored orange. On some vehicles, the high-voltage cables are enclosed in an orange shielding or casing; again the orange indicates high voltage. In addition, the high-voltage battery pack battery and other high-voltage components have "High Voltage" caution labels. It is important to remember that high voltage is also used to power some vehicle accessories (Figure 4-18).

With these high voltages come special precautions.

❏ If a repair or service is incorrectly done, an electrical shock, fire, or explosion can result. Always precisely follow the correct procedures.

❏ Always follow the manufacturer's procedures for disconnecting the power source before working on the high-voltage system (Figure 4-19).

❏ Systems may have a high-voltage capacitor that must discharge after the high-voltage system has been isolated. Make sure to wait the prescribed amount of time (about 10 minutes) before working on or around the high-voltage system.

❏ Wear insulating gloves when working on or around the high-voltage system. Make sure the gloves have no tears, holes, or cracks and that they are dry.

❏ After removing a high-voltage cable, cover the terminal with vinyl electrical tape.

❏ Always use insulated tools.

❏ Alert other technicians that you are working on the high-voltage systems with a warning sign such as "high voltage work: do not touch."

❏ Do not carry any metal objects, such as a mechanical pencil or a measuring tape, that could fall and cause a short circuit.

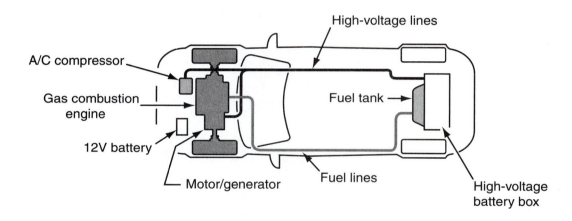

Figure 4-18 The A/C compressor is powered by the high-voltage system on the hybrid.

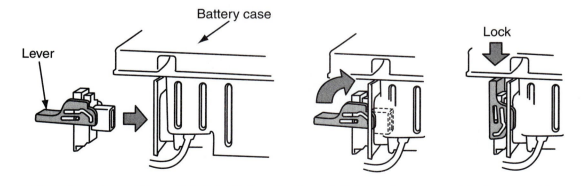

Figure 4-19 Removal of the service plug on a Toyota hybrid to disconnect the high-voltage circuits.

Lead-Acid Batteries

The most common automotive batteries are lead-acid designs. The wet cell, gel cell, absorbed glass mat (AGM), and valve regulated are versions of the lead-acid battery. The wet cell comes in two styles: serviceable and maintenance-free.

A lead-acid battery consists of grids, positive plates, negative plates, separators, elements, an electrolyte, a container, cell covers, vent plugs, and cell containers (Figure 4-20). The **grids** form the basic framework of the battery plates. Grids are the lead alloy framework that supports the active material of the plate. Plates are typically flat, rectangular components that are either positive or negative, depending on the active material they hold.

The positive plate has a grid filled with lead peroxide as its active material. Lead peroxide (PbO_2) is a dark brown, crystalline material. The material pasted onto the grids of the negative plates is sponge lead (Pb). Both plates are very porous and allow the liquid electrolyte to penetrate freely.

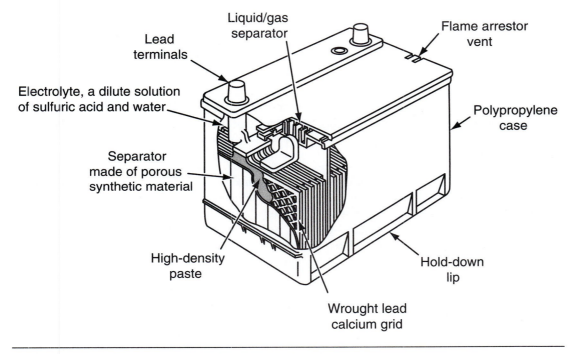

Figure 4-20 The construction of a typical lead-acid battery.

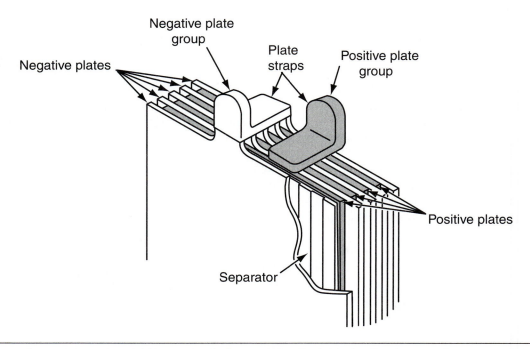

Figure 4-21 A battery is comprised of a number of individual cells. This is an illustration of the positive and negative plates, insulators, and straps that make up one cell in a lead-acid battery.

Basic Construction

Each battery contains a number of elements. An **element** is a group of positive and negative plates. The plates are formed into a plate group, which holds a number of plates of the same polarity. The like-charged plates are welded to a lead alloy post or plate strap. The plate groups are then alternated within the battery—positive, negative, positive, negative, and so on. There is usually one extra set of negative plates to balance the charge. To prevent the different plate groups from touching each other, separators are inserted between them. Separators are porous plastic, electrically insulating sheets that allow for a transfer of ions between plates.

When the element is placed inside the battery case and immersed in electrolyte, it becomes a cell (Figure 4-21). A 12-volt battery has six cells connected in series. Each cell has an open circuit voltage of approximately 2.1 volts; therefore, the total open circuit voltage of the battery is 12.6 volts.

The lead peroxide and sponge lead on the plates are the active materials in the battery. However, these materials are not active until they are immersed in electrolyte. The electrolyte is a solution of sulfuric acid and water. The sulfuric acid supplies sulfate, which chemically reacts with both the lead and lead peroxide to release electrical energy. In addition, the sulfuric acid is the carrier for the electrons as they move inside the battery.

To achieve the required chemical reaction, the electrolyte must be the correct mixture of water and sulfuric acid. At 12.6 volts, the desired solution is 65% water and 35% sulfuric acid. Available voltage decreases when the percentage of acid in the solution decreases.

Casing Design The container or shell of the battery is usually a one-piece, molded assembly of polypropylene, hard rubber, or plastic. The case has a number of individual cell compartments. Cell connectors are used to join all cells of a battery in series.

The top of the battery is encased by a cell cover. The cover may be a one-piece design, or the cells might have their own individual covers. The cover must have vent holes to allow hydrogen and oxygen gases to escape. These gases are formed during charging and discharging. Battery vents can be permanently fixed to the cover or be removable, depending on the design of the battery. Vent plugs or caps are used on some batteries to close the openings in the cell cover and to allow for the topping off the cells with electrolyte or water.

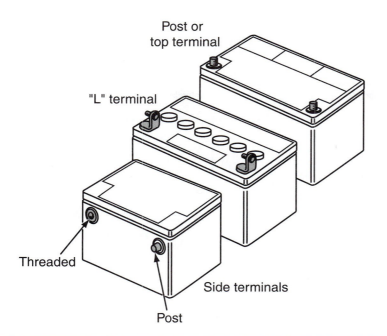

Post or
top terminal

"L" terminal

Threaded

Post

Side terminals

Figure 4-22 Various terminals for lead-acid batteries.

Terminals The battery has two external terminals: a positive (+) and a negative (-). These terminals are either two tapered posts, "L" terminals, threaded studs on top of the case, or two internally threaded connectors on the side (Figure 4-22). The terminals have either a positive (+) or a negative (–) marking, depending on which end of the series they represent.

The size of the tapered terminals is specified by standards set by the Battery Council International (BCI) and Society of Automotive Engineers (SAE). This means that all positive and negative cable clamps will fit any corresponding battery terminal, regardless of the battery's manufacturer. The positive terminal is slightly larger, usually around 11/16-inch in diameter at the top, while the negative terminal usually has a 5/8-inch diameter. This minimizes the danger of installing the battery cables in reverse polarity.

Side terminals are positioned near the top of the battery case. These terminals are threaded and require a special bolt to connect the cables. Some batteries are fitted with both top and side terminals to allow them to be used in many different vehicles.

Discharging and Charging

Remember, a chemical reaction between active materials on the positive and negative plates and the acid in the electrolyte releases electrical energy. When a battery discharges (Figure 4-23), lead in the lead peroxide of the positive plate combines with the sulfate radical (SO_4) to form lead sulfate ($PbSO_4$).

A similar reaction takes place at the negative plate. In this plate, lead (Pb) of the negative active material combines with sulfate radical (SO_4) to also form lead sulfate ($PbSO_4$), a neutral and inactive material. Thus, lead sulfate forms at both types of plates as the battery discharges.

As the chemical reaction occurs, the oxygen from the lead peroxide and the hydrogen from the sulfuric acid combine to form water (H_2O). As discharging takes place, the electrolyte becomes weaker and the positive and negative plates become like one another.

The recharging process (Figure 4-24) is the reverse of the discharging process. Electricity from an outside source, such as the vehicle's generator or a battery recharger, is forced into the battery. The lead sulfate ($PbSO_4$) on both plates separates into lead (Pb) and sulfate (SO_4). As the sulfate (SO_4) leaves both plates, it combines with hydrogen in the electrolyte to form sulfuric acid (H_2SO_4). At the same time, the oxygen (O_2) in the electrolyte combines with the lead (Pb)

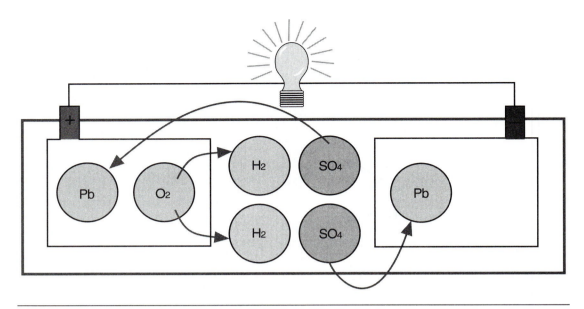

Figure 4-23 The chemical reaction during discharge of a lead-acid battery.

at the positive plate to form lead peroxide (PbO$_2$). As a result, the negative plate returns to its original form of lead (Pb), and the positive plate reverts to lead peroxide (PbO$_2$).

An unsealed battery gradually loses water due to its conversion into hydrogen and oxygen; these gases escape into the atmosphere through the vent caps. If the lost water is not replaced, the level of the electrolyte falls below the tops of the plates. This results in a high concentration of sulfuric acid in the electrolyte and permits the exposed material of the plates to dry and harden. In this situation, premature failure of the battery is certain. The electrolyte level in the battery must be checked frequently.

Temperature Batteries do not work well when they are cold. At 0°F (-17.8° C) a battery is only capable of working at 40% of its capacity. There is also the possibility of the battery freezing when it is very cold and its charge is low. When the battery is allowed to get too hot, the electrolyte can evaporate. Batteries used in hot climates should have their electrolyte level checked frequently.

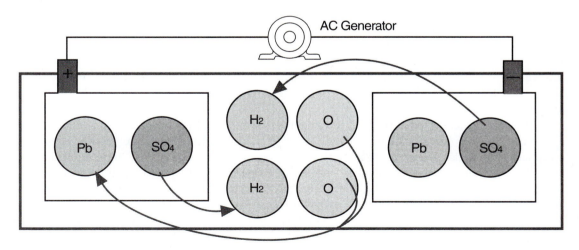

Figure 4-24 The chemical reaction during the recharging of a lead-acid battery.

Designs

A lead-acid battery can be designed as a starting battery or a deep-cycle battery. A deep-cycle battery has thicker and fewer plates than a starting battery. The exact chemical composition of lead-acid batteries also depends on the designed purpose of the battery. However, all lead-acid batteries are based on the reaction of lead and acid.

Maintenance-Free and Low-Maintenance Batteries The majority of batteries installed in today's vehicles are either low-maintenance or maintenance-free designs. A **low-maintenance battery** is a heavy-duty version of a normal lead-acid battery. Many of the components have thicker construction, and different, more durable materials are used. Similar in construction but made with different plate material, a **maintenance-free battery** experiences little gassing during discharge and charge cycles. Therefore, maintenance-free batteries do not have external holes or caps (Figure 4-25). They are equipped with small gas vents that prevent gas-pressure build-up in the case. Water is never added to maintenance-free batteries.

Low-maintenance batteries are still equipped with vent holes and caps, which allow water to be added to the cells. A low-maintenance battery will require additional water substantially less often than a conventional battery.

Hybrid Batteries A **hybrid battery** can withstand many deep cycles and still retain 100% of its original reserve capacity. The grid construction differs from other batteries in that the plates have a lug located near the center of the grid. In addition, the vertical and horizontal grid bars are arranged in a radial design. With this design, current has less resistance and a shorter path to follow. This means the battery is capable of providing more current at a faster rate.

The separators are constructed of glass covered with a resin or fiberglass. The glass separators offer low electrical resistance with high resistance to chemical contamination.

Recombination Batteries A recombination or **recombinant battery** is a completely sealed maintenance-free battery that uses an electrolyte in a gel form. In a gel cell battery, gassing is minimized and vents are not needed. During charging, the negative plates in a recombination battery never reach a fully charged condition and therefore cause little or no release of hydrogen. Oxygen is released at the positive plates, but it passes through the separators and recombines with the negative plates. Because the oxygen released by the electrolyte is forced to recombine with the negative plate, these batteries are called recombination batteries.

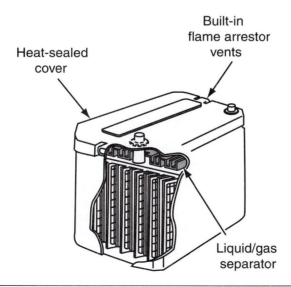

Figure 4-25 The construction of a maintenance-free battery.

Absorbed Glass Mat Batteries The electrolyte in **absorbed glass mat (AGM) batteries** is held in moistened fiberglass matting instead of existing as a liquid or gel. The matting is sandwiched between the battery's lead plates, where it doubles as a vibration dampener. AGM batteries are recombinant batteries.

Rolls of high-purity lead plates are tightly compressed into six cells (Figure 4-26). The plates are separated by acid-permeated vitreous separators. Vitreous separators absorb acid in the same way a paper towel absorbs water. Each of the cells is enclosed in its own cylinder within the battery case, forming a sealed, closed system that resembles a six-pack of soda. During normal use, hydrogen and oxygen within the battery are captured and recombined to form water within the electrolyte. This eliminates the need to ever add water to the battery.

The spiral rolled plates and fiberglass mats are virtually impervious to vibration and impact. AGM batteries will never leak, have short recharging times, and have low internal resistance, which provides increased output.

Valve-Regulated Batteries Valve-Regulated Lead-Acid (VRLA) batteries are similar to AGM batteries. They are recombinant batteries. The oxygen produced on the positive plates of this lead-acid battery is absorbed by the negative plate. That, in turn, decreases the amount of hydrogen produced at the negative plate. The combination of hydrogen and oxygen produces water, which is returned to the electrolyte. Therefore, this battery never needs to have water added to its electrolyte mixture.

One plate in a VRLA is comprised of a lead-tin-calcium alloy with porous lead dioxide; the other is also made of a lead-tin-calcium alloy but has spongy lead as the active material. The electrolyte is sulfuric acid that is absorbed into plate separators made of a glass-fiber fabric. The battery is equipped with a valve that opens to relieve any excessive pressure that builds up in the battery. At all other times, the valve is closed and the battery is totally sealed.

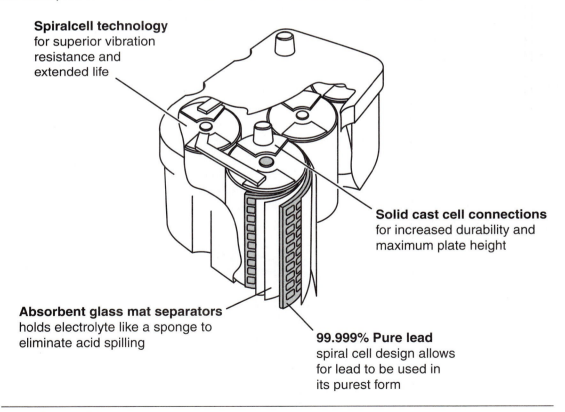

Spiralcell technology for superior vibration resistance and extended life

Solid cast cell connections for increased durability and maximum plate height

Absorbent glass mat separators holds electrolyte like a sponge to eliminate acid spilling

99.999% Pure lead spiral cell design allows for lead to be used in its purest form

Figure 4-26 An absorbed glass mat battery.

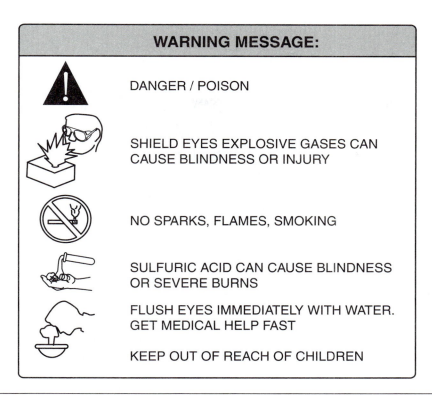

WARNING MESSAGE:

DANGER / POISON

SHIELD EYES EXPLOSIVE GASES CAN CAUSE BLINDNESS OR INJURY

NO SPARKS, FLAMES, SMOKING

SULFURIC ACID CAN CAUSE BLINDNESS OR SEVERE BURNS

FLUSH EYES IMMEDIATELY WITH WATER. GET MEDICAL HELP FAST

KEEP OUT OF REACH OF CHILDREN

Figure 4-27 The warning label on a lead-acid battery.

Service

Lead-acid batteries require proper maintenance. As the charge/discharge process takes place, the water in the electrolyte releases hydrogen into the air and the electrolyte becomes a stronger acid. This destroys the plates and causes the battery to quickly lose its efficiency. Therefore, most lead-acid batteries need to have their electrolyte fluid levels checked at periodical intervals. When refilling the electrolyte, only use mineral-free or distilled water.

A lead-acid battery should also be kept clean. Deposits on the outside of the battery case and around the battery terminals can cause current to flow to ground, which will serve as a constant discharge of the battery. The battery and its hold-downs and tray should be periodically cleaned with a couple of tablespoons of baking soda and mixed with a pint of water. Cable connections and terminals should also be kept clean and their connections tight.

Many of today's vehicles suffer from a problem called **parasitic drain**. Parasitic drain is a load put on a battery with the key off. This drain is not necessarily caused by a problem (although it certainly could be) but is typically caused by systems that operate when the engine is not running. Remember, the starting battery in most vehicles is designed to provide starting current and all other current drains are covered by the charging system. When the battery is faced with drains while the engine is not running, the battery's potential output is decreased and may not be able to maintain the systems running in the vehicle. This potential problem gets worse when there is a problem, such as dirt or corrosion. These conditions can drain the battery and make it incapable of providing the required energy when the engine is starting or running. A constant low or dead battery caused by excessive parasitic drain can shorten the life of a battery.

Testing

Testing a lead-acid battery should begin with a thorough inspection of the battery and its terminals (Figure 4-28). The following items should be checked:

1. Check the age of the battery by looking at the date code on the battery.
2. Check the condition of the case. A damaged battery should be replaced.

■ **CAUTION:**
Lead-acid batteries use sulfuric acid as the electrolyte (Figure 4-27). Sulfuric acid is poisonous, highly corrosive, and produces gases that can explode in high heat. Make sure you are wearing safety glasses (preferably a face shield) and protective clothing when working around and with batteries.

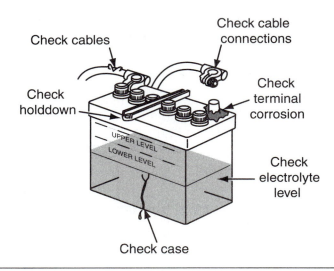

Figure 4-28 Inspect the battery and its cable connections and hold-downs.

3. If the battery is not sealed, check the electrolyte levels in all cells. If the level is low, use distilled water to bring the level up to one-half inch above the plates. If water was added, charge the battery before conducting any test on the battery.
4. Check the condition of the battery terminals and cables. Clean any corrosion from the cable ends and terminals. Make sure the cable ends are tightly fastened to the terminals.
5. Make sure the battery hold-downs are holding the battery securely in place.
6. If the battery has a built-in hydrometer, check its color. If the dot is green, the battery has a 65% charge. If the dot appears to be dark, the battery needs to be charged before testing. If the indicator appears to be clear, a low electrolyte level is indicated.

Prior to conducting any battery test, make sure the battery is fully charged. Also, remove the surface charge of the battery by turning on the headlights with the engine off. Keep the lights on for at least three minutes. Then with a voltmeter, measure the voltage across the battery terminals. This is called open terminal or circuit voltage. A fully charged lead-acid 12-volt battery should have a terminal voltage of 12.6 volts. However, sealed AGM and Gel-Cell batteries will have a slightly higher voltage (12.8-12.9 volts). If the voltage reading is 10.5 volts or lower and the battery is charged, this indicates that at least one cell is shorted.

Battery Hydrometer On unsealed batteries, the specific gravity of the electrolyte can be measured to give a fairly good indication of the battery's state of charge. A **hydrometer** is used to perform this test. A battery hydrometer consists of a glass tube or barrel, rubber bulb, rubber tube, and a glass float or hydrometer with a scale built into its upper stem. The glass tube encases the float and forms a reservoir for the test electrolyte. Squeezing the bulb pulls electrolyte into the reservoir.

When filled with test electrolyte, the sealed hydrometer float bobs in the electrolyte. The depth to which the glass float sinks in the test electrolyte indicates its relative weight compared to water. The reading is taken off the scale by sighting along the level of the electrolyte.

The electrolyte of a fully charged battery is usually about 64% water and 36% sulfuric acid. This corresponds to a specific gravity of 1.270. **Specific gravity** is the weight of a given volume of any liquid divided by the weight of an equal volume of water. Pure water has a specific gravity of 1.000, while battery electrolyte should have a specific gravity of 1.260 to 1.280 at 80°F (26.7°C). In other words, the electrolyte should be 1.260 to 1.280 times heavier than water.

The specific gravity of the electrolyte decreases as the battery discharges. This is why measuring the specific gravity of the electrolyte with a hydrometer can be a good indicator of

■ **CAUTION:**
Electrolyte is very corrosive. It can cause severe injuries if it comes in contact with your skin or eye. If electrolyte gets on you, immediately wash with baking soda and water. If the acid gets in your eyes, immediately flush them with cool water. Then get medical help.

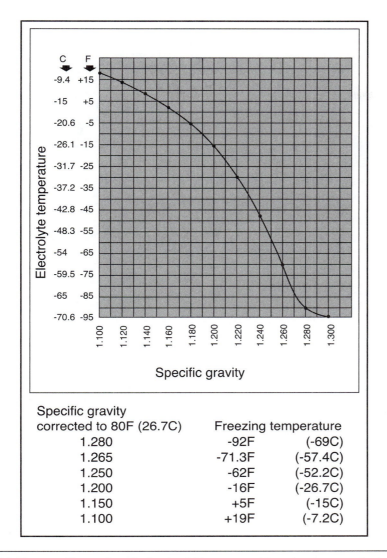

Specific gravity corrected to 80F (26.7C)	Freezing temperature	
1.280	-92F	(-69C)
1.265	-71.3F	(-57.4C)
1.250	-62F	(-52.2C)
1.200	-16F	(-26.7C)
1.150	+5F	(-15C)
1.100	+19F	(-7.2C)

Figure 4-29 The specific gravity of a lead-acid battery and how it relates to the battery's state-of-charge.

how much charge a battery has lost (Figure 4-29). Hydrometer readings should not vary more than .05 differences between cells.

Volt/Ampere Tester A volt/ampere tester (VAT) is used to test batteries, starting systems, and charging systems (Figure 4-30). The tester contains a voltmeter, ammeter, and carbon pile. The carbon pile is a variable resistor. When the tester is attached to the battery and turned on, the carbon pile will draw current out of the battery. The ammeter will read the amount of current draw. The maximum current draw from the battery, with acceptable voltage, is compared to the rating of the battery to see if the battery is okay. A VAT will also measure the current draw of the starter and current output from the charging system.

This test is commonly referred to as a **load test**. The load put on the battery during the test simulates the current draw of a starting motor. The amount of current draw is determined by the rating of the battery. Normally the load is equal to one-half of the CCA rating of the battery. For example, a battery rated at 600 CCA would load test with 300 amperes for 15 seconds. If the battery's voltage fell below 9.6 volts during the test, the battery should be replaced. Ampere-hour ratings can also be used to determine the test load. Normally the load is set to three times the ampere-hour rating and the voltage is observed for 15 seconds.

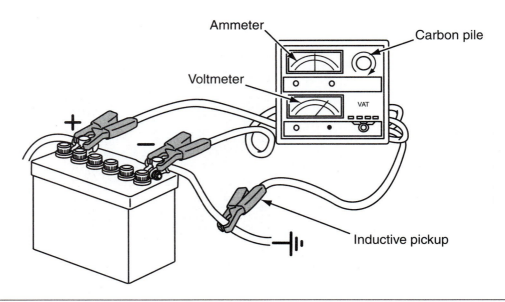

Figure 4-30 The typical connections for a VAT tester.

Battery Capacitance Test Many manufacturers recommend that a **capacitance** or **conductance test** be performed on batteries. Conductance describes a battery's ability to conduct current. It is a measurement of the plate surface available in a battery for chemical reaction. Measuring conductance provides a reliable indication of a battery's condition and is correlated to battery capacity. Conductance can be used to detect cell defects, shorts, normal aging, and open circuits, which can cause the battery to fail.

A fully charged new battery will have a high conductance reading, anywhere from 110% to 140% of its CCA rating. As a battery ages, the plate surface can sulfate or shed active material, which will lower its capacity and conductance.

When a battery has lost a significant percentage of its cranking ability, the conductance reading will fall well below its rating and the test decision will be to replace the battery. Because conductance measurements can track the life of the battery, they are also effective for predicting end of life before the battery fails.

To measure conductance, the tester creates a small signal that is sent through the battery, and then measures a portion of the AC current response. The tester displays the service condition of the battery. The tester will indicate that the battery is good, needs to be recharged and tested again, has failed, or will fail shortly.

Battery Safety

The potential dangers caused by the sulfuric acid in the electrolyte and the explosive gases generated during battery charging require that battery service and troubleshooting are conducted under absolute safe working conditions. Always wear safety glasses or goggles when working with batteries no matter how small the job.

Sulfuric acid can also cause severe skin burns. If electrolyte contacts your skin or eyes, flush the area with water for several minutes. When eye contact occurs, force your eyelid open. Always have a bottle of neutralizing eyewash on hand and flush the affected areas with it. Do not rub your eyes or skin. Receive prompt medical attention if electrolyte contacts your skin or eyes. Call a doctor immediately.

When a battery is charging or discharging, it gives off quantities of highly explosive hydrogen gas. Some hydrogen gas is present in the battery at all times. Any flame or spark can ignite this gas, causing the battery to explode violently, propelling the vent caps at a high velocity, and spraying acid in a wide area. To prevent this dangerous situation, take these precautions.

Figure 4-31 When working on a battery or when disconnecting it, always disconnect the negative cable first.

❏ Never smoke near the top of a battery and never use a lighter or match as a flashlight.
❏ Remove wristwatches and rings before servicing any part of the electrical system. This helps to prevent the possibility of electrical arcing and burns.
❏ Even sealed, maintenance-free batteries have vents and can produce dangerous quantities of hydrogen if severely overcharged.
❏ When removing a battery from a vehicle, always disconnect the battery ground cable first (Figure 4-31). When installing a battery, connect the ground cable last.
❏ Always disconnect the battery's ground cable when working on the electrical system or engine. This prevents sparks from short circuits and prevents accidental starting of the engine.
❏ Always operate charging equipment in well-ventilated areas. A battery that has been overworked should be allowed to cool down and air should be allowed to circulate around it before attempting to jump-start the vehicle. Most batteries have flame arresters in the caps to help prevent explosions, so make sure that the caps are tightly in place.
❏ Never connect or disconnect charger leads when the charger is turned on. This generates a dangerous spark.
❏ Never lay metal tools or other objects on the battery, because a short circuit across the terminals can result.
❏ Always disconnect the battery ground cable before fast charging the battery on the vehicle. Improper connection of charger cables to the battery can reverse the current flow and damage the generator.
❏ Never attempt to use a fast charger as a boost to start the engine.
❏ As a battery gets closer to being fully discharged, the acidity of the electrolyte is reduced, and the electrolyte starts to behave more like pure water. A dead battery may freeze at temperatures near zero degrees Fahrenheit. Never try to charge a battery that has ice in the cells. Passing current through a frozen battery can cause it to rupture or explode. If ice or slush is visible or the electrolyte level cannot be seen, allow the battery to thaw at room temperature before servicing. Do not take chances with sealed batteries. If there is any doubt, allow them to warm to room temperature before servicing.
❏ As batteries get old, especially in warm climates and especially with lead-calcium cells, the grids start to grow. The chemistry is rather involved, but the point is that plates can grow to the point where they touch, producing a shorted cell.

- ❑ Always use a battery carrier or lifting strap to make moving and handling batteries easier and safer.
- ❑ Acid from the battery damages a vehicle's paint and metal surfaces and harms shop equipment. Neutralize any electrolyte spills during servicing.

Nickel-Based Batteries

Two designs of nickel-based rechargeable batteries are commonly used: the **Nickel Metal Hydride (NiMH)** and **Nickel-Cadmium (NiCad)**. Except for the materials used as the anode, a NiMH cell is constructed in the same way as a NiCad cell. Both designs are found in the same types of equipment such as laptop computers, digital cameras, and electric vehicles. The cell voltage from both designs is 1.2 volts, which makes them potentially interchangeable. Most of today's hybrid vehicles use NiMH batteries.

NiMH batteries have replaced NiCad batteries in many applications because of their higher energy densities (more energy is available for a given amount of space) and they use environmentally friendly metals. Cadmium is harmful to the environment.

NiCad batteries also suffer from something called the "memory effect." This problem occurs when a battery is not fully discharged and then recharged. If the battery is consistently being recharged after it is only partially discharged, 50% for example, the battery will eventually only accept and hold a 50% charge. NiMH batteries tend not to be affected by the memory effect.

Nickel/Metal Hydride Cells

NiMH cells are available in the cylindrical and prismatic designs. Both are used in today's hybrid vehicles. The prismatic design requires less storage space but had less energy density than the cylindrical design. NiMH cells are currently the battery of choice for hybrids (Figure 4-32) because of their capacity.

These cells are still being studied and developed to overcome some of their weaknesses and limitations. They have a relatively short service life; however, most batteries used in HEVs have an eight-year warranty. Service life is reduced by subjecting the battery to many deep cycles of charging and discharging. NiMH cells generate more heat while being charged and they require slightly longer charge times than NiCad batteries.

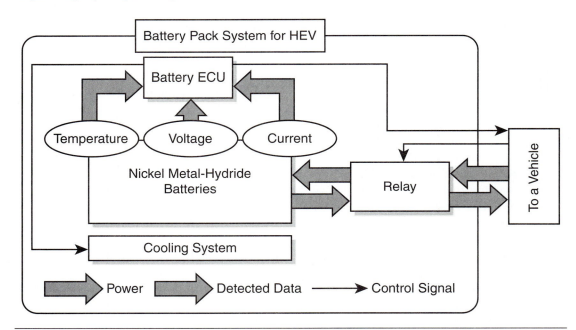

Figure 4-32 A diagram showing how a NiMH battery fits into the operation of a hybrid vehicle.

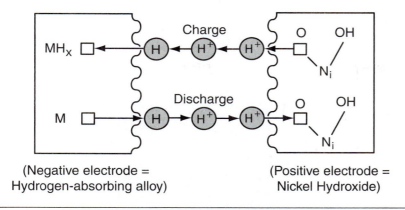

Charge

MH$_X$ ← H ← H$^+$ ← H$^+$ ← □ O OH
                                    \
                                     N$_i$

Discharge

M → H → H$^+$ → H$^+$ → □ O OH
                              \
                               N$_i$

(Negative electrode =
Hydrogen-absorbing alloy)

(Positive electrode =
Nickel Hydroxide)

Figure 4-33 The chemical action inside a NiMH cell when it is supplying electrical energy.

Chemical Reactions Nickel-metal hydride batteries have a positive plate that contains nickel hydroxide. The negative plate is made of hydrogen-absorbing metal alloys. The plates are separated by a sheet of fine fibers saturated with an aqueous and alkaline electrolyte—potassium hydroxide. The components of the cell are typically placed in a metal housing and the unit sealed. There is a safety vent that allows high pressures to escape, if needed.

The most commonly used hydrogen-absorbing alloys are compounds of titanium, vanadium, zirconium, nickel, cobalt, manganese, and aluminum. An alloy formed by the combination of two or three of these metals has the ability to absorb and store hydrogen. The amount of hydrogen that can be stored is many times greater than the actual volume of the alloy.

When a NiMH cell discharges, hydrogen moves from the negative to the positive plate (Figure 4-33). The electrolyte has no active role in the chemical reaction; therefore, the electrolyte level is not changed by the reaction. When the cell is recharged, hydrogen moves from the positive to negative electrode.

Nickel-Cadmium Cells

NiCad (also referred to as NiCd) cells are commonly used in many appliances and may have a future in electric drive vehicles. They have a number of advantages over NiMH cells, one of which is the fact they can withstand many deep cycles—about three times as many as NiMH cells. NiCad batteries can be produced at relatively low costs and have a long shelf and service life. NiCad batteries are especially good performers when high energy boosts are required. This is one feature that has lead to further development of NiCad cell use in electric drive vehicles. Most NiCad batteries are of the cylindrical design.

On the downside, NiCad batteries have toxic metals, suffer from the memory effect, have low energy densities, and require charging after they have been unused for a while.

Chemical Reactions NiCad batteries (Figure 4-34) use a nickel hydroxide positive electrode. This plate is made of a fiber mesh. The anode or negative plate is also made of fiber. The plate is covered with cadmium. An alkaline electrolyte, aqueous potassium hydroxide (KOH) serves as a conductor of ions and has only a slight role in the chemical reaction. During discharge, ions leave the anode and move to the cathode. During charging, the opposite occurs.

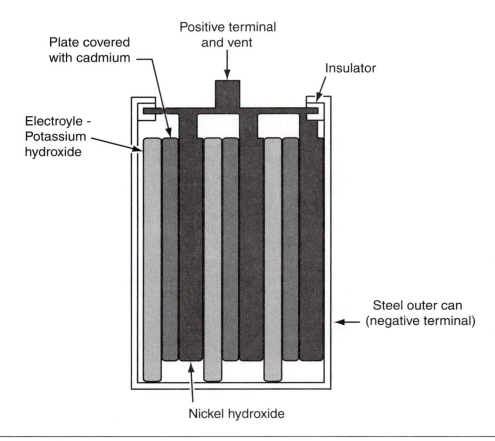

Figure 4-34 Construction of a typical NiCad cell.

Lithium-Based Batteries

Rechargeable lithium-based batteries are very similar in construction to nickel-based batteries and cells. They also have high energy density, do not suffer from the memory effect, and are environmentally more friendly. There are two major types of lithium-based cells: lithium-ion and lithium polymer.

Lithium is the lightest metal and provides the highest energy density of all known metals. Lithium is considered an alkali metal and is used for many purposes, including as the base for many medicines and as a base for heat transfer alloys. Lithium oxidizes very rapidly in air and water and therefore is highly flammable and slightly explosive when exposed to air and water. Lithium metal is also corrosive.

Lithium Ion Battery

Battery cells do not use lithium metal due to safety issues; instead they use lithium compounds. A variety of lithium compounds is used. The term **lithium ion (Li-Ion)** applies to all batteries that use lithium regardless of the materials mixed with the lithium. These compounds have been the focus of much research during the development of lithium-ion batteries for electric drive vehicles. Recently, a manganese lithium ion battery has been developed that may last twice as long as a NiMH battery.

Li-Ion cells have a high-voltage output of 3.6 volts. The compounds used in the cell determine the energy density of the cell. However, the higher density designs are more dangerous. Safety issues also surface when connecting Li-Ion cells, in series or parallel, to form a battery pack. Not all lithium ion cells are designed to be used in a battery pack; only cells that meet tight voltage and capacity tolerances should be used. If the connected cells do not have the same output and capacity, the battery pack can be overcharged and cause a fire.

Lithium ion batteries are expensive to produce. This is mostly due to the required protection circuit. Because lithium metal is very reactive and can explode, the cycling of the battery must be monitored. The protection circuit limits the peak voltage of each cell during charging and prevents the voltage from dropping too low during discharge. The temperature of the battery pack is also monitored and charge and discharge activity is controlled to prevent high temperatures. The circuit also contains electronics and/or fuses to prevent polarity reversal.

Chemical Reactions As with most other rechargeable cells, ions move from the anode to the cathode when the cell is providing electrical energy and during recharging, the ions are moved back from the cathode to the anode (Figure 4-35). The anode is made of graphite, a form of carbon. The cathode is mostly comprised of graphite and a lithium alloy oxide. The construction of the cathode is one of the areas that researchers are working on to produce a safer and stronger Li-Ion battery.

The electrolyte is also the target of much research. The basic electrolyte is a lithium salt mixed in a liquid. Polyethylene membranes are used to separate the plates inside the cells and, in effect, separate the ions from the electrons. The membranes have extremely small pores that allow the ions to move within the cell.

Charging Li-Ion batteries have a nominal voltage of 3.6 volts and a typical charging voltage of 4.2 volts. The cells should be charged with a constant voltage. Charging current should change in response to the voltage of the cell. In other words, as the voltage of the cell increases, the charging current should decrease. Lithium ion batteries should not be fast charged.

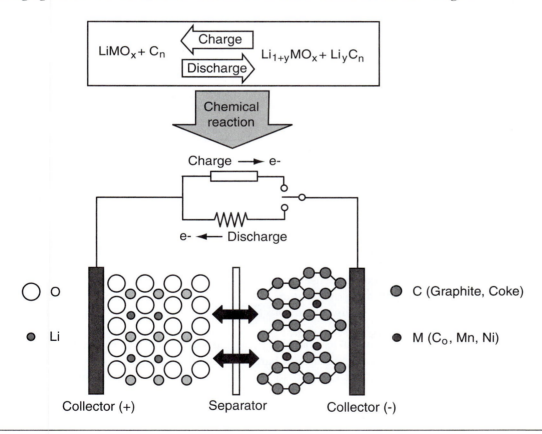

Figure 4-35 The chemical action inside a Li-Ion cell when it is supplying electrical energy and when it is being charged.

Due to the explosive nature of lithium, certain precautions must be followed to safely discharge and charge Li-Ion batteries:

❏ Never connect cells in parallel and/or series that are not designed to be connected or do not have identical output voltages.
❏ Never charge or discharge the battery if it is not connected to its protection circuit.
❏ If the protection circuit does not have a temperature sensor, carefully monitor the battery's temperature while charging and discharging.
❏ Never charge a Li-Ion battery that is physically damaged.

Lithium Polymer Batteries

The **lithium polymer (Li-Poly or LiPo) battery** was developed through the continuous research on the Li-Ion battery. The electrolyte used in Li-Poly cells is not a liquid; rather, the lithium salt is held in a solid polymer composite (such as polyacrylonitrile). The polymer electrolyte is not flammable and these batteries are less hazardous than Li-Ion batteries.

The dry polymer electrolyte does not conduct electricity. Instead, it allows ions to move between the anode and cathode. The polymer electrolyte also serves as the separator between the plates. The dry electrode has very high resistance and therefore cannot provide bursts of current for heavy loads. Heating the cell to 140°F (60°C) or higher increases its efficiency. The voltage of a Li-Poly cell is about 4.23 volts when fully charged. The cells must be protected to prevent overcharging. However, these cells are more resistant to overcharging than Li-Ion cells and there is much less of a chance for electrolyte leakage.

Li-Poly cells are expensive to manufacture and have a much higher cost-to-energy ratio than lithium-ion cells. However, since Li-Poly cells do not use a metal case, they are lighter and can be packaged in many ways. As a result of the flexible packaging, Li-Poly cells have a much higher energy density than Li-Ion, NiMH, and NiCad batteries.

In some Li-Poly cells, a gelled electrolyte has been added to enhance ion conductivity. These are called lithium-ion-polymer cells.

Ultra-Capacitors

Ultra- or super-capacitors are used in many current hybrid vehicles and in some experimental fuel cell electric vehicles. Before discussing what an ultra-capacitor is or how it works, we must take a look at conventional capacitors.

Capacitors

A **capacitor** is an electrical device used to store and release electrical energy. Capacitors can be used to smooth out current fluctuations, store and release a high voltage, or block DC voltage. Capacitors are sometimes called condensers.

Although a battery and a capacitor store electrical energy, the battery stores the energy chemically. A capacitor stores energy in an electrostatic field created between a pair of electrodes (Figure 4-36).

A capacitor can release all of its charged energy in an instant, whereas a battery slowly releases its charge. A capacitor is quick to discharge and quick to charge. A battery needs some time to discharge and charge, but can provide continuous power. A capacitor only provides power in bursts.

A capacitor, like a battery, has a positive and a negative terminal. Each terminal is connected to a thin electrode or plate (usually made of metal). The plates are placed in parallel to each other and are separated by an insulating material called a dielectric. The dielectric can be paper, plastic, glass, or anything that does not conduct electricity. Placing a dielectric between the plates allows the plates to be placed close to each other without allowing them to touch.

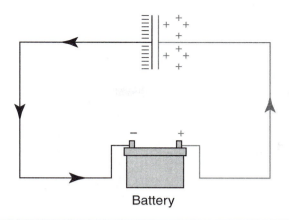

Figure 4-36 A capacitor stores energy in an electrostatic field created between a pair of electrodes.

When voltage is applied to a capacitor, the two electrodes receive equal but opposite charges (Figure 4-37). The plate in the capacitor that is connected to the negative terminal of the battery, or other power source, accepts electrons and stores them on its surface. The other plate in the capacitor loses electrons to the power source. This action charges the capacitor. Once the capacitor is charged, it has the same voltage as the power source. This energy is stored statically until the two terminals are connected together.

The ability of a capacitor to store an electric charge is called **capacitance**. The standard measure of capacitance is the **farad (F)**. A one-farad capacitor can store one coulomb of charge at one volt. A coulomb is 6.25 <u>billion</u> <u>billion</u> electrons. One ampere equals the flow of one coulomb of electrons per second, so a one-farad capacitor can hold one ampere-second of electrons at one volt. Most capacitors have a capacitance rating of much less than a farad and their values are expressed in one of these terms:

❏ microfarads: μF (1 μF = 10^{-6} F)
❏ nanofarads: nF (1 nF = 10^{-9} F)
❏ picofarads: pF (1 pF = 10^{-12} F) </BL>

Three major factors determine how much capacitance a capacitor (Figure 4-38) has: the insulating qualities of the dielectric, the surface area of the electrodes, and the distance between the electrodes. The amount of capacitance is directly proportional to the surface areas of the plates and the non-conduciveness of the dielectric, and is inversely proportional to the distance between the plates.

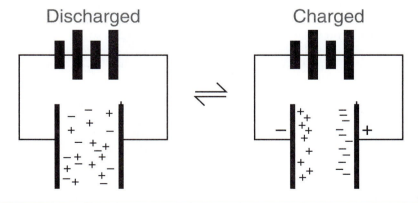

Figure 4-37 When voltage is applied to a capacitor, the two electrodes receive equal but opposite charges.

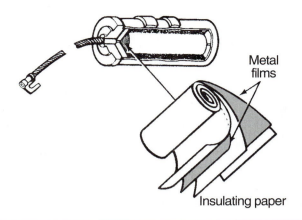

Figure 4-38 The construction of a typical capacitor (condenser).

Ultra-Capacitors

Ultra-capacitors are capacitors with a large electrode surface area and a very small distance between the electrodes. These features give them very high capacitance, which is why they are called ultra- or super-capacitors. Some ultra-capacitors are rated at 5000 farads.

Ultra-capacitors use an electrolyte rather than a dielectric and store electrical energy at the boundary between the electrodes and the electrolyte. Although an ultra-capacitor is an electro-chemical device, no chemical reactions are involved in the storing of electrical energy. As a result, they have no negative impact on the environment.

Ultra-capacitors are maintenance-free devices. They can withstand an infinite number of charge/discharge cycles without degrading and have a long service life. They also are very good at capturing the large amounts of energy from regenerative braking, and can deliver power for acceleration and heavy loads quickly. Also, because they charge very quickly, they have energy available shortly after they have been discharged.

Ultra-capacitors, however, cannot store as much total energy as batteries and they are expensive to manufacture. To provide the required high voltages for electric vehicles, several capacitors must be connected in series. Each cell of an ultra-capacitor can only store between 2 to 5 volts. Up to 500 cells are required to meet the needs of a typical electric drive vehicle.

Construction A regular capacitor is made of conductive foils and a dry separator (Figure 4-39). An ultra-capacitor has two special electrodes and some electrolyte, much like a battery cell. The electrodes are typically made of carbon but can be made from a metal oxide or conducting polymers. The carbon surface of the electrodes is very coarse, with thousands of microscopic peaks and valleys. These irregularities increase the electrodes' surface area. In fact, an ounce (28.35 grams) of carbon provides nearly 13,500 square feet (1250 sq. m) of surface area.

The plates are immersed in an electrolyte. The electrolyte is normally boric acid or sodium borate mixed in water and ethylene glycol or sugars to reduce the chances of evaporation. When voltage is applied across the capacitor, the electrolyte becomes polarized. The charge of the positive electrode attracts the negative ions in the electrolyte and the charge of the negative electrode attracts the positive ions. When the positively charged ions form a layer on the surface of the negative electrode, electrons within the electrode but beneath the surface move to match up with them. The same occurs on the positive electrode and these two layers of separated charges form a strong static charge.

A porous, dielectric separator is placed between the two electrodes to prevent the charges from moving between them. This separator is ultra-thin; in fact, it is about half the size of the ions in the electrolyte. This small separator and the immense amount of surface area is what allow an ultra-capacitor to have high capacitance. However, the thin insulator is also the reason cell voltage must be kept low. High voltages would easily cause arcing across the plates.

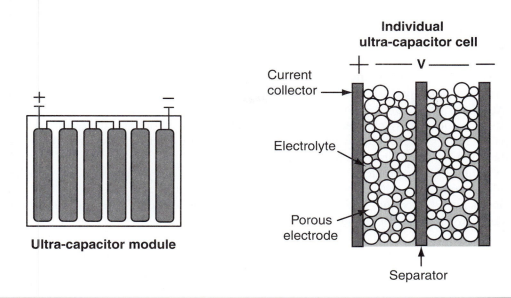

Figure 4-39 An ultra-capacitor cell.

Applications The Toyota Prius was the first automobile to use a bank of ultra-capacitors. The ultra-capacitors store energy captured during deceleration and braking and release that energy to assist the engine during acceleration (Figure 4-40). The energy in the capacitors is also used to restart the engine during the stop/start sequence. Other full hybrid vehicles have also included ultra-capacitors in their power platform.

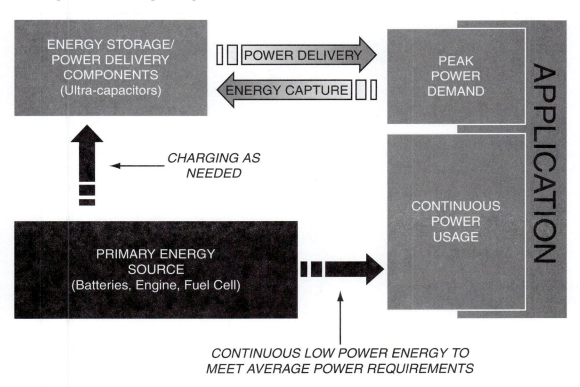

Figure 4-40 A diagram showing how an ultra-capacitor fits into the power train of a hybrid vehicle.

Mild hybrids, those with regenerative braking, a starter/generator, and the stop/start feature, can also benefit from the use of ultra-capacitors. These systems are 42-volt systems. In city driving when the vehicle stops, the engine shuts down. When it is time to accelerate, the engine starts again. In very heavy traffic, this cycle can occur several times within a short period of time. The cycling is very hard on batteries. An ultra-capacitor can be used to provide the power needed to start the engine. Recharged by regenerative braking and/or the generator, the ultra-capacitor can be quickly charged and is capable of providing enough energy for two engine restarts after it is charged. Supplying only 42 volts, the ultra-capacitor pack is much less expensive than those required in a full hybrid.

A bank of ultra-capacitors can also improve the performance of pure electric vehicles and fuel cell electric vehicles. Ultra-capacitors can allow a battery to be discharged and charged at a continuous rate. This would allow the use of simpler battery designs, ones that do not need to provide bursts of power or absorb bursts of energy from braking.

Charging An ultra-capacitor is normally placed in parallel with a battery pack and is recharged by regenerative barking. An ultra-capacitor can also be charged with a battery charger. A typical ultra-capacitor needs about 10 seconds to be fully recharged. The charging process is much the same as that of a battery—the initial charge takes little time and current should be limited during the final stages of charging. The ultra-capacitor is fully charged when it reaches the voltage of whatever is charging it. Ultra-capacitors accept only as much energy as needed and, therefore, there is no possibility of overcharging. Once they are charged, they will not accept further charging. Ultra-capacitors can be recharged and discharged an unlimited number of times.

Review Questions

1. Describe the difference between a cylindrical and prismatic battery cell.
2. How can you identify the high-voltage system in most hybrid vehicles?
3. List five reasons ultra-capacitors are and can be effectively used in electric drive vehicles.
4. List the factors that determine the capacitance of a capacitor.
5. Which of the following cells are NOT considered environmentally friendly?
   A. NiCad
   B. NiMH
   C. Li-Ion
   D. Ultra-capacitors
6. Which of the following statements about NiMH cells is true?
   A. The cylindrical design requires less storage space but has less energy density than the prismatic design.
   B. They have a relatively long service life.
   C. Service life can be extended by frequent deep cycles.
   D. They do not respond well to overcharging; they should be trickle charged.
7. Which of the following statements about battery ratings is true?
   A. The ampere-hour rating is defined as the amount of steady current that a fully charged battery can supply for 1 hour at 80°F (26.7°C) without the cell voltage falling below a predetermined voltage.

   B. The cold cranking amps rating represents the number of amps that a fully charged battery can deliver at 0°F (-17.7°C) for 30 seconds while maintaining a voltage above 9.6 volts for a 12V battery.
   C. The cranking amp rating expresses the number of amperes a battery can deliver at 32°F (0°C) for 30 seconds and maintain at least 1.2 volts per cell.
   D. The reserve capacity rating expresses the number of amperes a fully charged battery at 80°F can supply before the battery's voltage falls below 10.5 volts.
8. True or False. Heating a lithium polymer cell increases its efficiency.
9. Which of the following statements about NiMH cells is NOT true?
   A. When a NiMH cell discharges, hydrogen moves from the negative to the positive plate.
   B. Nickel-metal hydride batteries have a negative plate that contains nickel hydroxide.
   C. The alkaline electrolyte has no active role in the chemical reaction.
   D. The plates are separated by a sheet of fine fibers saturated with potassium hydroxide.
10. True or False. All electrochemical batteries have three major parts: an anode, a cathode, and an electrolyte.

THE BASICS OF A BATTERY-OPERATED ELECTRIC VEHICLE

CHAPTER 5

After reading and studying this chapter, you should be able to:

❑ Describe the basic differences between a battery-operated electric vehicle and an internal combustion engine vehicle.

❑ List some of the advantages of driving an electric vehicle.

❑ List some of the disadvantages of driving an electric vehicle.

❑ Describe the various types of emissions that result from the use of fossil-fueled motor vehicles.

❑ Describe the major systems that make up a BEV.

❑ Compare motor specifications with ICE specifications.

❑ Describe the purpose and function of a battery control system.

❑ Explain the advantages and disadvantages of having an AC motor rather than a DC motor in an electric vehicle.

❑ Describe the purpose of a motor controller.

❑ Describe the purpose of an inverter and a converter.

❑ Explain how regenerative braking works.

❑ Describe the different methods used to recharge the batteries in a BEV.

❑ Explain the differences between conductive and inductive battery charging.

❑ Describe the differences in the operation of accessories and auxiliary systems in a BEV and an ICEV.

❑ Describe some of the unique things about driving an electric vehicle.

❑ Explain how most electric vehicle's problems are diagnosed.

❑ Describe some precautions that should be followed when troubleshooting and repairing an electric vehicle.

❑ Describe what needs to be done to convert an ICEV to a BEV.

Introduction

Today, there are no pure battery electric vehicles (BEVs) manufactured by the major automobile companies. Although many different models of EVs were available during the 1990s, none were sold in large numbers because of their high price and limited driving range. Altogether, only about 5,000 EVs were sold or leased between 1990 and 2003, and none thereafter. Table 5-1 compares one example of an EV and the gasoline-powered version of the same vehicle. The disadvantages of the EV are quite apparent when comparing the two.

Some features of an EV are greatly accepted by the public: reduced air pollution and reduced fuel cost. But more popular has been hybrid electric vehicles. HEVs have long driving ranges, are affordable, and are very fuel efficient. In fact, on any list of the most fuel-efficient vehicles, the first seven will be different hybrid automobiles.

If HEVs are popular and EVs have been discontinued, why do we want to study pure electric vehicles? There are four answers: (1) the electric vehicles of the 1990s provided the technology that is now used in today's hybrids, (2) new technologies may be developed that make BEVs practical in the future, (3) fuel cell vehicles are based on BEV platforms, and (4) there are still a large number of pure electric vehicles in use today. About 500 electric buses and trams, over a million electric forklifts and factory/utility vehicles, and over 5 million electric golf carts are in operation in the United States today. All of these are "electric vehicles" and share the same basic propulsion system.

Key Terms

Conductive charging
Data Link Connector (DLC)
Diagnostic Trouble Codes (DTCs)
Electro-hydraulic brake systems
Electro-hydraulic steering systems
Flywheel energy storage
Flywheel power storage
Heat pump
Inductive charging
Insulated (Isolated) Gate BiPolar Transistors (IGBTs)

Key Terms (continued)

Malfunction Indicator Light (MIL)

Pulse width modulation (PWM)

State of charge (SOC) meter

Table 5-1 COMPARISON OF GASOLINE AND ELECTRIC VERSIONS OF THE RAV4 SUV MANUFACTURED BY TOYOTA.

Characteristic	Propulsion System	
	Gasoline	Electric
Driving range without refueling or recharge	336 miles	126 miles
Top speed in mph	109 mph	79 mph
Acceleration time from 0 to 60 mph	10.5 sec.	18 sec.
Price	$24,000	$40,000

Advantages

There are three basic types of electric drive vehicles: battery electrical vehicles, hybrid electric vehicles, and fuel cell electric vehicles. BEVs use the electrical energy stored in batteries to power the drive or traction motors (Figure 5-1). BEVs have zero emissions, but are unable to travel far between battery recharges, and their batteries need replacement after a few hundred discharge-charge cycles. HEVs can travel far between fuel fill-ups and there is no need to stop and recharge the batteries, as they are charged by the vehicle's ICE-driven generators and byre-generative braking. The engine in a hybrid still requires fossil fuels and has undesirable emissions. However, HEVs have emissions levels much lower than pure ICE vehicles. Fuel cell vehicles use an expensive in-vehicle fuel cell to generate the electrical energy required to power the vehicle's traction motors. FCEVs are zero-emission vehicles but they rely on hydrogen as the fuel. There is no infrastructure for dispensing hydrogen, although fuel reformers can be used to extract hydrogen from other fuels. Electric drive vehicles help reduce our dependency on oil and the amount of pollutants we release into our environment.

With a BEV, there is the convenience of being able to "fill up" at home, eliminating the need to go a service station. There are also some remote charge stations available in a few states. The cost to refuel is very low, typically a recharge costs less than $3. Electrical sources, such solar and wind generation systems, have zero emissions and can further reduce the cost and further decrease the dependency on natural resources to generate electricity.

Because of the limited range, BEVs are ideal for commuting or traveling within a limited area. Studies have shown that 80% of commuters travel fewer than 40 miles per day; this is well within the range of most BEVs.

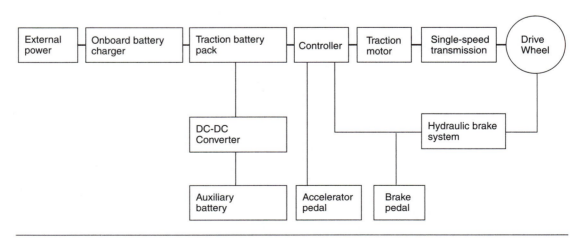

Figure 5-1 Major components of a battery-operated electric vehicle.

Cost

The initial cost of a BEV tends to be higher than a comparable ICE vehicle. This is due to the limited availability of BEVs and the cost of the battery packs. New batteries developed to extend the range of a BEV are, unfortunately, more expensive. However, as more BEVs are produced and sold, their cost should decrease.

The batteries` in some BEVs may need to be replaced after a few years. Golf car batteries, for example, can only experience a few hundred cycles. Once that number is reached, the battery pack must be replaced. However, this cost is offset by the low cost to operate a BEV on a daily basis. Toyota's RAV4-EV had an equivalent fuel economy rating of 112 miles per gallon.

The true cost of driving a BEV depends on the cost of electricity per kilowatt/hour (kWh) and the efficiency of the vehicle. Actual operating costs are reduced by making the cars lighter, more aerodynamic, and with less rolling resistance.

BEV vs. ICEV

To move and operate, all vehicles must have a way to store energy, a way to control the input and output of energy, and a way of changing the energy into a rotary motion to rotate the drive wheels (see Table 5-2). In an electric drive vehicle, the battery serves the same purpose as the fuel tank in an ICEV, storing energy until it is needed. To regulate the speed and acceleration of the vehicle, an ICEV uses a fuel injection system to control the flow of energy. In a BEV, the flow of energy is regulated by a controller. The controller provides electrical energy to the motor at the required rate. Like in an ICEV, the rate is adjusted according to the position of the accelerator pedal. In both types of vehicles, the power output is used to rotate the drive wheels.

An EV can look like a conventional car without an exhaust pipe (Figure 5-2). Internally, however, a BEV is quite a bit different. In many EVs, there is no transmission because the rotary motion of the motor can be applied directly to the differential gears. A motor is capable of providing enough torque throughout its speed range to move the vehicle without torque multiplication from transmission gears. The motor can be positioned to provide front-wheel or rear-wheel drive.

An engine loses much energy through heat (Figure 5-3). Much of the energy produced during combustion merely heats up the engine or goes out of the exhaust, rather than serving as energy to power the vehicle. Also, some of the energy from the engine is used to drive accessories; this also decreases the efficiency of the engine. BEVs generate much less heat and, therefore, much less energy is wasted. BEVs can also reclaim or capture energy through regenerative braking. In BEVs, most accessories are driven by power from the battery and not by the motor.

Table 5-2 A COMPARISON OF THE MAJOR COMPONENTS OF AN ICEV AND A BEV.

Major Components of a Gasoline Vehicle	Functions of the Components	Major Components of an Electric Vehicle
Gasoline tank	Stores the energy to run the vehicle	Battery
Gasoline pump station	Replaces the energy to run the vehicle	Charger
Gasoline engine	Provides the force to move the vehicle	Electric motor
Fuel injection system	Controls acceleration and speed	Controller
Generator/Alternator	Provides power to accessories	DC/DC converter
	Converts DC to AC to power AC motor	DC/AC converter
Emissions controls	Lowers the toxicity of exhaust gases	Not needed

Figure 5-2 This RAV4-EV looks just like its ICE-equipped brother. *Courtesy of Dewhurst Photography and Toyota Motor Sales, U.S.A., Inc.*

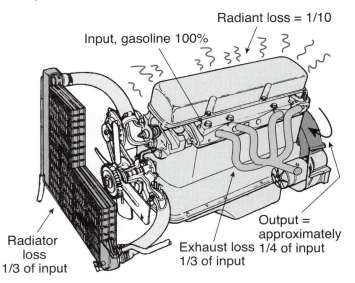

Radiant loss = 1/10

Input, gasoline 100%

Radiator loss 1/3 of input

Exhaust loss 1/3 of input

Output = approximately 1/4 of input

Figure 5-3 In an ICE, the energy produced during combustion merely heats up the engine or goes out of the exhaust, rather than serving as energy to power the vehicle.

An EV has few moving parts. The armature or rotor of the motor is the only moving part in the power plant system. An engine has hundreds of moving parts, each requiring clean lubrication and subject to wear. The rotor in a motor is normally mounted on sealed bearings and requires little, if any, additional lubrication throughout its life. The controller and battery charger are electronic units with no moving parts, and require little or no maintenance. The batteries are sealed and maintenance free. In summary, an EV requires less periodic maintenance and is more reliable than an ICE.

Emissions

BEVs produce zero emissions. The only emissions related to a BEV are those released when coal or natural gas are used in power plants to generate the electrical energy required to recharge the batteries. The use of hydroelectric, wind, sunlight, or other renewable sources to generate electricity would eliminate all emissions associated with EVs. It is impossible to have zero emissions from an ICE. Other than operating cost advantages, emissions, or the lack of, are the primary justification for having a BEV. There are many chemicals and substances related to the emissions of an ICE, none of which are emitted by an electric vehicle:

❑ Carbon monoxide (CO) is a byproduct of combustion and is a deadly gas.
❑ Sulfur dioxides (SO_x) are produced by combustion of coal, fuel oil, and gasoline, because these fuels contain sulfur. SO_x combined with water vapor in the air becomes the major contributor to acid rain.
❑ Hydrocarbons (HC) are the basic building blocks for all fossil fuels. An ICE releases HCs through its exhaust and its fuel storage system. HCs are best thought of as unburned fuel. These fumes contribute to the formation of ozone and HCs can cause cancer or irritate mucous membranes.
❑ Nitrogen oxides (NO_x) are released from the exhaust of ICEs and the burning of coal, oil, or natural gas. When nitrous oxides combine with the oxygen in the air, a poisonous gas is formed that can damage lung tissue.
❑ Volatile organic compounds (VOCs) contribute to ozone and smog formation. VOCs are emitted from transportation and industrial sources, such as automobile exhaust, gasoline/oil storage and transfer, chemical manufacturing, dry cleaners, paint shops, and other facilities using solvents.
❑ Ozone (O_3) is a toxic gas and is seen as a white haze (smog) over many cities. Ground-level ozone is formed by a chemical reaction between VOCs and NO_x, in the presence of sunlight. ICEVs are major sources of NO_x and VOCs.
❑ Carbon dioxide (CO_2) is a product of combustion. Some CO_2 is needed because vegetation needs it to survive. However, a high concentration of CO_2 traps heat and warms the atmosphere, which causes the "greenhouse effect" and global warming.

Disadvantages

Although there are many advantages to using an electric vehicle for transportation, there are some major disadvantages. Perhaps the biggest disadvantage is the very limited range with current battery technologies (Figure 5-4). The driving range between recharging with the batteries currently available is between 50 and 150 miles. Although some new battery designs have extended this range, long travel in an electric vehicle is still not practical. Long recharge times are also a problem with current battery technologies. Recharging the batteries takes much longer than filling the fuel tank at a service station.

In addition to the recharge times, there is a problem of where they can be recharged. If the owner is at home, the charger can be connected to electrical system of the house. If the vehicle is on the road, some sort of electrical hookup must be made.

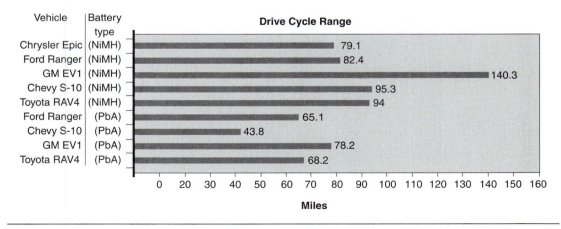

Figure 5-4 The drive cycle ranges for common BEVs.

The Toyota RAV4-EV used 24 12-volt NiMH batteries and had a range of 80 to 125 miles before recharging. It took approximately 6 hours to recharge the battery pack. The range, of course, depended on the vehicle's speed and traffic conditions, so much planning was required if the driver desired to cover many miles.

The EV1 was an all-electric vehicle offered by General Motors from 1996 to 2004. The lead-acid batteries used in the early EV1s required 5 to 6 hours to recharge and the vehicle had a driving range of 55 to 95 miles. Later EV1s had a nickel-metal hydride battery pack that required a 6- to 8-hour charge and provided a driving range of 75 to 130 miles.

Many individuals convert ICEVs to electric power by removing the engine and related equipment and installing a motor and battery pack, as well as other necessary equipment. The range of most of these conversions is fewer than 100 miles.

These driving ranges are adequate if the vehicle is not the only one in a household or if long trips are not desired. Given current battery technology, these ranges seem to be the limit. Adding more batteries to the battery pack would seem to extend the range, but the extra batteries add weight to the vehicle. The weight would then decrease the range. Remember, the smallest, lightest, and most aerodynamic electric vehicles will provide the longest range (see Table 5-3).

Another challenge facing BEV owners is the availability and access to skilled service technicians to service and maintain their vehicles. Training for current technicians must be readily available and there must be programs developed in schools to prepare individuals for this technology before BEVs are practical.

The Past, Present, and Future

Throughout the history of the automobile, many electric vehicles have been produced. A majority of automobiles sold in 1900 were non-gasoline powered. As the internal combustion engine became more practical and consumer friendly, electric vehicles all but disappeared. Due to environmental and oil dependency concerns, a resurgence in the development of electric drive vehicles reoccurred in the 1990s. Many manufacturers produced BEVs during that time.

General Motors invested much time and money in the development of the EV1 (Figure 5-5) only to drop the program and reclaim the 800 or so vehicles they had leased to customers. Although financially this project was a serious failure, they did gain. The propulsion system used in the EV1 is used as the basis of GM's advanced technology vehicle programs, from hybrids to fuel cells. When deciding to drop the EV1 program, GM cited the following reasons:

❏ A restricted driving range.
❏ The vehicle was a sporty commuter car with seating for only two people.
❏ Most consumers will not buy a vehicle that does not closely match most of their needs and lifestyles, no matter how advanced or environmentally friendly it may be.

Figure 5-5 General Motors's EV1. *Courtesy of Georgia Power Company*

Table 5-3 A COMPARISON OF VARIOUS MANUFACTURER PURPOSE-BUILT ELECTRIC VEHICLES. NOTE HOW THEIR BASIC CONSTRUCTION AFFECTS THEIR RANGE AND OTHER CHARACTERISTICS.

This list is arranged according to driving range (the lowest and highest values in each category are highlighted).

Base vehicle	Batteries	System voltage	Curb weight	0-50 mph (80 km/h)	Driving cycle range	Time to recharge
1999 GM EV1	NiMh	**343 V**	**2848 lbs (1292 kg)**	**6.3 sec**	**140.3 mi (226 km)**	6 hrs 58 min
1998 Chev S-10 E	NiMh	**343 V**	4230 lbs (1919 kg)	9.9 sec	95.3 mi (153 km)	**8 hrs 54 min**
1998 Toyota Rav4	NiMh	**288 V**	3507 lbs (1591 kg)	12.8 sec	94.0 mi (151 km)	6 hrs 47 min
1999 Ford Ranger EV	NiMh	300 V	4144 lbs (1880 kg)	10.3 sec	82.4 mi (133 km)	8 hrs 13 min
1999 Epic Dodge Caravan	NiMh	336 V	**4878 lbs (2213 kg)**	12.3 sec	79.1 mi (127 km)	8 hrs 45 min
1997 GM EV1	VLRA	312 V	2922 lbs (1325 kg)	**6.3 sec**	78.2 mi (126km)	5 hrs 18 min
1996 Toyota Rav4	VLRA	**288 V**	3364 lbs (1526 kg)	**13.15 sec**	68.2 mi (110 km)	8 hrs 29 min
1998 Ford Ranger EV	VLRA	312 V	4731 lbs (2146 kg)	11.6 sec	65.1 mi (105 km)	8 hrs 51 min
1997 Chev S-10 E	VLRA	312 V	4199 lbs (1905 kg)	9.75 sec	**43.8 mi (71 km)**	**5 hrs 15 min**

More than 100 EV1s remain in service. GM has donated some to schools and museums. Schools that received the EV1s are expected to use the vehicles as part of their engineering or design curricula. GM has also given more than 100 to GM employees and some state and local government agencies in Massachusetts and New York. The drivers of these vehicles are collecting performance data that will be used in the development of GM's hybrid and fuel cell vehicles.

Ford, Honda, and Toyota also manufactured BEVs. All have been discontinued. Improvements in battery technology, especially with lithium ion batteries, may extend the range of an electric vehicle enough to get the manufacturers interested in producing BEVs again.

Recently, a panel of technical experts representing the National Academy of Sciences claimed that government and industry researchers should develop battery electric vehicles as an alternative to vehicles powered by fuel cells. The panel reviewed government-funded research regarding the switching from gasoline to hydrogen. This panel came to the conclusion that fuel cells present many manufacturing and safety issues, such as on-vehicle storage of hydrogen. Based on this, the panel strongly recommended that more research be done on BEVs. GM, who spent over $1 billion on the research and development of the EV1, answered this by stating that fuel cells are closer to effective use than batteries in range, cost, and practicality. Obviously, the future of BEVs is hard to predict.

Conversions Although the major manufacturers have been concentrating on hybrids and fuel cells, BEVs are still being produced. These vehicles are being made by enthusiasts and small companies. There are also companies that specialize in providing the components needed to convert an ICEV to a BEV. Most of these converted vehicles use lead-acid batteries and have a very short driving range.

Flywheel Energy Storage Another technology for storing electrical energy is being studied: the use of **flywheel energy storage**, also called **flywheel power storage** (Figure 5-6). This system has promise for the future of electric vehicles. Flywheel power storage systems have storage capacities comparable to batteries but with faster discharge rates. Temperature does not affect the efficiency of the flywheel nor is there is a limit as to how much energy they can store. A typical flywheel storage system is composed of a large, heavy flywheel suspended by magnetic bearings and connected to a combination electric motor/generator. The flywheel is placed inside a sealed vacuum chamber to reduce friction. Kinetic energy is stored as the flywheel spins with

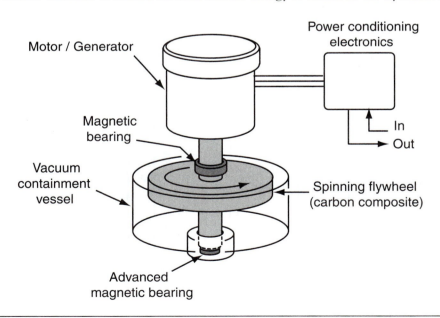

Figure 5-6 Basic construction of a flywheel energy storage system.

the motor. The momentum of the flywheel keeps it rotating when the motor stops. The rotating flywheel then spins the generator to produce electrical energy.

Major Parts

The basic systems in a BEV are a high-voltage battery pack, battery management system, the propulsion system, 12-volt system, converter and/or inverter, and the driver's displays and controls (Figure 5-7).

The battery pack consists of several battery cells connected together to provide the required voltage and current for the system. The battery management system monitors and controls the discharge and recharge activity of the batteries. It can also serve many other functions such as causing the motor to turn into a generator for regenerative braking. The battery management system may include an onboard charging system that uses an outside source of 110 or 220 volts to recharge the batteries while the vehicle is not in use. The propulsion system is made up of two primary components: the traction motor that provides the power to rotate the drive wheels and the controller that controls the power output of the motor. Two types of electric motors are used in electric vehicles: a direct current (DC) motor or an alternating current (AC) motor. The 12-volt system supplies the electrical power for vehicle's accessories, such as the radio and lights. An inverter is required to convert AC electricity to DC electricity. BEVs that use a DC motor do not need an inverter; BEVs with AC motors need the inverter to change the DC from the batteries into AC for the motor. A converter is used to reduce the system's high voltage in order to charge the 12-volt battery and to provide power to the low-voltage systems.

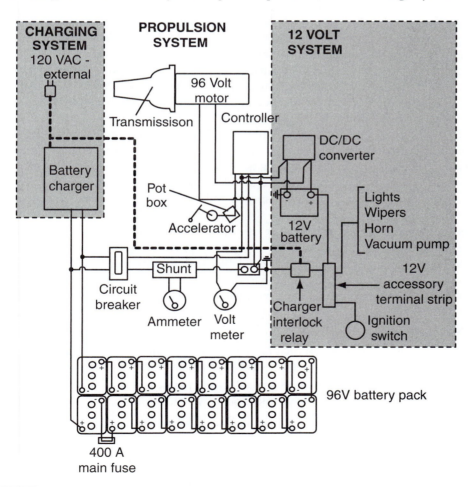

Figure 5-7 A basic wiring diagram of an electric vehicle.

Table 5-4 COMPARISON OF HYPOTHETICAL ICEV AND BEV SPECIFICATIONS.

Specification	ICEV	BEV
Engine or motor power	150 horsepower (hp)	150 kilowatts (KW) = 200 hp
Torque @ rpm	110 Ft Lbs @4500 rpm (Honda Civic)	165 Ft Lbs @ 0 rpm to 5000 rpm (AC Propulsion VW Golf)
Energy storage	20-gallon (gal) gas tank	20 kilowatt-hr (KWH) battery
Electrical system (for accessories)	12 volts	12 volts
Electrical system (for propulsion)	(none)	300 volts
Transmission	4-speed automatic	(none)

Component Specifications

Common automotive specifications include ratings of horsepower (hp) and gallons of gasoline, but in electric vehicles, different specifications are used. A comparison of EV and ICEV specifications is shown in the hypothetical example in Table 5-4. Note that power is stated in kilowatts. A kilowatt (kW) is the international unit to measure power (not only electrical); a kilowatt is 1000 watts. One kW equals 1.34 horsepower and 746 watts equals 1 horsepower. Stored electrical energy is stated in kilowatt-hours. A kilowatt-hour (kWh) is a unit of electrical energy or work, equal to what is accomplished by one kilowatt acting for one hour.

Because 1 horsepower equals 746 watts and 200 hp equals approximately 149,200 watts or 149 kW, a 149 kW motor may seem like a reasonable substitute for a 200 hp ICE. This, however, is not true! The power of ICE is always specified at the maximum horsepower it can produce at a specific speed (engine rpm). It is important to realize that the only time the engine produces this power is when it is at that speed. There is less available horsepower below and above that engine speed. The available torque from the engine also changes with engine speed; typically the maximum torque is available at an engine speed lower than the speed at which maximum horsepower is available. The available torque allows the vehicle to accelerate and pull loads. When the vehicle is maintaining highway speeds, only 10 to 20 horsepower are needed to maintain the cruising speed.

Electric motors are rated at maximum continuous (not on-demand) power output. An electric motor rated at 40 kW (approximately 54 hp) can provide a peak power output of at least 200 kW, which is five times its rating. This available power can be used for acceleration or passing. Therefore, a 30 kW motor may be able to provide the same performance as a 200 hp ICE in the same vehicle. In addition, an electric motor develops its maximum torque instantly.

Batteries

The performance of a BEV depends on a number of things, one of which is the system's voltage. Remember that higher voltages provide more power but require larger battery packs. Beyond available voltage, there are other factors that affect the performance of BEVs.

GM's EV1 was extremely aerodynamic and very light for its size. This resulted in a vehicle that accelerated very well and offered more than a respectable driving range. The EV1 was available with two different battery technologies: an advanced, high-capacity valve-regulated lead-acid and a Nickel Metal Hydride. The lead-acid pack of 26 12-volt lead-acid batteries (312 volts) stored 17.7 kWh of energy and provided an estimated driving range of 55 to 95 miles (90 to 150 km) per charge. This pack was rated at 18.7 kWh and weighed 1310 pounds (594 kg). The NiMH battery pack (343 volts) was composed of 26 13.2-volt cells capable of storing 26.4 kWh of energy with an estimated driving range of 75 to 130 miles (120 to 210 km) and weighed 1147 pounds (520 kg).

Chevrolet's S-10 Electric pickup was similarly equipped, but was quite heavy and not as aerodynamic. It was equipped with 26 lead acid batteries (312 volts) and had a driving range of about 42 miles (68 km); and required 2 to 3 hours to charge using an inductive charging station. An unusual feature of this pickup was it was modified to be front-wheel drive with the electric motor directly driving the front wheels through a differential.

Toyota's RAV4-EV was equipped with a Nickel Metal Hydride battery pack consisting of 24 individual 12-volt modules (288 volts) contained in a sealed housing (Figure 5-8). The battery pack is rated for 1,200+ recharging cycles. The battery pack was located under the floor to give the vehicle a low center of gravity for secure handling (Figure 5-9). To fully recharge the battery pack required 6 to 8 hours using a small-paddle inductive charging system.

Th!nk City (Figure 5-10) was an electric vehicle built in Norway and offered for sale by Ford Motor Company. The car was designed for commuter use, held two adults, and looked like a large golf car. It weighed only 2116 pounds (960 kg) and had a driving range of approximately 53 miles (85 km). This vehicle was equipped with a liquid-cooled three-phase AC induction motor and 19 water-cooled nickel-cadmium (NiCad) batteries (114-volts). The battery pack provided approximately 11.5 kilowatt-hours of energy and required about 8 hours to recharge.

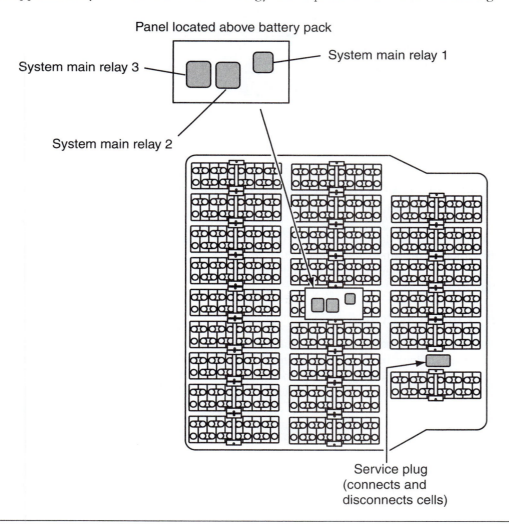

Figure 5-8 The battery pack in a RAV4-EV is composed of 24 individual batteries.

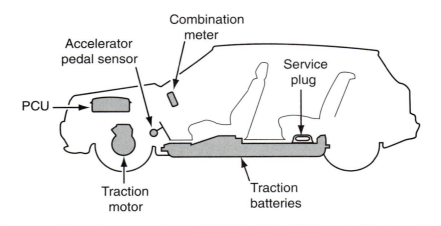

Figure 5-9 The sealed battery pack is located in the floor of a RAV4-EV for safety reasons.

Figure 5-10 Th!nk City was built in Norway and offered for sale by Ford Motor Company. *Courtesy of Georgia Power Company*

During the same time period, Ford offered an electric Ranger pickup. Unlike Chevrolet's S-10, this pickup remained rear-wheel drive and was fitted with 39 8-volt lead acid batteries (312 volts). It had a range of approximately 50 miles (80 km) and took 6 to 8 hours to recharge the batteries.

Battery Control System

The battery control or management system monitors the discharging and charging of the battery pack and the auxiliary battery. The system is tied into the charging system (both internal and external) to keep the charging voltages and currents at levels that are best for the batteries. The control system monitors the batteries' state of charge, temperature, and other vital conditions. It also controls discharge rates to optimize driving range. Many systems also have a driver alert warning system or prevention mechanism that keeps the traction motor circuit open during charging so the driver cannot inadvertently drive off when an external charger is connected to the vehicle.

Often the system is tied into the instrument panel to keep the driver aware of several conditions. As an example, Toyota's RAV4-EV has a combined gauge cluster with a **state of charge**

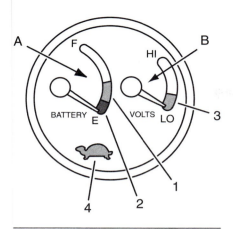

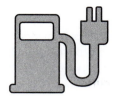

Figure 5-12 Warning lamp symbol when a recharge is needed.

Figure 5-11 (A) SOC (state of charge) meter (1) yellow zone, (2) red zone **(B)** traction battery voltmeter (3) yellow zone, and (4) the output control warning light.

(SOC) meter and traction battery voltmeter (Figure 5-11). The SOC meter indicates the battery charging capacity when the motor switch is "on" or when the batteries are being charged. Low state of charge is indicated when the gauge's needle reaches the lower part of the meter. This area is divided into two areas: the yellow zone and the red zone. The yellow zone tells the driver the batteries should be recharged. Along with the gauge reading yellow, a warning lamp (Figure 5-12) lights to further alert the driver. The warning lamp flashes when the SOC meter moves to the red zone. This tells the driver to immediately recharge the batteries. At this point, the driver may not be able to continue driving because there is not enough power.

The traction battery voltmeter displays the voltage level of the high-voltage batteries. When there is an increase in energy consumption, such as during acceleration, heavy use of accessories, or moving heavy loads, the voltage reading on the meter will decrease. When the load is decreased, the voltmeter may read higher, indicating less consumption. The traction battery voltmeter also has a yellow zone. This zone indicates heavy energy consumption and in order to have a longer driving range, drivers should drive in such a way that keeps the meter out of the yellow zone. Also, as the ambient temperature becomes colder, the voltmeter tends to enter the yellow zone because lower temperatures affect battery performance.

When the voltmeter is in the yellow zone, the output control warning light (Figure 5-13) comes on. When the battery voltage drops further, a buzzer sounds. The warning lamp alerts the driver of heavy energy consumption, high or low ambient temperatures, or low traction battery voltage. During these times, the amount of power available for the motor is limited, acceleration may be uneven, and the vehicle's maximum speed may be decreased. The warning lamp instructs the driver to drive at moderate speeds in order to reduce energy consumption.

(indicator and buzzer)

Figure 5-13 The symbol for the output control warning light. This tells the driver to drive at moderate speeds.

Traction Motors

Propulsion of a BEV is provided by traction motors. The motors are either AC or DC motors specifically designed for this use. The motors can be liquid- or air-cooled and are normally lubricated for life. Most production BEVs used AC motors and many conversion EVs use a DC motor. The latter is a consequence of cost. DC motors can be powered directly by the batteries, whereas AC motors require converters and inverters to change the DC voltage stored in the batteries into the AC required by the motors (Figure 5-14).

DC electric motors are quite reliable, but the brushes and commutators present some durability concerns. The carbon brushes spark, wear out, and the spring tension on the brushes must be kept within specifications. Excessive spring tension causes excessive friction and wear of the brushes and commutator. When the spring tension is too low, sparking occurs between the brush and the commutator causing damage to both and the brushes can bounce which breaks the circuit. Both of these problems can result in overheating the motor and a decrease in reliability. Although brushless DC motors do not have this problem, most cannot provide enough power to move a vehicle.

DC motors are typically more expensive than comparable AC motors. In addition, the available torque from a DC motor is at its peak when the rotor or armature is not turning. The available torque decreases from that peak as armature speed increases. DC motors also tend to run hotter; therefore, they need proper cooling. DC motors also do not provide for regenerative braking unless they are separately excited (Sep-Ex) DC motors fitted with a special controller that allows for efficient switching of the motor to a generator and back to a motor.

AC motors are lighter than comparable DC motors. They are also very reliable. Because they have only one moving part, the shaft, they should last the life of the vehicle with little or no maintenance. In an AC motor, there is no commutator to distort or burn and no brushes to wear or spark.

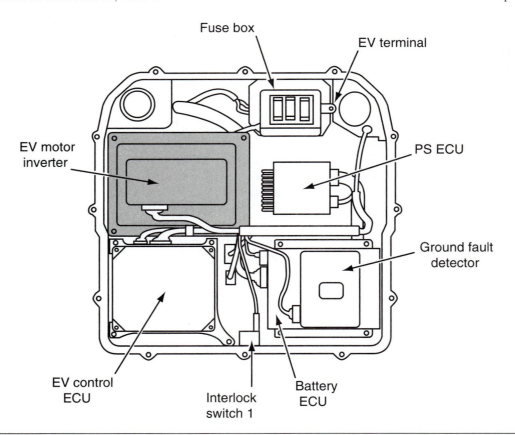

Figure 5-14 The power control unit (PCU) in a RAV4-EV contains several control units, including an inverter.

AC systems typically operate at higher voltage and lower current than DC systems with the same power output. In an AC induction motor, the torque output is constant through a wide range of speeds. This provides even acceleration and often allows driving without the need of a transmission and different speed gears. There is no need for additional controllers or electronics to have regenerative braking.

The primary disadvantage of an AC motor is, again, the cost of the electronic systems required to convert (invert) the battery's DC to AC for the motor. However, most electronic equipment is becoming less expensive (as well as better) and this disadvantage has less merit than it did a few years ago and will be less of a consideration in the future. Manufacturers used AC motors simply because of their efficiency and lighter weight.

Applications General Motors's EV1 was a front-wheel drive vehicle. It had direct drive to the differential gears, which provided a final drive gear ratio of 10.946:1. The three-phase AC induction motor was liquid-cooled and was rated at electric 102 kilowatts (137 horsepower) at 7,000 rpm. The available torque rating was 110 lb-ft (150 Nm) at 0-7,000 rpm. The EV1 was able to accelerate from 0 to 60 mph (0 to 96.6 km/h) in less than 9 seconds and had a governed top speed of 80 mph (129 km/h). In 1994, a specially modified EV1 prototype set a land-speed record of 183 mph (294.5 km/h).

Chevrolet's S-10 Electric pickup was similarly equipped. It also was front-wheel drive with direct drive and a three-phase AC induction motor rated at 85 kW (114 hp). The design of the motor is slightly different from that used in the EV1. The copper rotor used in the EV1 was replaced with an aluminum one to save weight. In addition, the size and the number of turns in the stator were changed to meet the truck's additional torque requirements.

Toyota's RAV4-EV was also a single-speed front-wheel drive vehicle. The RAV4-EV was powered by a permanent magnet AC motor (Figure 5-15) that produces 50 kW (67 hp) from 3,100 to 4,600 rpm and 140 lb/ft (190 Nm) of torque from 0 to 1,500 rpm. Its 140 lb/ft of peak torque is immediately available from 0 to 1,500 rpm, making the RAV4-EV quite responsive in city driving.

The two electric vehicles offered by Ford Motor Company were equipped with three-phase AC induction motors. The TH!NK City had direct front-wheel drive and a 27 kW motor. The Ranger-EV had a single-speed transmission with rear-wheel drive and a 90 hp motor.

In-Wheel Motors Continued research with BEVs has led to the use of different battery types, as well as different motor types. Mitsubishi has been developing a four-wheel-drive performance electric car, the Lancer Evolution MIEV. This car combines in-wheel motors with lithium-ion

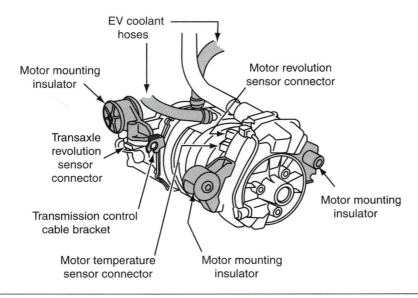

Figure 5-15 The traction motor and transaxle assembly in a RAV4-EV.

battery technology. The AC motors have a hollow donut construction that puts the rotor outside the stator. This is the opposite of conventional motor designs; however, it is less complex, provides substantial weight savings, and overcomes steering problems associated with normal wheel-mounted motors. The Lancer Evolution MIEV produces 67 hp and 382 lb/ft of torque, and has a curb weight of 3505 pounds.

Controller

A controller for a BEV is a device used to control the voltage and current to the traction motor in response to the driver's input on the accelerator (throttle) pedal. The controller also responds to movement of the brake pedal when the vehicle has regenerative braking and to stop power to the motor. The controller may also reverse the current flow to the motor when reverse gear is selected.

In electric vehicles with DC motors, a simple variable-resistor-type (Figure 5-16) controller can be used to regulate the speed of the motor. With this type of controller, full current and power is drawn from the battery all of the time. At slow speeds, when full power is not needed, a high resistance in the resistor reduces current flow to the motor. With this type of system, a large percentage of the energy from the battery is wasted as an energy loss (heat) at the resistor. The only time all of the available power is used was at high speeds.

Modern controllers adjust motor speed through **pulse width modulation (PWM)**. Pulse width is the length of time, in milliseconds, that a component is energized. Controllers rely on transistors to rapidly interrupt the flow of electricity to the motor. High electrical power (during high speed, acceleration, and/or heavy loads) is available when the intervals that the current is stopped are short. During slow speeds, little power is needed and the intervals of no current flow are longer (Figure 5-17).

Inverter/Converter

An AC power inverter converts the traction battery's DC voltage into three-phase AC voltage to power the traction motor. The output voltage varies according to the demands of the driver and the vehicle. Normally the power inverter is controlled by an electronic control module. GM called this control module the Propulsion Control Module (PCM). The output from a typical inverter is constantly being calculated using input signals from the accelerator pedal, the motor's shaft speed sensor, the motor's direction sensor, and the brake pedal. The inverter is essential to the operation of the motor and, if it fails, the motor cannot run. Electric vehicles that use DC motors do not need an inverter.

General Motors's PCM is located inside the inverter and controls six power modules that distribute the power to the motor. Each of these power modules contains six **Insulated** (or sometimes called Isolated) **Gate Bipolar Transistors (IGBTs)**. An IGBT is a combination of transistor designs; it is capable of controlling its input, high-speed switching, and increasing voltage and

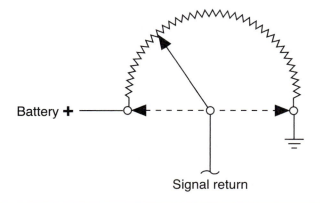

Figure 5-16 In electric vehicles with DC motors, a simple variable-resistor-type (potentiometer) controller was used to regulate the acceleration and speed of the vehicle.

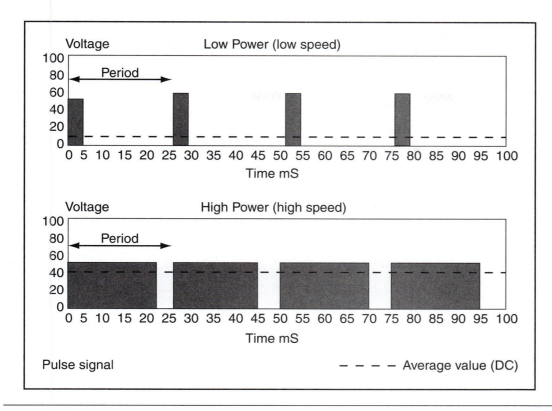

Figure 5-17 An explanation of pulse width modulation at low and high speeds.

current output. All of these actions result in a high-voltage, three-phase alternating current to power the motors. The inverter is liquid-cooled and the heat from the inverter can be used to supplement the heat from the heat pump to save energy. This is done automatically whenever the controls are set for heat.

In DC systems, the voltage and current from the battery pack merely needs to be controlled. The actual current flow to the motor is regulated by the controller and the subsequent CEMF in the motor. Remember a DC motor draws a maximum amount of current and produces its maximum torque when it has zero speed. If a DC converter fails, it is possible for a motor to receive maximum current and the vehicle can suddenly move with the highest possible torque from the motor. This could be very dangerous and is one of the primary reasons major manufacturers use AC systems in their vehicles.

A DC/DC converter reduces the voltage from the main battery pack to provide power for the 12-volt accessories, such as the head and taillights, wipers, radio, windows, power steering pump, and so on. The DC/DC converter also keeps the 12-volt auxiliary battery charged. The auxiliary battery may be used as an emergency power source if the main converter fails. Instantaneous power demands can be provided by an ultra-capacitor wired in parallel to the converter's output. The ultra-capacitor takes care of power demands for a fraction of a second.

Regenerative Braking

The controllers on most vehicles also have a system for regenerative braking. Regenerative braking is one of the most important differences between ICEVs and EVs. In ICEVs, the energy flow in the propulsion system is in only one direction: from the gas tank to the drive wheels. In EVs, the energy flow can be in both directions: from the battery to the wheels during acceleration and cruise, and from the wheels to the battery during braking or coasting. This reverse flow of electricity causes the effect known as regenerative braking, which slows the vehicle and

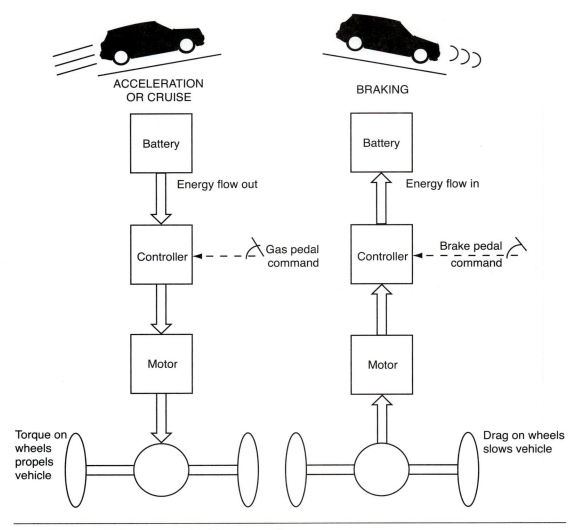

Figure 5-18 Regenerative braking energy flow.

partially recharges the battery. This is possible because electric motors can also act as generators. Figure 5-18 shows the operation of the regenerative braking system.

During the regenerative braking, some of the vehicle's kinetic energy that is normally absorbed by the brakes and turned into heat is converted to electricity by the traction motor. The energy then passes back through the controller to the battery, where it is stored, ready to be used again. The controller regulates how fast this energy is converted, and thus regulates how fast the braking occurs. The controller gets its signal from the brake pedal, telling it how much regenerative braking to apply and how much hydraulic braking will be used.

Regenerative braking increases an EV's range potential, especially when the vehicle's speed is changing, like in city traffic where the brakes are used frequently. By using regenerative braking, the driving range of a BEV can be increased by 25% (compared to relying only on the batteries). Regenerative braking not only increases the range but it also decreases brake wear and reduces maintenance costs.

Some vehicles are equipped with a switch or lever that allows the driver to activate the regenerative braking system during coasting or deceleration without depressing the brake pedal. Doing this slows down the vehicle and adds some charge to the battery. Regenerative braking is generally found on more expensive electric vehicles and all hybrid-electric vehicles (HEVs). It is not currently used on most small, simple, low-priced EVs such as golf cars.

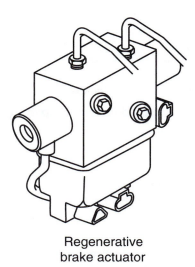

Regenerative
brake actuator

Figure 5-19 This unit senses pressure on the brake pedal and sends a signal to the controller to activate regenerative braking.

Regenerative braking works better when the generator can spin quickly. It works poorly at low speed; therefore, the controller must calculate the amount of regenerative braking based on the speed of the vehicle and the pressure applied to the brake pedal. The controller must also apply the regenerative braking smoothly so the vehicle does not jerk when it is engaged.

In most BEVs, the basic brake system is very similar to the hydraulic system used in conventional vehicles. Regenerative braking systems are not a physical part of the brake system. The hydraulic brake system is directly activated by the brake pedal and most EVs had anti-lock brakes. To save weight, GM's EV1 had electrically applied rear drum brakes, but had conventional hydraulic front disc brakes. The motor controller responds to signals from the brake pedal, accelerator, and/or hydraulic pressure sensors (Figure 5-19) and changes the motor to a generator. The torque required to turn the generator's shaft is what helps slow and stop the vehicle.

Battery Charging

Refueling a BEV simply means charging the batteries. Recharging involves connecting a battery charger to a source of electricity and connecting the charger to the battery pack. Battery chargers (Figure 5-20) may be internal or onboard (in the vehicle) or external or offboard (at a fixed location). There are advantages and disadvantages to both. An onboard charger allows the batteries to be recharged wherever there is an electrical outlet. The disadvantage of onboard chargers is their added weight and bulk. To minimize this, manufacturers normally equip the vehicles with low power chargers that require long charge times. Offboard chargers, however, force the driver to charge the batteries at specific locations but offer more power and decrease the time required to charge the batteries. Some manufactured BEVs with offboard chargers also have a convenience charger. These onboard chargers plug into standard 110-volt outlets and allow the driver to recharge batteries wherever there is electricity available.

It normally takes several hours to recharge the battery pack. The required time varies with the size and type of battery pack and battery charger (Figure 5-21). Much research and development of battery chargers has taken place. New designs of chargers have been able to recharge a battery pack in less than 20 minutes. These chargers use sophisticated electronics to monitor the cells and regulate the charging voltage and current. Being able to quickly charge the batteries would certainly make an electric vehicle more practical.

The connections between the battery charger and the power outlet can be an ordinary plug (as used for golf cars) or a specialized connector to improve safety (as used for electric

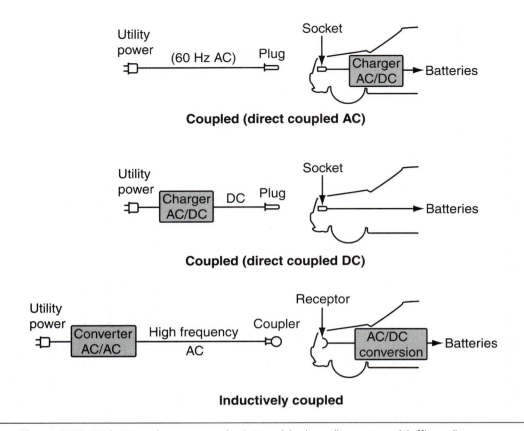

Figure 5-20 EV battery chargers may be internal (onboard) or external (offboard).

automobiles). These specialized connections contain ground-fault interrupters that break the circuit if any electrical current leakage to ground is detected, such as when charging the vehicle when it is wet. A block diagram of the golf car and electric automobile charging systems are shown in Figure 5-22.

There are two basic ways a BEV is connected to an external source of electricity for charging. One is the traditional plug, called a conductive coupling. The coupling is plugged into a receptacle on the vehicle where it connects into the wiring for the batteries. The other end of the coupling is connected to a 110-volt or 220-volt outlet or a battery charger. The other type of

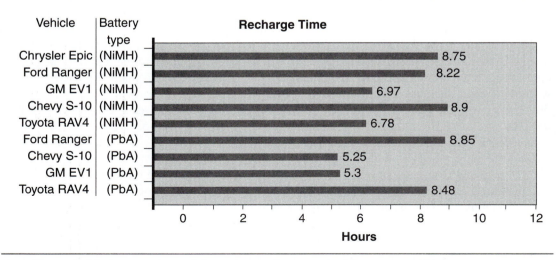

Figure 5-21 A comparison of the recharge times required for common EVs.

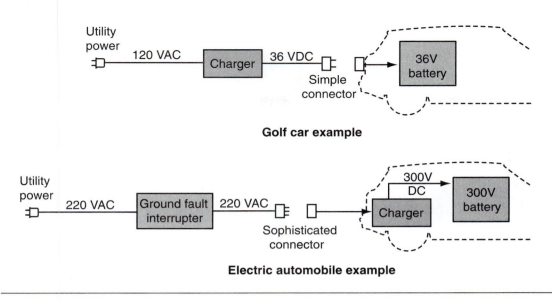

Golf car example

Electric automobile example

Figure 5-22 A block diagram of the golf car and electric automobile charging systems.

coupling is called an inductive coupling. This coupling uses a paddle that fits into a socket on the vehicle. Rather than transferring the power by a direct wire connection, power is transferred by induction. This is actually a magnetic coupling between the windings of two separate coils, one in the paddle and the other mounted in the vehicle. Because there is no exposed metal on the insulated magnetic coils, this is a very safe connection.

Charge Levels

Battery chargers are classified by the level of power they can provide to the battery pack:

- ❏ Level 1—Level 1 chargers use the standard household 3-prong electrical plug. They are usually portable and have ratings of up to 120 VAC and 15 amps.
- ❏ Level 2—Typically an onboard charger with ratings of up to 240 VAC and 60 amps.
- ❏ Level 3—An onboard charger with ratings of greater than 240 VAC and 60 amps.

Fast chargers are rated as Level 3 chargers. However, not all Level 3 chargers are fast chargers. A charger can be considered a fast charger if it is capable of charging an average battery pack in 30 minutes or less.

Conductive Charging

Conductive charging is an 110 or 220 V recharging method. AC electricity from the local utility or other source is transformed to the voltage required for the battery pack, converted into DC, and fed to the batteries via conductive, metal-to-metal contact. With a conductive charger, a connector, such as the AVCON (Figure 5-23), safely makes the link between the power supply and the vehicle's charge port. The connector makes a weatherproof direct electrical connection to the vehicle's internal charge port. This type of charging is used with most onboard chargers. Some offboard chargers also use a conductive coupling.

The connector has multiple pins that carry data. This data is used to control the action of the charger based on the conditions of the battery pack. AVCON connectors have been installed in many applications, including some Fords and Hondas. External chargers are available in many different sizes and can be wall or pedestal mounted (Figure 5-24) with an AVCON connector.

The Chrysler EPIC EV and early Toyota RAV4-EV used a different design of connector. Conductive charging is accomplished with a fuel nozzle looking connector called the ODU (Figure 5-25). The connector had many round male pins that mate to female ends in the vehicle. Similar to adding fuel to the vehicle, the connector is placed into an opening on the vehicle and refueling or recharging can take place.

Figure 5-23 An AVCON charging coupler. *Courtesy of Avcon Corporation*

Figure 5-24 An AVCON EV power pack charging system. *Courtesy of Avcon Corporation*

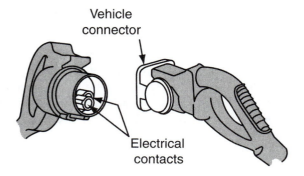

Figure 5-25 On some vehicles, conductive charging is accomplished with a fuel nozzle looking connector that has round male pins that mate to female ends in the vehicle.

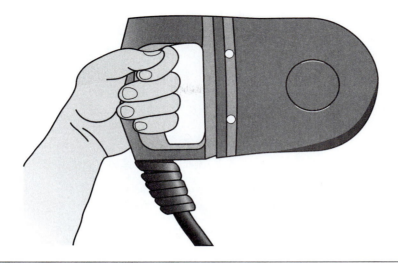

Figure 5-26 A paddle-type inductive charging coupler.

Inductive Charging

Inductive charging is a 220 VAC recharging system that transfers electricity from a charger to the vehicle using magnetic principles. To charge the batteries, a weatherproof paddle is inserted into the vehicle's charge port (Figure 5-26). The paddle and charge port form a magnetic coupling. The external charging unit sends current through the primary winding inside the paddle. The resulting magnetic flux induces an alternating current in the secondary winding, which is in the charge port. The connection is basically a transformer with the primary winding in the paddle and the secondary winding in the vehicle. The induced AC is then converted to DC (within the vehicle) to recharge the batteries. There is no metal-to-metal contact between the charge paddle and the charge port of the vehicle. This system provides a safe and easy-to-use way to recharge the batteries.

Inserting the paddle begins the charging process. The insertion of the paddle completes a communication link between the charger and the vehicle. The charger displays what percent of charge remains in the batteries and an estimate the time needed to fully charge the batteries. This link also allows the charging unit to enter into self-diagnostics and prevents the vehicle from being driven while the paddle is inserted in its port. If the charging cable becomes damaged or cut, power will shut off within milliseconds. The charging process ends immediately after the paddle is removed from the port.

The EV1 was one of the first vehicles to use inductive charging systems. Nissan and Toyota, as well as GM's S-10 EV, followed and used the technique. GM included a 220-volt MagneCharge™ inductive charger with each EV1 it leased. Drivers of the EV1 had the inductive charging station installed in their homes. Using an external charger reduces the weight of the vehicle and simplifies the charging process. In addition to the home charger, GM formed partnerships with utility companies to build an electric car charging infrastructure (Figure 5-27).

Charge Times

There are three primary things that affect the required time to recharge the batteries: the current state of charge of the battery, the chemicals used in form the cells of the battery, and the type of charger used.

Using the EV1 as an example, the estimated time required to take the sealed lead-acid battery pack with 0% charge to 100% SOC is 5.5 to 6 hours using the 220-volt (6.6kW) charger, or 22 to 24 hours using the 110-volt (1.2kW) convenience charger. The recharge times for the Nickel -Metal Hydride battery pack is slightly different. They require 6 to 8 hours using the 220-volt (6.6kW) charger and only 2 to 3 hours to charge with the inductive charger.

Figure 5-27 This charging station was released by GM and Southern California Edison. It was called the Magne Charger® and was capable of filling a battery pack from almost empty to almost full in about 12 minutes. *Courtesy of Southern California Edison*

Charging Procedures

Each vehicle has a specific charging procedure. These procedures vary with the type of charger, charger coupling, and battery. Always follow the procedure for the vehicle being worked on. The following are some general guidelines to follow:

❏ Make sure the gear selector is in the park position and the parking brake is applied before charging.

❏ Before charging, make sure the motor switch is off and the key is removed.

❏ To reduce the likelihood of explosion or fire, charge the batteries in a well-ventilated area. Flammable gas may be produced by the batteries during charging.

❏ To avoid getting an electric shock, never operate the charger with wet hands.

❏ Avoid charging under high temperatures or direct sunlight.

❏ When the ambient temperature is high, charge indoors, in the shade, or at night.

❏ When the ambient temperature is low, it is recommended to charge indoors.

❏ Never touch the terminals of the conductive terminals on the vehicle or coupler; you may get an electric shock.

❏ Do not modify the charge coupler.

❏ The charge coupler should be firmly installed without any tension on the cable.

❏ If the charge coupler is damaged, repair or replace it as soon as possible. The use of a damaged coupler can cause burns or electrical shock.

❏ Make sure water, dirt, or other foreign objects do not enter the charge port on the vehicle. This can cause a failure of the equipment or create an unsafe condition.

❏ When charging the batteries, always apply a full charge.

❏ Do not disconnect the charge coupler until the batteries are fully charged, unless it is necessary to prematurely stop charging.

❏ The auxiliary battery can discharge if the charge coupler is left inside of the charge port for a long period of time after charging is completed.

Accessories

As with ICEVs, electric vehicles have a number of auxiliary systems and quite a few accessories. For example, the RAV4-EV was available with power mirrors, power windows and door locks, rear window defroster and wiper, heated front seats, heated windshield, traction control, and audio system. In addition, it had a four-wheel anti-lock brake system (ABS) and dual front airbags, in addition to the normal light, horn, wiper, and other safety-related systems.

A major difference between BEVs and ICEVs is the operation of these systems. In an ICEV, all of this equipment is powered by the gasoline engine. Belts drive the power steering, air conditioning, and 12-volt generator; there is engine vacuum for the power brakes; and there is hot engine coolant for heating. In a BEV, the traction motor is not used to rotate belts and pulleys and there is no engine vacuum. Therefore, all of these auxiliaries and accessories are powered directly by the high-voltage traction battery or the auxiliary 12-volt battery (see Table 5-5). The 12-volt battery operates the lights, radio, and other 12-volt systems.

Some systems, such as the radio, lights, and horn, operate the same way as they do in a conventional vehicle. Other systems, such as the power steering and power brakes, require additional small electric motors, which have an impact on the vehicle's range. The range is especially affected by the air conditioning and heating systems. Because all accessories and auxiliaries systems operate on electricity, their electrical power needs reduces the capacity of the battery. Table 5-6 shows how the major systems affect the driving range of a typical BEV.

Table 5-5 COMPARISON OF ACCESSORIES OPERATION.

	Vehicle Type	
Accessory	ICEV	BEV
Power steering	Fan-belt drives power-steering hydraulic pump	Electric motor drives power steering directly, or a hydraulic pump in some models
Power brakes	Engine provides vacuum for brake power assist	Electric motor operates vacuum pump
Air conditioning	Fan-belt drives compressor	Electric motor drives compressor
Heating of passenger compartment	Hot water from engine cooling system	Electric heater or heat pump
12-volt accessories	Fan-belt drives 12-volt generator to charge battery	DC/DC converter reduces high voltage to 12 volts

■ **CAUTION:**
When working on BEVs, remember that the auxiliaries and accessories may be powered by high voltage. Never attempt to work on these components (or the main propulsion system components) without thorough training that includes all safety procedures. Normally, the high-voltage components are housed in the controller or another safety housing, and the high-voltage cables are identified by their orange color.

Table 5-6 THE IMPACT OF ACCESSORIES ON BEV RANGE.

Accessory	Range Impact	Comments
Air conditioning	Up to 30%	Highly dependent on use, ambient temperature, cabin temperature, and air volume
Heating	Up to 35%	Highly dependent on use and ambient and cabin temperatures
Power steering	Up to 5%	
Power brakes	Up to 5%	
Defroster	Up to 5%	Depending on use
Others, such as lights, stereo, power-assisted seats, windows, door locks	Up to 5%	Depending on use

HVAC

To meet federal safety standards, all vehicles must be equipped with passenger compartment heating and windshield defrosting systems. In an ICEV, these systems use the heat of the engine's coolant. In a BEV, there is no engine and therefore there is no direct source for heat. The heat must be provided by an auxiliary heating system. Some electric vehicles use an electric resistance heater with a fan, which requires a significant amount of electrical energy and can reduce the driving range by as much as 35%.

Other BEVs have liquid heaters. Water, or a mixture of water mixed with ethylene glycol, is held in a tank. The liquid in the tank is kept heated by a resistive heating element submerged in the tank. When the driver turns on the heating system, a small pump circulates the heated liquid through a heater core in the passenger compartment. A fan moves air over the core to provide heated air.

BEV air conditioning systems also have a significant impact on the driving range. In many cases, the air conditioning system uses a high-voltage motor to rotate the compressor. Obviously, the energy used to power the air conditioning puts a drain on the battery pack. The amount of energy consumed by the air conditioning system depends on how often it is used, the outside temperature, and the selected temperature for the passenger compartment.

Many manufactured BEVs were equipped with an electrically driven heat pump. A **heat pump** functions as either a heater or an air conditioner (Figure 5-28). They also use much less energy. A heat pump does not produce heat; rather, it transfers heat. This is why heat pumps require less energy to operate.

Air, regardless of temperature, has heat. Naturally, the colder the air is, the less heat it has. When a heat pump is functioning as a heater, it takes heat from the outside air and transfers it to the passenger compartment. When it is functioning as an air conditioner, heat from inside the passenger compartment is transferred to the outside air.

Most vehicles equipped with a heat pump have auxiliary heating systems that provide additional heat when the outside temperatures are very low. These heating systems can be electrical heating elements or diesel fuel-fired heaters.

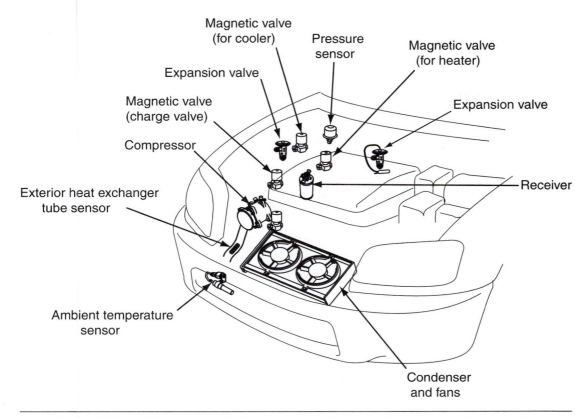

Figure 5-28 A heat pump system looks very much like a traditional air conditioning system.

Some vehicles equipped with heat pumps have a feature that allows for warming or cooling of the passenger compartment while the batteries are being recharged. Because the charger is replacing the energy consumed by the heat pump, the batteries can still be fully charged.

Power Brakes

Many power brake systems use engine vacuum and atmospheric pressure to multiply the effort applied to the brake pedal during braking. Because there is no engine in a BEV, there is no direct vacuum source. However, normal vacuum assist power brake systems can be used if fitted with an electrically powered vacuum pump. These pumps are similar to those used on diesel-engined vehicles. The pump may be connected to a storage tank. The tank reduces the time the pump needs to operate and therefore minimizes the effect the pump has on driving range.

Another type of power brake system uses hydraulic pressure, from a pump, to reduce the pedal effort required to apply the brakes. Some BEVs use an electric pump to provide the necessary hydraulic pressure (Figure 5-29). These systems are called **electro-hydraulic brake systems**.

Because both types of power brake systems for BEVs operate on electrical power, brake boost is available at all times. The rest of the brake system in nearly all electric vehicles is the same as that found on any vehicle.

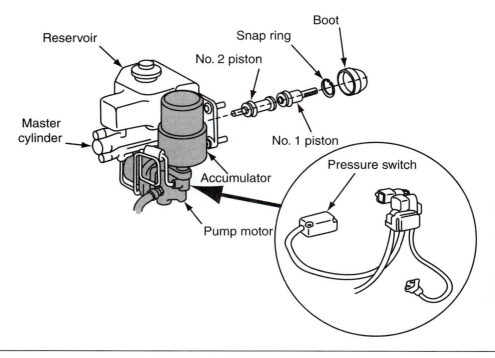

Figure 5-29 The pump assembly for an electro-hydraulic brake system.

Power Steering

Most vehicles use hydraulic pressure to reduce steering effort. In conventional vehicles, the power steering pump is driven by the engine. This pump can be driven by an electric motor, which is how many BEVs are equipped (Figure 5-30). The control for the pump can be programmed to provide more assist at lower speeds, and less at higher speeds. The system can also be programmed to only run the pump when it is needed; this reduces the effect power steering has on the driving range. These systems are called **electro-hydraulic steering systems**.

Some power steering systems are purely electrical and mechanical systems. An electric motor moves the steering linkage. These systems are programmable and the energy consumed by the motor depends on the amount the steering wheel is turned. While driving straight, the motor may not run. However, when the steering wheel is fully turned, the motor is drawing its maximum current.

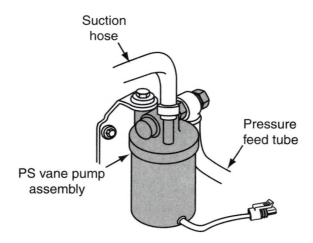

Figure 5-30 A electrically driven power steering pump.

Driving an EV

Driving a BEV is like driving any other vehicle but with some notable exceptions. There is still a steering wheel, a brake pedal, and an accelerator pedal. However, there is little, if any noise, and no emissions. A BEV typically has adequate acceleration and can travel at highway speeds. The biggest difference for the driver is that attention must be paid to the consumption of energy. Failure to minimize consumption and carefully plan travel routes can lead to reduced power and a need to recharge the batteries at inconvenient locations or times. If the batteries are not charged, the vehicle will not move.

Starting

The biggest adjustment a driver needs to make when preparing to drive a BEV is starting it or getting it ready for action. A BEV has no noise or vibration when it is ready to go. The driver must look at the instrument cluster to determine it is ready. Like conventional vehicles, the ignition (motor) switch has several positions. One is "lock," during which the traction motor is off and the steering wheel is locked. The key can be removed only at this position. "Accessories" allows some accessories to work but the traction motor is off. "START" actually gets the traction motor ready to work, and "ON" is the normal position for driving. Never leave the switch in the on position when the vehicle is not in use. Doing so can discharge the auxiliary battery and damage the traction motor.

Before starting, make sure the charge coupler is not connected to the vehicle. Always check that charging is completed and then disconnect the charge coupler. The traction motor will not run with the coupler in place. Make sure the gear lever is in the "park" position and that the parking brakes are on. The accelerator should never be depressed during starting.

To turn on the traction motor, turn and hold the motor switch to START with the brake pedal depressed until the READY light in the instrument cluster comes on (Figure 5-31). On some vehicles, a buzzer will sound when this happens. Once the READY lamp is lit, the motor switch can be released to allow it to move to the ON position. At this point, the traction motor will run when the accelerator is depressed and all accessories are ready to operate. If the READY light does not illuminate during the start process, there is a problem with the traction motor or its circuit, or the auxiliary battery is discharged.

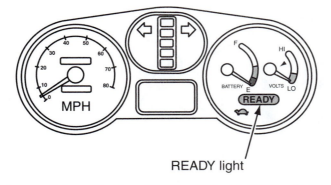

READY light

Figure 5-31 The READY light in the instrument cluster of a RAV4-EV.

Driving and Braking

Most BEVs have a single-speed automatic transmission and the gearshift lever has five positions (Figure 5-32):

❑ **P** — Position for parking, engine starting, and key removal
❑ **R** — Reverse position
❑ **N** — Neutral position
❑ **D** — Normal driving position
❑ **B** — Position for engine braking (regenerative braking)

In addition to these positions, the vehicle may have an "engine brake (EB)" button or switch. When this switch is on, regenerative braking will slow the vehicle when the accelerator is released. When the shift lever is in the "B" position, more regenerative braking will take place.

Typically, the shift lever can only be shifted out of "P" when the motor switch is in the "ON" position. When moving out of "P" into "D" or "R," the brake pedal must be depressed. It is important that the accelerator not be depressed when shifting gears. Doing this can cause the vehicle to unsafely and quickly move and can cause damage to the motor. Once the shift lever has been moved and with the brake pedal still depressed, the parking brake can be released.

To begin moving, press the accelerator. Drive normally with the realization that because the transmission is a single speed, the accelerator is the only thing that controls vehicle speed. When the accelerator pedal is released while driving in the EB mode, vehicle speed will decrease because the wheels are now turning the motor that just became a generator. If more vehicle slowing and/or more regeneration are needed, the shift lever can be moved down to the "B" position. Once coasting has been completed, the lever can be moved back into the "D" position.

To back up, bring the vehicle to a complete stop. Then depress the brake pedal and move the shift lever into the "R" position. It is important to keep in mind that a BEV can accelerate just as quickly in reverse as it does in drive. However, it is more difficult to steer any vehicle in reverse therefore the accelerator should be gently pressed when backing up.

To park and shut down the vehicle, come to a complete stop. Then apply the parking brake. While depressing the brake pedal, move the shift lever to the "P" position. Now turn the motor switch to the "LOCK" position and remove the key.

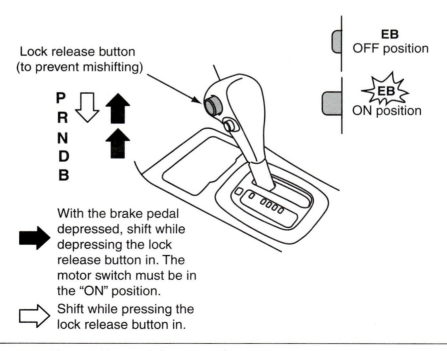

Figure 5-32 The possible gearshift positions for a single speed transmission.

Maximizing Range

The driving range of a BEV is reduced by high driving speeds, stop-and-go driving, hills, cold weather (requiring use of heater), warm weather (requiring use of the air conditioner), and the age of the battery. If these conditions are minimized, driving range can be extended. There are certain things a driver can do to extend the range and the life of the batteries.

- ❏ Keep the voltmeter out of the heavy consumption or yellow zone.
- ❏ Avoid high-speed driving. Maintain a moderate speed on highways.
- ❏ Avoid driving up inclines.
- ❏ Avoid frequent speed increases or decreases. Attempt to drive at a steady pace.
- ❏ Avoid unnecessary stopping and braking.
- ❏ Avoid heavy stop-and-go traffic whenever possible.
- ❏ Avoid full throttle acceleration.
- ❏ Accelerate slowly and smoothly.
- ❏ The vehicle should be well maintained, including proper tire inflation pressures.
- ❏ Unnecessary weight in the vehicle will shorten the driving range.

Basic Diagnosis

Diagnosis of BEV concerns can be simpler than diagnosing concerns on a conventional vehicle because there are fewer components. However, most manufactured BEVs have complex electronics that are unique and require a solid understanding of how the vehicle's systems operate. Fortunately, most BEVs have self-diagnostics with retrievable trouble codes.

Manufacturer-supplied checklists are especially helpful when deciding what should be known about a particular problem and repair (Figure 5-33). In the vehicle's service manual, there may be symptom-based diagnostic aids. These can guide you through a systematic process. As you answer the questions given at each step, you are guided to the next step.

When these diagnostic aids are not available or prove to be ineffective, good technicians conduct a visual inspection and then take a logical approach to solving the problem. Logical diagnosis follows these steps:

1. Gather information about the customer's concern. Find out when and where it happens and what exactly happens.
2. Verify that the problem exists. Take the vehicle for a road test and try to duplicate the problem, if possible.
3. Thoroughly define what the problem is and when it occurs. Pay strict attention to the conditions present when the problem happens. Also pay attention to the entire vehicle; another problem may be evident to you that is not evident to the customer.
4. Research all available information and knowledge to determine the possible causes of the problem. Try to match the exact problem with a symptoms chart or think about what is happening and match a system or some components to the problem.
5. Isolate the problem by testing. Narrow the probable causes of the problem by checking the obvious or easy to check items.
6. Continue testing to pinpoint the cause of the problem. Once you know where the problem should be, test until you find it!
7. Locate and repair the problem, then verify the repair. Never assume that your work solved the original problem. Make sure the problem is resolved before returning it to the customer.

CUSTOMER PROBLEM ANALYSIS CHECK

EV CONTROL SYSTEM Check Sheet	Inspector's Name

Customer's Name		Model	
Driver's Name		Model Year	
Data Vehicle Brought in		Frame No.	
License No.		Odometer Reading	km / miles

Problem Symptoms	☐ READY does not turn ON ☐ Vehicle does not move ☐ Poor acceleration ☐ Noise ☐ Vibration ☐ Harshness ☐ Smoke is rising ☐ Smell of or the likes burn ☐ Other _____

Data Problem Occurred		
Problem Frequency		☐ Constant ☐ Sometimes(times per day/month) ☐ Once only ☐ Other _____
Condition When Problem Occurred	Weather	☐ Fine ☐ Cloudy ☐ Rainy ☐ Snowy ☐ Various/Other _____
	Outdoor Temperature	☐ Hot ☐ Warm ☐ Cool ☐ Cold(approx. ____°F/____°C)
	Place	☐ Highway ☐ Suburbs ☐ Inner City ☐ Uphill ☐ Downhill ☐ Rough road ☐ Other _____
	Traction Motor	☐ Just after starting vehicle(min.) ☐ Standing with READY ON ☐ Driving ☐ Constant speed ☐ Acceleration ☐ Deceleration ☐ Other _____

Condition of MIL	☐ Remains on ☐ Sometimes lights up ☐ Does not light up
DTC Inspection	☐ Normal ☐ Malfunction code(s) (code) ☐ Freeze frame data ( )

Figure 5-33 A checklist for inspecting and road testing a BEV. *Courtesy of Toyota Motor Sales, U.S.A., Inc.*

Precautions

During diagnosis and repair of a BEV always keep in mind that the vehicle has very high voltage. This voltage can kill you! Therefore, always adhere to the safety guidelines given by the manufacturer. Here are a few of things that should be done to prevent being shocked by the vehicle's electrical system:

❏ Wear dry and undamaged insulated gloves while working on the vehicle.
❏ Disable or disconnect the high-voltage system. Do this according to the procedures given by the manufacturer.
❏ After the high-voltage system is disconnected, wait the prescribed time before handling any part of the system.
❏ Always use insulated tools.
❏ When disconnecting electrical connectors, do not pull on the wires. When reconnecting the connectors, make sure they are securely connected.
❏ Do not leave tools or parts anywhere in the vehicle.
❏ Do not wear metallic objects such as rings and necklaces.

Self-Diagnostics

The vehicle's control unit or computer may have a built-in self-diagnostic system. In these systems, a malfunction in the computer or in components of the propulsion system can be detected. When a fault is detected, the computer will store that information and may illuminate the **Malfunction Indicator Light (MIL)** on the instrument panel. The faults held in the computer's memory can be retrieved as **Diagnostic Trouble Codes (DTCs)**.

To retrieve these codes, connect the hand-held scan tool to the appropriate **Data Link Connector (DLC)** on the vehicle (Figure 5-34). The scan tool will also be able to display other operational data. Many scan tools also have a freeze frame feature. With this feature, the tool records the conditions that were present when a particular malfunction was detected.

Before connecting the scan tool to the vehicle, measure the voltage of the auxiliary battery. If the voltage is lower than specifications, recharge it before continuing with your tests. Also, inspect all fuses, fusible links, wiring harness, connectors, and ground in the low-voltage circuit. Repair them as necessary.

Turn the motor switch to the ON position and make sure the MIL is lit. If the MIL does not light, check for a burnt bulb, a bad circuit fuse, or an opening in the circuit. Again, correct the problem before proceeding. The MIL should go off when the READY lamp lights. If the MIL stays on, the computer has found a problem and related information is stored in its memory. Turn the motor switch to the OFF position.

Make sure the scan tool is set up for the vehicle being tested. Then connect it securely to the DLC. Turn the motor switch to the ON position and turn the scan tool on. Check for DTCs and freeze frame data and record all codes and data displayed on the scan tool. Refer to the manufacturer's reference to determine what the DTCs indicate.

Following the correct procedures, verify the trouble and repair the problem. After completing any repair of the motor or related parts, erase the DTCs retained in the computer's memory with the scan tool. Then test it again to make sure the fault is no longer present.

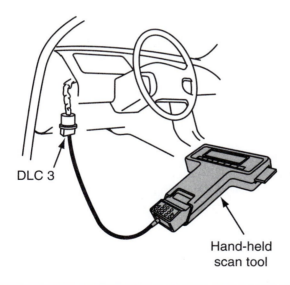

DLC 3

Hand-held
scan tool

Figure 5-34 A scan tool connected to the DLC for the electric drive on a BEV.

Building or Converting

Many individuals who want an electric vehicle build one themselves. They do this because the major manufacturers no longer produce any and they want the challenging but rewarding experience of making their own. Of course, they want an electric vehicle for the obvious reasons: no emissions, no gasoline, less maintenance, and so on.

Most homemade BEVs are based on a conventional vehicle. There are companies that specialize in EV conversions and can provide guidance for making this change. What follows is a brief outline of what is involved. The exact steps that should be followed depend on many factors: your driving needs, driving desires, the vehicle to be converted, and budget. The Internet has much information available that can help in the planning and building process.

The conversion process follows these basic steps:

1. Determine exactly what you expect from your electric vehicle and what you can afford.
2. Do some research to find conversion kits for your specific vehicle. Most conversion kits include everything you need, except for the batteries.
3. If a kit is not available for that vehicle, locate sources for the following components:

   ❏ Motor—most conversions use a DC motor to keep the expenses low
   ❏ Adaptor plate—needed to mount the motor to the transmission
   ❏ Motor controller—must be matched to the motor
   ❏ Potbox (Potentiometer)—this is the accelerator
   ❏ Battery pack—voltage must be matched to the needs of the motor
   ❏ Auxiliary battery—can be the same battery as was in the vehicle
   ❏ Main contactor— an ignition switch that matches the voltage of the new system
   ❏ Switching circuit breaker—allows for quick shutdown of the high-voltage circuit in the case of an emergency
   ❏ Main fuse—this fuse is for the high-voltage system and must be matched to the voltage of the system
   ❏ Voltmeter—match the range of the meter to the voltage range of the vehicle
   ❏ Charger interlock relay—prevents power to the motor when the batteries are being charged
   ❏ DC/DC converter—lowers system voltage to charge the auxiliary battery

4. Determine and secure all tools and supplies that will be needed.
5. List all of the safety precautions that should be followed during and after the conversion.
6. Remove all of the ICE components on the vehicle.
7. Install the motor, components, battery box, and batteries.
8. Route and connect the wiring for propulsion system, auxiliary 12-volt system, charging system, meters, and controls.
9. Check all wiring and connections.
10. Connect the batteries.
11. Test the battery charger.
12. Test the vehicle.

Review Questions

1. Why are carbon dioxide emissions a concern?
2. List five things a driver of a BEV can do to extend the driving range of the vehicle.
3. Which of the following statements about the use of heat pumps in BEVs is NOT correct?
   - **A.** A heat pump is very efficient at producing heat.
   - **B.** When a heat pump is functioning as a heater, it takes heat from the outside air and transfers it to the passenger compartment.
   - **C.** When it is functioning as an air conditioner, heat from inside the passenger compartment is transferred to the outside air.
   - **D.** Most vehicles equipped with a heat pump have auxiliary heating systems that provide additional heat when the outside temperatures are very low.
4. What makes up the propulsion system in a BEV?
5. What basic factors affect the required time to recharge the battery pack in a BEV?
6. There are two basic ways a BEV is connected to an external source of electricity for charging: conductive and inductive. What is the difference between the two?
7. List three advantages for using an AC motor rather than a DC motor in a BEV.
8. Explain what the ratings of kilowatts and kilowatt-hours mean.
9. What is the purpose of an AC power inverter?
10. List some of the components that must be purchased in order to convert an ICEV to a BEV.

HEV BASICS

After reading and studying this chapter, you should be able to:

❏ Explain why a hybrid vehicle is more efficient than a vehicle powered only by an internal combustion engine.

❏ Describe the basic difference between series and parallel hybrid configurations.

❏ Explain why hybrid vehicles are more expensive to manufacture.

❏ Describe the importance of electronics in the operation of a hybrid vehicle.

❏ Explain how the stop-start feature operates.

❏ Explain how regenerative brakes work.

❏ Describe how the operation of accessories and auxiliary systems in a HEV differ from those in an ICEV and a BEV.

❏ Explain the differences between a full, assist, and mild hybrid.

❏ Describe the primary advantage of plug-in hybrid vehicles.

❏ Describe the basic operation of a hydraulic hybrid.

Key Terms

Assist hybrid
Full hybrid
Hydraulic hybrids
Mild hybrid
Plug-in hybrid electric vehicles (PHEVs)

Introduction

Hybrid electric vehicles (HEVs) are the only electric drive vehicles manufactured for highway use today. Hybrid vehicles combine the technologies of battery-operated electric vehicles (BEVs) and internal combustion engine vehicles (ICEVs). HEVs are designed to take advantage of the positives of ICEVs and the positives of BEVs. They also suffer from some of the disadvantages of each, however these are minimized.

This chapter covers the basics of hybrid technology. It describes the various configurations and equipment used to construct a current hybrid vehicle. It also discusses some of the primary benefits and features of the typical hybrid vehicle. Later chapters detail each of the configurations and equipment options. This chapter also explores some of the other hybrid designs currently in development including plug-in hybrids, hydraulic hybrids, and various powerplant combinations.

What Is a Hybrid Vehicle?

A hybrid vehicle is one that has at least two different types of power or propulsion systems (Figure 6-1). Ideally, each of them work together to improve the efficiency and performance of each other while minimizing the disadvantages of each other. Today's hybrid vehicles have an internal combustion engine and a battery-powered electric motor (Some vehicles have more than one electric motor.).

The logic for using two power sources is simple. A typical ICEV has more available power than it needs for most driving situations. Most ICEs can produce more than 150 horsepower, however only 20 to 40 horsepower are typically needed to maintain a cruising speed. The rest of the horsepower is needed only for acceleration and overcoming loads, such as climbing a hill. These high-output engines use quite a bit of gasoline when they are providing the power to accelerate. An electric motor consumes no fuel and can provide near instantaneous power. Hybrid vehicles typically use a smaller ICE and an electric motor to provide additional power for acceleration and overcoming loads.

Hybrid vehicles use much less fuel in city driving than comparable ICEVs. This is because the engine does not need to supply all of the power required for stop and go traffic. The power from the electric motor supplements the engine's power. There is also improvement in highway

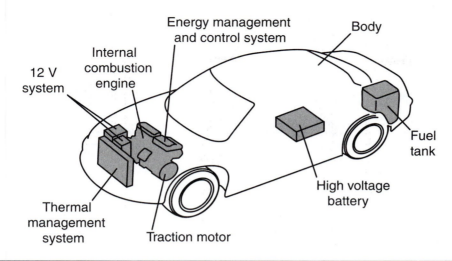

Figure 6-1 The main components of a hybrid electric vehicle.

fuel mileage, because of the use of smaller and more efficient engines. These advanced engines, however, cannot produce the power needed for reasonable acceleration by themselves.

The overall efficiency of a hybrid can be, and in most cases is, enhanced by a number of other features. One of these is the stop-start system. When a hybrid vehicle is stopped in traffic, the engine is temporarily shut off. It restarts automatically when the driver presses the accelerator pedal, releases the brake pedal, or shifts the vehicle into a gear. In addition, to reduce the required energy to drive the generator, hybrids have regenerative braking. Rotated by the vehicle's wheels, the electric drive motor acts as a generator to charge the batteries when the vehicle is slowing down or braking. This feature recaptures part of the vehicle's kinetic energy that would otherwise be lost as heat in a conventional vehicle.

No current hybrid needs to be plugged in to recharge the batteries. However, there is much research being done on "plug-in" hybrids. It is felt that externally charging the batteries, in addition to regenerative braking and the ICE-driven generator, would decrease the use of gasoline and significantly increase the time the vehicle could be driven by electricity only.

Most hybrids use transmissions specifically designed to keep the ICE operating at its most efficient speed. Efficiency can also be increased by the use of low-rolling resistance (LLR) tires, which are stiff and narrow to minimize the amount of energy required to turn them. Hybrids may also be designed to minimize aerodynamic drag and made lighter.

Types

Depending on the system used, the engine may power the vehicle by itself, or it may drive a generator, or both. The electric motor drives the wheels, sometimes alone, or sometimes the engine and motor work together to provide the required power. Often, the motor is used as the propulsion unit when the vehicle is traveling at low speeds. Once a specific speed or load is reached, the engine takes over. The electric motor may also provide a power boost when there is a heavy load on the engine. The motor is powered by the batteries and/or ultra-capacitors, both of which are charged by the generator and regenerative braking. The engine may use gasoline, diesel, ethanol, methanol, compressed natural gas, hydrogen, or another alternative fuel.

Often hybrids are categorized as series or parallel types (Figure 6-2). In a series hybrid, the engine never directly powers the vehicle. Rather it drives a generator, and the generator either charges the batteries or directly powers the electric motor that drives the wheels. Currently, no true series hybrid automobiles are manufactured. In a parallel hybrid, the engine, the motor, or both can power the drive wheels. The engine also drives the generator to charge the battery pack.

There are many variations of hybrid systems used by manufacturers today. Some vehicles labeled as hybrids do not fit into either category. These are called mild or micro hybrids and will be discussed later in this chapter, as will other classifications used to describe a hybrid.

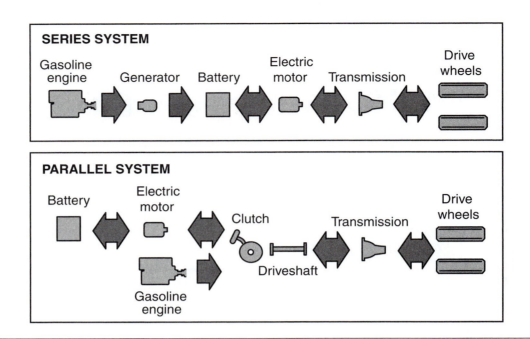

Figure 6-2 Basic layout of series and parallel hybrid powertrains.

Fuel Cell Hybrids Not all hybrids have an engine and electric motor. By definition, a HEV is a vehicle powered by two separate energy sources. Some hybrids being developed combine a fuel cell with batteries to power electric motors. The electricity produced by the fuel cell can be used to charge the batteries and power the motor. Hydrogen fuel cells generate electric energy without combustion and therefore have zero emissions. Also, fuel cells produce DC voltage that can be used to power motors and lights, and charge batteries without additional electronics. Nearly all fuel cell vehicles have a series hybrid configuration.

Benefits of a Hybrid

Hybrid vehicles have the potential to be two to three times more fuel-efficient than conventional vehicles, with much lower emission levels. Also, the combination of an engine and electric motor can provide increased power and/or additional auxiliary power for electrical devices. A hybrid vehicle has two main advantages over a battery-operated electric vehicle:

1. Drivers are more comfortable with HEVs because there are no battery range limitations (therefore no fear of being stranded) and the vehicles are refueled in the same way as any gasoline-powered vehicle.
2. A hybrid requires a much smaller battery than a pure EV, which dramatically reduces the vehicle's price. The battery cost in a pure EV can be as much as $30,000, whereas the battery cost in hybrids is less than $5000.

When comparing the pollution and fuel consumption of EVs and HEVs, it is important to remember that EVs are not completely pollution-free, nor oil-independent. Many of the electric power plants that generate the electricity to charge EVs burn oil, natural gas, or coal. Only electric power produced by hydroelectric dams, wind turbines, solar, or geothermal plants are pollution-free and are renewable (they do not consume any type of fossil fuel). These clean sources, however, produce only a small fraction of the electricity in the United States today, and most electricity is generated by burning a fuel.

Although HEVs still burn gasoline and emit exhaust pollutants, the amount of pollution produced, and gasoline consumed per mile are substantially less than in a typical gasoline-powered vehicle because:

❏ The gasoline engine can be smaller for the same level of vehicle performance because there is electric motor assist.

❏ The propulsion system can be designed to allow the engine to run at its most fuel-efficient speed.

❏ The engine can be shut off while the car is stopped at traffic lights, decelerating, or moving at low speed.

❏ Regenerative braking captures and recycles much of the energy used to accelerate the vehicle.

Fuel Economy

Hybrids consume significantly less fuel than vehicles powered by gasoline alone. Therefore, they can reduce the country's dependence on fossil fuels and foreign oil. (Fuel economy ratings for hybrid and conventional vehicles can found at *www.FuelEconomy.gov*.) Table 6-1 compares the performance of a standard Honda Civic to a Civic HEV. As you can see, there is a substantial difference in fuel consumption. Also, notice, although the acceleration times are slightly different, other performance measurements are near identical.

HEVs typically have the same or greater range than conventional vehicles. For example, Honda's Insight has a range of about 700 miles on a tank of fuel. This extended range certainly offsets the disadvantage of a BEV, which has been impaired by having very short driving ranges.

Air Pollution

Hybrids can have more than 90% fewer emissions than the cleanest conventional vehicles. HEVs also produce significantly lower total fuel-cycle ("well to wheel") emissions when compared to equivalently sized conventional vehicles.

The low emissions result from the use of smaller and more efficient internal combustion engines. The engine's power is boosted by electric motors that produce zero emissions. Also, the engine can be shut down when it is not needed. In addition, many HEVs can move in an

Table 6-1 COMPARING TWO HONDA CIVICS. *COURTESY OF THE U.S. DEPARTMENT OF ENERGY*

What's the Difference?	Civic EX	Civic HEV
On-Road Fuel Economy (mpg)	40.1	47.8
EPA Fuel Economy (city/highway; mpg)	31/38 (auto)	48/47 (CVT)*
Acceleration (0–60 mph)	10.84	12.88
Slalom (cones every 50 ft; average mph)	33.25	33.39
Skid Pad (average G clockwise)	0.69	0.66
Braking (ft from 60–0 mph)	140	136
Passenger/Luggage Volume (ft³)	88.1/12.9	91.4/10.1
Curb Weight (lb)	2,615 (auto)	2,732 (CVT)

* Continuously Variable Transmission; mileage based on 2003 estimates.

Source: **U.S. Department of Energy Technology Snapshot — Featuring the Honda Civic Hybrid** from their website

electric-only mode. In this mode, the vehicle has no emissions. Clean electricity is also used to power many accessories and other equipment that typically are driven by the ICE. This means the engine has less work to do and therefore will use less fuel and emit fewer pollutants.

Hybrids will never be zero-emission vehicles, because they rely on an engine for much of its power. However, most are rated as being close to zero emission vehicles.

Global-Warming Pollution Using ICEs and electric motors, HEVs also have very low carbon dioxide emissions. Hybrids emit one-third to one-half less carbon dioxide than conventional vehicles.

Cost

HEVS have a higher initial cost than comparable conventional vehicles and tend to be heavier. The added weight decreases fuel economy, which is why some larger hybrid SUVs see little gain in fuel mileage. The additional weight results from the addition of large battery packs and electric motor/generators. These same items contribute to the higher cost, as well. However, the cost of some of these items will decrease as the volume of production increases. Currently, a HEV costs nearly $4000 more to produce (See Table 6-2).

Many believe the increased fuel economy of HEVs offsets their higher initial cost. As seen in Table 6-3, consumers pay more for a HEV but typically gain a substantial increase in fuel economy. The initial cost can also be partially offset by tax incentives. A federal tax deduction for the purchase of a hybrid vehicle may be available on some vehicles, and some states may provide additional incentives or tax credits for purchasing a hybrid vehicle.

Not all automotive experts believe the savings on fuel does, indeed, offset the higher initial cost of a HEV. These opinions are based on the service life of a HEV. According to *www.edmunds.com*, gasoline would have to cost $5.60 a gallon for a Ford Escape Hybrid owner and $10.10 a gallon for a Prius owner to financially justify the higher initial cost of a hybrid. The debate will undoubtedly continue for quite some time. However, it is safe to say that most who buy hybrids are doing so for other reasons than using less fuel. These reasons are based on the benefits of hybrid vehicles such as lower emissions and decreasing the country's dependence on fossil fuels. Plus, some feel owning a hybrid makes a statement they want to make!

Table 6-2 A QUICK LOOK AT SOME OF THE ADDITIONAL PRODUCTION COSTS FOR BUILDING A TYPICAL HEV

Why HEVs cost more to manufacture:

Add	
Batteries, cooling system, and battery controller	$1,400
Electronic controls and inverter	$1,400
Electric motor (50 kW)	$600
Harness, safety circuitry, and AC-DC converter	$600
Added cost over standard vehicle	$4000

Source: Energy and Environmental Analysis Inc.

Table 6-3 COMPARISON OF THE COST AND EPA FUEL ECONOMY RATINGS FOR VARIOUS 2006 MODELS OF THE HONDA CIVIC SEDAN

Sedan Models	Transmission	MSRP $	City/Hwy.
Civic DX Sedan	5-Speed Manual	14,560	30/38
	5-Speed Automatic	15,360	30/40
Civic LX Sedan	5-Speed Manual	16,510	30/38
	5-Speed Automatic	17,310	30/40
Civic EX Sedan	5-Speed Manual	18,260	30/38
	5-Speed Automatic	19,060	30/40
Civic EX Sedan with Navi	5-Speed Manual	19,760	30/38
	5-Speed Automatic	20,560	30/40
Civic **Hybrid** Sedan	Continuously Variable Transmission (CVT)	21,850	49/51
Civic **Hybrid** Sedan with Navi	Continuously Variable Transmission (CVT)	23,350	49/51

Availability

Hybrid vehicles are available in many shapes and sizes. Manufacturers are building or are planning to build a great variety of hybrid models. The public seems to have accepted hybrid vehicles, judging by the success of a few models. Initially, hybrid vehicles were designed to address the disadvantages of pure electric vehicles. BEVs never were generally accepted, mostly because of their limited driving range. A hybrid vehicle is much like a conventional vehicle, except it uses less fuel, emits fewer emissions, and operates quietly (sometimes). Most modern hybrids even look like conventional vehicles.

This was not the case when the first HEV was released for sale in North America. The Honda Insight was a small and different looking car (Figure 6-3). It was also rather expensive. Honda did not sell many of these vehicles, but the Insight was a success for the automotive industry. It opened the consumer's eye to this new technology.

Figure 6-3 The look of a Honda Insight. *Courtesy of American Honda Motor Co.*

Shortly after the introduction of the Insight, Toyota brought the Prius to this country. The original Prius was also a small and different looking car. Nevertheless, it was large enough to be deemed more practical and sold quite well. The Prius has since been redesigned and made larger. Today, Honda and several other major manufacturers either offer HEVs or plan to do so in the near future. Newer designs of hybrids are considerably more conventional looking. Many look exactly the same, except for badging, as their non-hybrid counterparts. To consumers, today's HEVs are practical while providing improved fuel economy and reduced emissions.

Today's hybrids are not just about increased fuel economy. Many larger hybrids use the supplemental power from their electric motor to improve performance. Some hybrid sedans offer nearly the same performance as cars typically called "performance sedans."

Sales

Modern HEVs became available in this country during the 2000 model year. Sales numbers were low but the acceptance of the technology grew. In 2002, about 38,000 hybrids were sold and that number increased to about 54,000 in 2003. These numbers certainly do not indicate a high demand for the vehicles, especially when one considers the number of vehicles sold annually in the United States (approximately 17 million). However, hybrid sales are expected to triple over the next two years, to more than 235,000 in 2007 in the United States. It has been estimated that 535,000 to 3.5 million hybrids will be sold in the United States in 2010.

The estimates vary because some include only full hybrids and others include all types of hybrids, including mild ones. Mild hybrids use an electric motor to assist the gasoline-powered engine or feature the stop-start function, but the electric motor alone cannot drive the vehicle.

When looking at sales numbers, one must look at the lack of availability. The battery packs and supporting electronics are not readily available in large quantities; therefore, auto manufacturers cannot provide enough vehicles to meet the demands of the public. It seems a sure bet that if more hybrids were available, the sales numbers would be much greater than they are.

Technology

Today's hybrids are rolling examples of modern technology. The switching between the electric motor and gasoline engine is controlled by computers, as are other features of vehicle. The control systems are extremely complex. They have very fast processing speeds and real-time operating systems. The individual computers in the control system are linked together and communicate with each other by high-speed "communication buses." CAN (Controller Area Network) communications take place between many control systems (Figure 6-4): the electric motor controller, engine controller, battery management system, brake system controller, transmission controller, electrical grid controller, and some systems have 12- or 42-volt components that also must be controlled.

The basic components of a hybrid vehicle include batteries, fuel tank, transmission, electric motor, and internal combustion engine. Both the batteries and fuel tank store the energy needed to run the primary parts of the drive train. Obviously, the batteries store energy for the electric motor. The fuel tank is the energy storage device for the internal combustion engine. The battery pack must be larger and heavier than a typical fuel tank because batteries have a much lower energy density than gasoline. To store as much energy as a gallon (3.8 liters) of gasoline, which would weigh about 7 pounds (3.2 kg), a battery would need to weigh approximately 1000 pounds (454 kg).

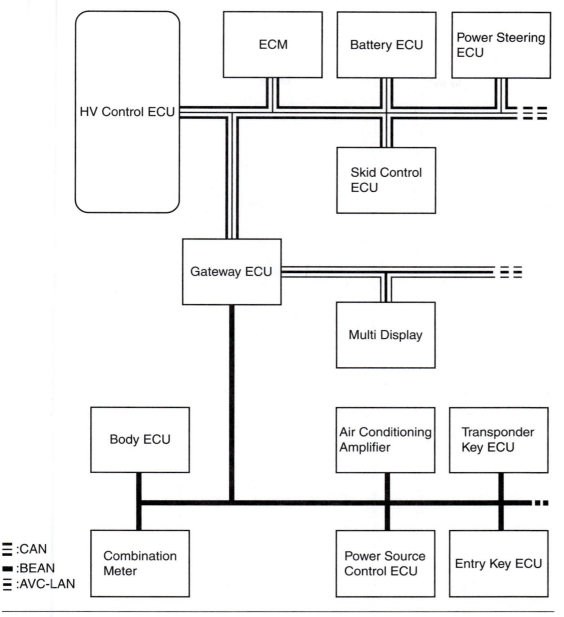

Figure 6-4 The individual computers in the control system are linked together and communicate with each other by high-speed "communication buses," known as CAN (Controller Area Network). *Courtesy of Toyota Motor Company*

The transmission used in a HEV can be a normal transmission or one especially designed for the vehicle. In either case, the gear ratios are designed to allow the engine to run at its most efficient speed during normal operating conditions. Often a continuously variable transmission (CVT) is used. These transmissions do not have fixed gear ratios, rather the overall drive ratio changes according to load (Figure 6-5).

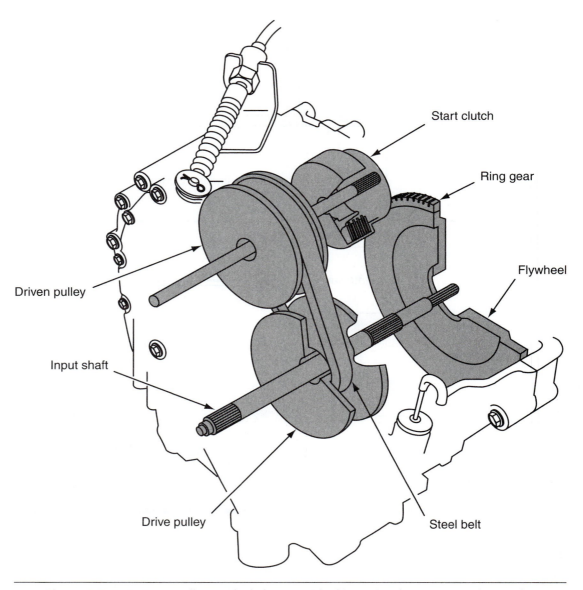

Start clutch

Ring gear

Flywheel

Driven pulley

Input shaft

Drive pulley

Steel belt

Figure 6-5 A CVT uses pulleys and a belt to provide drive ratios that vary according to the needs of the vehicle.

The addition of an electric motor or motors to a conventional vehicle is what makes a hybrid fuel efficient. The motor provides full torque at low speeds, which helps with good acceleration and uses zero fuel when it is operating on its own. Although the engine is designed for fuel economy, it alone cannot provide adequate power for safe acceleration and overtaking loads. The motors also have low production costs, low noise, and high efficiency. Advanced electronics allow the motor to act as a generator, when its power is not needed. The electric motor can be configured, through electronics, to assist the engine during acceleration, passing, overtaking heavy loads, or hill climbing. The intervention of the electric motor allows

Figure 6-6 The electric assist motor is placed between the engine and the transaxle is this hybrid powertrain.

for the use of a smaller, more efficient engine (Figure 6-6). The hybrid controller or computer control system coordinates or synchronizes the action of both.

Internal Combustion Engine The engine found in most hybrids is a four-stroke cycle engine that burns gasoline. These engines are very similar to those used in conventional vehicles; however, they tend to be smaller and deliver less power. The power from the electric motor supplements the engine's power so the vehicles drive much like conventional ones. The engine in a hybrid also uses advanced technologies to reduce emissions and increase overall efficiency.

Diesel Engines The engine in a hybrid does not need to run on gasoline. In other countries where diesel fuel is commonly used, hybrids are being tested with Direct Injection Diesel Engines. Diesel engines have the highest thermal efficiency of any internal combustion engine. Because of this efficiency, diesel hybrids can achieve outstanding fuel economy. Diesel engines do have drawbacks and these define the reasons diesels automobiles are not common in the United States. These disadvantages include particulate matter and nitrogen oxides in the exhaust, noise, and vibration. Diesel engines also have the capabilities of running on biodiesel fuel, which means a diesel hybrid would not use fossil fuels.

Stop-Start Feature

Stop-start systems automatically shut down the engine when the driver applies the brakes and brings the vehicle to a complete stop. This prevents wasting energy while the engine is idling and can increase fuel economy by more than 5%. The benefit of these systems is most evident during stop-and-go driving. With the engine off, the vehicle's heating and air conditioning systems and its electrical systems continue to run using the battery power. The engine is restarted automatically when the drivers releases the brake pedal or when the control system senses the need.

When the engine is restarted is a defining element of mild and full hybrids. Electric components turn off the engine when the vehicle is stopped and is quickly restarted when the brakes are released. Most mild hybrids have a belt-driven starter-generator (Figure 6-7).

Full hybrids can be powered by only the electric motor, the gasoline engine alone, or both. They have the stop-start feature but the engine does not automatically restart when the brakes are released. Because full hybrid can accelerate with only the electric motor, the engine is not restarted until the engine's power is needed. This is determined by the control system and is based on the vehicle's load and the batteries' state of charge.

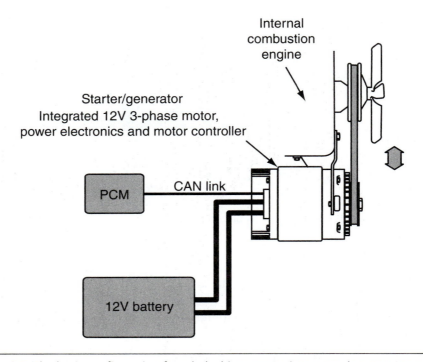

Figure 6-7 The basic configuration for a belt-driven starter/generator in a stop-start system.

Regenerative Brakes

Regenerative braking is the process that allows a vehicle to recapture and store part of the kinetic energy that would ordinarily be lost during braking. A vehicle has more kinetic energy when it is moving fast, therefore, regenerative braking is more efficient at higher speeds (Figure 6-8). When the brakes are applied in a conventional vehicle, friction at the wheel brakes converts the vehicle's kinetic energy into heat. With regenerative braking, that energy is used to recharge the batteries. Regenerative braking can capture approximately 30% of the energy normally lost during braking in conventional vehicles. It is claimed that the electric energy resulting from regenerative braking supplies 20% of the energy used by a Toyota Prius.

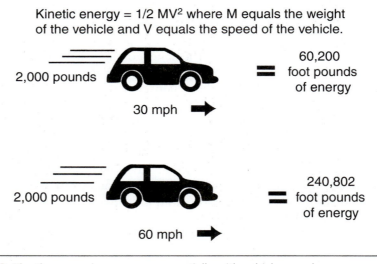

Figure 6-8 Kinetic energy increases exponentially with vehicle speed.

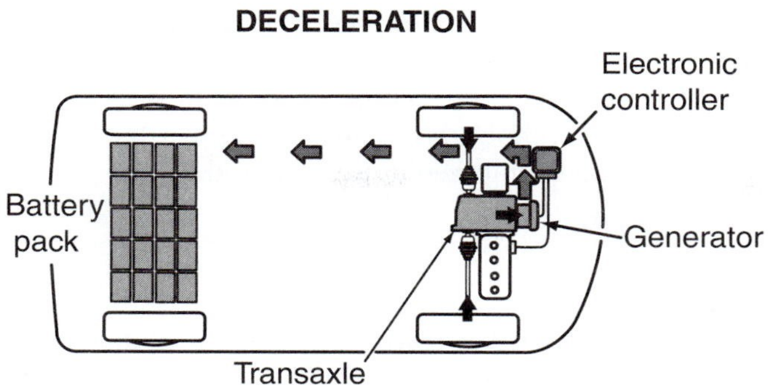

DECELERATION

Figure 6-9 During deceleration, kinetic energy is used to drive a generator, which recharges the batteries.

In a regenerative braking system, the rotor of the generator is turned by the vehicle's wheels as the vehicle is slowing down. The activation of the generator applies resistance to the drivetrain causing the wheels to slow down. The kinetic energy of the vehicle is changed to electrical energy until the vehicle is stopped (Figure 6-9). At that point, there is no kinetic energy.

In many hybrids, there is no separate generator for the braking system. Rather the control system changes the circuitry at the motor and it acts as a generator. The motor now converts motion into electricity rather than converting electricity into motion. The captured energy is sent to the batteries or to an ultra-capacitor. Some hybrids use ultra capacitors because they can capture the energy quickly and can release it quickly during acceleration.

Regenerative braking is not used to completely stop the vehicle. A combination of conventional hydraulic brakes and regenerative braking is used. Hydraulic, friction-based brakes must be used when sudden and hard braking is needed.

The amount of energy captured by a regenerative braking system depends on many things, such as the state of change of the battery, the speed at which the generator's rotor is spinning, and how many wheels are part of the regenerative braking system. Most current HEVs are front-wheel drive; therefore, energy can only be reclaimed at the front wheels. The rear brakes still produce heat that is wasted.

Brake-By-Wire Systems By-wire systems are those that are controlled and operated by electricity and have no mechanical or hydraulic connections. Traditionally, brake systems use both mechanical and hydraulic systems. They rely on hydraulic pressure forced through brake lines to stop the vehicle. With regenerative braking, part of the braking is done by the interaction of magnetic fields in the motors; the rest is done through hydraulics. However, some hybrid vehicles will be fitted with electromechanical brakes at two wheels (both wheels are on the same axle), in addition to a regenerative braking system.

When the brake pedal is depressed, a signal is sent to the motors, which apply brake pads at the individual wheels. By eliminating the hydraulic system, the weight and space required for the brake system is reduced. These systems also provide for electric parking brakes, which can be controlled to prevent their application any time the vehicle is moving.

Accessories

Hybrids can have a number of auxiliary systems and quite a few accessories. In a conventional vehicle, nearly all of these are powered by the engine. In a HEV, there is an engine and a high-voltage system. The accessories are powered by either, depending on the model. Some systems, such as the radio, lights, and horn, operate the same way as they do in a conventional vehicle.

Other systems, such as the power steering and power brakes, may be operated by small electric motors. It must be remembered that when working on HEVs, these auxiliaries and accessories may be powered by high voltage. Never attempt to work on these components (or the main propulsion system components) without thorough training that includes all safety procedures. Normally, most of the high-voltage components are clearly identified and the high-voltage cables are orange.

HVAC

All vehicles are equipped with passenger compartment heating and windshield defrosting systems. In a HEV, the engine can be used to supply the heat so heating and defrosting systems are similar to those used in conventional vehicles. Some hybrids however have additional electrical heaters. These keep the passenger compartment warm when the engine is off.

HEV air conditioning systems are typically identical to those used in a conventional vehicle, except a high-voltage motor may be used to rotate the compressor (Figure 6-10). This increases the efficiency of the engine and allows for conditioned air when the engine is off.

Power Brakes

Many power brake systems use engine vacuum and atmospheric pressure to multiply the effort applied to the brake pedal during braking. Because there is an engine in a HEV, there is a natural vacuum source. Some HEVs have an electrically powered vacuum pump fitted to the vacuum assist power brake system. This allows for power boost when the engine is off. Some hybrids have an electro-hydraulic brake system. An electric pump provides the necessary hydraulic pressure for a hydraulic brake booster.

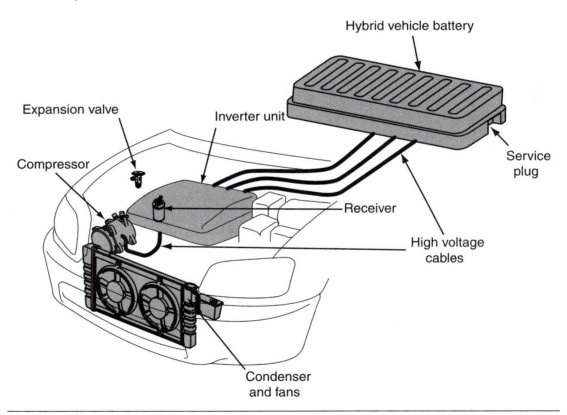

Figure 6-10 HEV air conditioning systems are typically driven by a high-voltage motor rather than the engine's crankshaft.

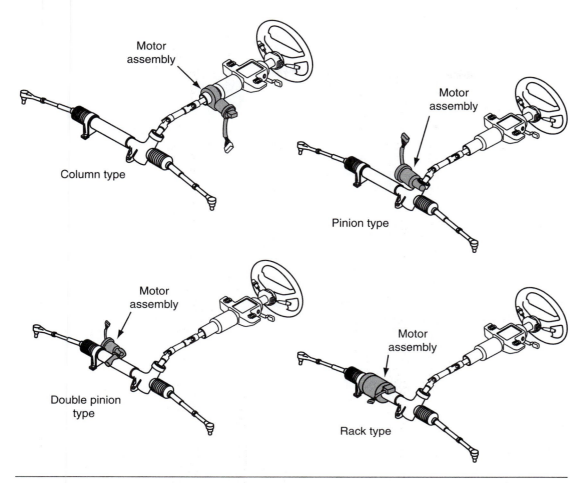

Figure 6-11 The different designs of vehicle-speed sensing electric power rack and pinion steering systems used in hybrid vehicles.

Power Steering

Power steering systems in HEVs are typically pure electrical and mechanical systems (Figure 6-11). An electric motor directly moves the steering linkage. These systems are also very programmable and the energy consumed by the motor depends on the amount the steering wheel is turned. While driving straight, the motor may not run. However, when the steering wheel is fully turned, the motor is drawing its maximum current.

Types Of HEVS

A series-type HEV is moved by the power of a single source, the electric motor. The engine never drives the vehicle; rather it drives a generator (Figure 6-12). The generator, in turn, charges the batteries. No true series hybrids are being manufactured today. In a parallel hybrid, the electric motor and ICE work together (in parallel) to power the vehicle (Figure 6-13).

Many current hybrids use a combination of the series and parallel hybrid configurations. In this design, the vehicle can be driven by either the engine or the electric motor (Figure 6-14). The hybrid control system shuts off the engine when there is ample power from the motor to move the vehicle. The engine is turned on when extra power is needed or when the batteries need to be recharged.

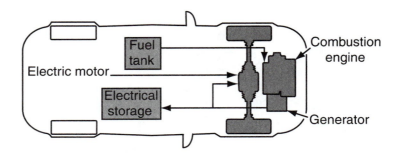

Figure 6-12 A series hybrid configuration.

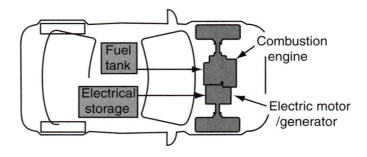

Figure 6-13 A parallel hybrid configuration.

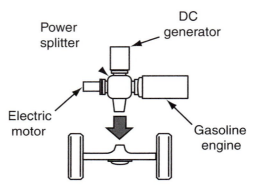

Figure 6-14 The power splitter allows the ICE and/or the electric motor to power the vehicle.

There are two other variations of the series and parallel configuration: the fuel-cell hybrid and the "roadway-coupled" hybrid. In the fuel cell hybrid, the fuel tank is replaced by a hydrogen tank, and the engine and generator are replaced by the fuel cell. Because the output of the fuel cell is only electricity (there is no mechanical motion), fuel cells can <u>only</u> be used in a series configuration. In the roadway-coupled hybrid, the engine and electric motor are not connected within the vehicle. Instead, one set of wheels is powered by the engine, and the other set is powered by the motor. The coupling between the two is made by the road through the tires. This system is shown in Figure 6-15 and is often referred to as a split hybrid design.

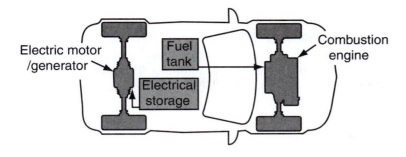

Figure 6-15 A split hybrid configuration.

Hybrid configurations are further defined by the role of the electric motor. Naturally, the more the electric motor powers the vehicle, the less time the ICE must run (Figure 6-16). A **full hybrid** is a vehicle that can run on just the engine, just the batteries, or a combination of the two. The Prius and the Escape are examples of full hybrids. An **assist hybrid** cannot be powered only by the electric motor. The electric motor helps or assists the engine to overcome increased load. At all other times, the vehicle is powered by the engine. A micro or **mild hybrid** is a vehicle equipped with stop-start technology combined with regenerative braking. The electric motor/generator never drives the wheels or adds power to the drivetrain.

Many different types of hybrids have been released or are about to be released. The manufacturers call all of them "hybrids" but they differ greatly in design and purpose. This has led some to categorize them by feature and purpose, rather than configuration. The currently available hybrids can be classified as:

❏ <u>Conventional vehicle (not really a hybrid)</u>, which has stop-start and possibly regenerative braking.
❏ <u>Mild hybrid</u>, which has stop-start, regenerative braking, electric motor assist, and an operating voltage greater than 60 volts.
❏ <u>Full hybrid</u>, which has stop-start, regenerative braking, electric motor assist, an operating voltage greater than 60 volts, and can be power by only electricity.
❏ <u>Performance hybrid</u> (Some call these "muscle hybrids"), which is a full hybrid designed for improved acceleration rather than fuel economy.
❏ <u>Plug-in hybrid</u>, which are full hybrids that can use an external electrical source to charge the batteries, thereby extending the electric-only driving range.

	Fuel economy improvement				Driving performance	
	Idling stop	Energy recovery	High-efficiency operation control	Total efficiency	Acceleration	Continuous high output
Series	●	◎	●	●	○	○
Parallel	●	●	○	●	●	○
Series/ parallel	◎	◎	◎	◎	●	●

◎ Excellent   ● Superior   ○ Somewhat unfavorable

Figure 6-16 A comparison of the fuel efficiencies of the various hybrid configurations.

Plug-in Hybrids (PHEVs)

Plug-in hybrid electric vehicles (PHEVs) are full hybrids with larger batteries and the ability to recharge from an electric power grid. They are equipped with a power socket that allows the batteries to be recharged when the engine is not running. The socket can be plugged into a normal 120-volt outlet. Charged overnight, PHEVs can drive up to 60 miles without the engine ever turning on. When the batteries run low, the engine starts and powers the vehicle and the generator to charge the batteries.

The biggest advantage of plug-in hybrids is they can be driven in an electric-only mode for a much greater distance. During that time, the vehicle consumes no fuel. Under normal conditions, a plug-in hybrid can be twice as fuel-efficient as a regular hybrid. A fully charged PHEV will produce half the emissions of a normal HEV. This is simply due to the fact there are no emissions when the engine is not running. Table 6-4 compares the various hybrid configurations and the resultant fuel economy.

The manufacturing costs of a PHEV are about 20% higher than a regular HEV. The increase in cost is mainly due to the price of the larger batteries. Of course, as battery technology advances and more "high-tech" batteries are produced, the cost will decrease. It is projected that consumers will accept PHEVs because they have a driving range that is equal to or greater than that of a conventional vehicle with very low emissions and improved fuel economy.

Table 6-4 ESTIMATED FUEL ECONOMY POTENTIAL FOR VARIOUS HYBRID CLASSIFICATIONS

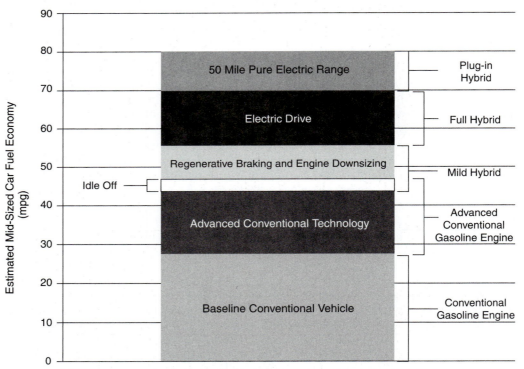

NOTES: Hybrid fuel economy levels assume specific engine and battery motor sizing in a mid-sized vehicle parallel hybrid driveline configuration, altering that sizing, the driveline configuration, or the vehicle type will affect the fuel economy to some degree. This should only be used as a general guide.

Mercedes-Benz Sprinter Daimler Chrysler has built a Mercedes-Benz Sprinter equipped with a diesel engine and an electric motor. This utility van is a plug-in hybrid. The electric motor is positioned between the transmission and clutch. The motor uses energy from a nickel/metal hydride (NiMH) battery pack and has an electric-only driving range of up to 20 miles (30 km). The battery requires approximately six hours recharging when plugged into a conventional electrical outlet. The battery is also recharged by the engine and through regenerative braking.

Hydraulic Hybrids

Hydraulic hybrids function in the same way as hybrid electric vehicles, except energy for the alternative power source is stored in tanks of hydraulic fluid under pressure rather than in batteries. Also, rather than being fitted with an electric motor, these vehicles have a hydraulic propulsion system, which can power the vehicle by itself.

The SHEP (Stored Hydraulic Energy Propulsion) system captures energy during braking and uses that energy when the vehicle is accelerating from a stop. Like regenerative braking systems, the system captures a large percentage of the energy normally lost during braking. That energy is stored in hydraulic tanks attached to the vehicle's chassis (Figure 6-17). The system also aids in the halting of the vehicle. When the brake pedal is depressed, the control unit opens solenoids that send fluid from the low-pressure tank to the pump at the drive shaft. As the drive shaft turns, it turns the pump and fluid pressure increases. This causes the pump and drive shaft to slow down. The fluid under pressure now moves to the high-pressure tank for storage.

During acceleration, the system's computer instructs the pump to send the stored high-pressure fluid back to the drive shaft and to the low-pressure tank. At this point, the vehicle moves without power from the engine and without burning any fuel. Once the computer senses that the energy stored in the tanks has been used, the engine will start and take over the operation of the vehicle. The energy is restored in the tanks during the next brake application.

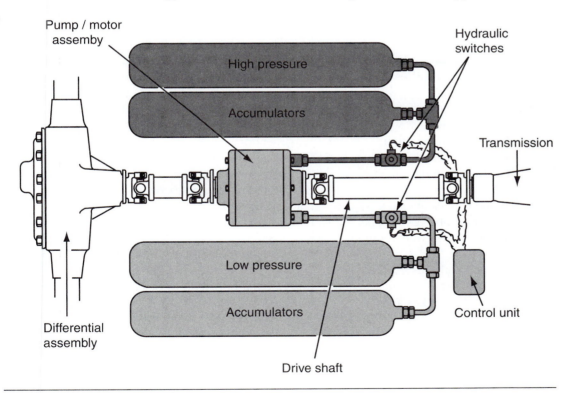

Figure 6-17 The layout of a full series hydraulic hybrid in an urban delivery vehicle.

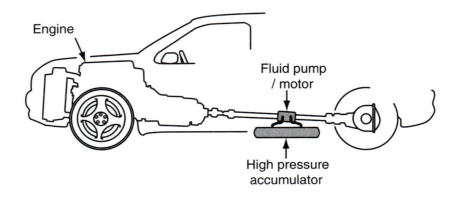

Engine

Fluid pump / motor

High pressure accumulator

Figure 6-18 The location of the major components of a hydraulic hybrid system.

The basic components of the hybrid propulsion system (Figure 6-18) are:

❏ **Hydraulic Storage:** Two hydraulic tanks (accumulators) are installed under the vehicle to store the kinetic energy captured during braking.

❏ **Pump/Motor Assembly:** A variable displacement hydraulic pump is used to transfer the energy stored in the tanks to the vehicle's drive shaft. This also serves as a motor to rotate the drive shaft.

❏ **Electronic Control System:** Monitors and controls the entire system.

Hydraulic hybrid technology is being developed for use in heavy vehicles like buses, trucks, and military vehicles. It is projected that hydraulic hybrids will use 60–70% less fuel than conventional delivery vehicles. The trucks will also have lower emissions because the engine is not used during acceleration.

Review Questions

1. Which of the following types of hybrid vehicles are NOT currently being manufactured?
   A. Full hybrid with a parallel configuration
   B. Mild hybrid with a series configuration
   C. Full hybrid with a series/parallel configuration
   D. Mild hybrid with a parallel configuration
2. Describe the importance of electronics in the operation of a hybrid vehicle.
3. Explain why a hybrid vehicle is more efficient than a vehicle powered only by an internal combustion engine.
4. Which of the following statements is NOT true about stop-start systems?
   A. Only mild hybrids are equipped with this feature.
   B. Stop-start systems automatically shut down the engine when the driver applies the brakes and brings the vehicle to a complete stop.
   C. The engine is restarted automatically when the driver releases the brake pedal or when the hybrid control system senses the need.
   D. Most mild hybrids have a belt-driven starter-generator.
5. Describe the basic difference between series and parallel hybrid configurations.
6. Explain the differences between a full, assist, and mild hybrid.
7. Explain why hybrid vehicles are more expensive to manufacture.
8. Describe the basic operation of a hydraulic hybrid.
9. Explain how regenerative brakes work.
10. Describe the primary advantage of plug-in hybrid vehicles.

MILD AND ASSIST HYBRIDS

After reading and studying this chapter, you should be able to:

❏ Describe the difference between a mild and an assist hybrid.

❏ Identify the advantages of the stop-start feature in hybrids.

❏ Describe the operation of a flywheel alternator starter hybrid system.

❏ Explain how a belt alternator starter system works.

❏ Explain why high voltage is needed in assist-type hybrids.

❏ Describe the basic operation of the hybrid system used in Honda's Insight.

❏ Describe the differences between the first, second, and third generations of Honda's IMA system.

❏ Describe the basic electronic control system used in Honda's hybrids.

❏ Explain the advantages and disadvantages of using a lean-burn engine.

❏ Describe the operation of a cylinder idling system.

Introduction

This chapter covers the common hybrid designs that are not considered "full" hybrids. The two primary classifications of these less-than-full designs are mild and assist hybrids. Although some would move some assist designs into the mild category, this chapter will refer to all hybrids that do not have the ability to use the electric motor in propulsion as mild hybrids. The discussion of assist hybrids includes those hybrids that use an electric motor to assist the engine but do not have the ability to move the vehicle solely by battery power (Table 7-1).

Nearly all **mild hybrid** vehicles are fitted with a flywheel/alternator/starter or a belt-driven alternator/starter hybrid system. These systems provide for stop-start. Strong motors are used to spin the engine fast enough to provide for quick engine restarts. Because most of the vehicle's accessories are powered by the battery, they continue to run when the engine is off. Regenerative braking may be used to supplement the recharging capability of the generator. Fuel comsumption is decreased because the engine does not run when it is not needed.

Assist hybrids typically have an electrical motor connected in series with the engine. The motor adds power to the output of the engine when needed. When the motor is not assisting the engine, it may serve as a generator. The motor also provides for stop-start. Fuel consumption is decreased because of the stop-start feature and, with the assist of the motor, smaller and more efficient engines are used.

Key Terms

Active Control Engine Mount (ACM)
Active Noise Control (ANC)
Assist hybrids
Belt Alternator Starter (BAS)
Continuously variable transmission
Cylinder idling system
Electro-hydraulic power steering (EHPS)
Integrated Motor Assist (IMA)
Integrated starter alternator damper (ISAD)
Intelligent Power Unit (IPU)
Linear Air-Fuel (LAF) sensor
Mild hybrid
Motor Control Module (MCM)

Table 7-1 COMPARISON OF TYPICAL MILD AND ASSIST HYBRIDS.

FEATURE	MILD HYBRID	ASSIST HYBRID
Stop-start	yes	yes
Battery charging with motor	yes	yes
Regenerative braking	yes	yes
Engine assist	no	yes
Pure electric propulsion	no	no
Nominal voltage	42V	144V

Flywheel Alternator Starter Hybrid System

The flywheel/alternator/starter assembly replaces the starter and generator on a conventional engine. The assembly is sometimes called an **integrated starter alternator damper (ISAD)**. In most applications, the ISAD is positioned between the engine and the transmission (Figure 7-1), although it can be mounted to the side of the transmission.

This unit costs about $1,000 less than a full hybrid system. The compact assembly does not require the very high voltages required by other hybrid systems. Most ISAD systems rely on 42-volt power sources. The ISAD allows for regenerative braking.

General Motors

General Motors Corp.'s hybrid Chevrolet Silverado and GMC Sierra full-sized pickup trucks are equipped with the ISAD system. The system is designed to provide:

- ❏ Quick and quiet engine starting
- ❏ Stop-start technology
- ❏ Dampening of driveline vibrations
- ❏ Charging voltages for the batteries
- ❏ Regenerative braking
- ❏ Generation of electricity for auxiliary electrical outlets

The ISAD system replaces the conventional starter, generator, and flywheel with an electronically controlled compact AC asynchronous induction electric motor. This unit is housed in the transmission's bell housing. The stator of the starter/generator is mounted to the engine block. The rotor is attached to the end of the engine's crankshaft (Figure 7-2). As the crankshaft rotates, so does the rotor, or vice versa. Current is sent to the stator when the unit is functioning as a motor. When functioning as a generator, current flows from the stator. When working as a generator, the electrical motor can provide up to 14,000 watts of continuous electric power.

The electricity generated by the system is used to recharge the 12- and 42-volt battery packs, both of which are used to power the various vehicle systems. The electricity can also be used to run power tools or home appliances (Figure 7-3). These trucks have four 120-volt, 20-amp AC power outlets. The auxiliary outlets can power up to four accessories while driving or when the vehicle is parked, as long as the engine is running. It is claimed the generators can power tools or appliances for up to 32 hours on a full tank of gasoline. The system is also designed to turn off the engine when the fuel level gets low. This feature allows the vehicle to be driven somewhere to be refueled. All power supply circuits are protected by a ground fault detection system to prevent overloads and short circuits.

The overall performance of the trucks is the same as non-hybrid models. Both models use the same 5.3-liter Vortec V-8 and 4-speed automatic. Because the electric motor provides no power assist, their power output ratings are also identical. This means these hybrid trucks have the same towing capacity and payload ratings as a conventional pickup.

Figure 7-1 An ISAD assembly. *Courtesy of Continental Automotive Systems*

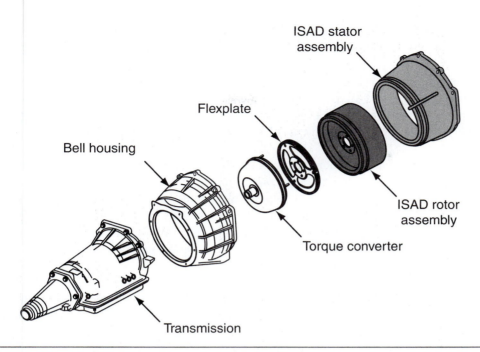

Figure 7-2 The basic layout of an ISAD assembly.

The reduction in fuel consumption of the hybrid pickup is mostly noticeable during city driving, like most hybrids. The EPA mileage ratings for a Sierra 4WD hybrid with an automatic transmission are 17 miles per gallon (mpg) city and 19 mpg highway. The ratings for a similar non-hybrid model are 15 mpg city and 18 highway. This amounts to an approximate gain of 12% during city driving.

The hybrid system improves fuel economy and reduces emissions because it has three primary features. The system allows for regenerative braking, which decreases the amount of load placed on the engine to turn the generator. However, the major contributor to fuel efficiency is the stop-start feature. An additional feature to save fuel is the shutdown of the engine's fuel injection system whenever the vehicle is coasting, decelerating, and braking.

The transmission in the hybrid models has been slightly modified to meet the demands of the system. The torque converter is smaller and has a stronger lockup clutch. The transmission

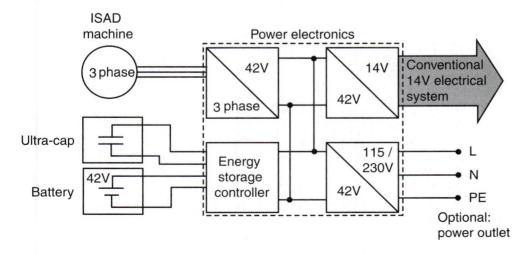

Figure 7-3 A look at a simplified electrical circuit for an ISAD system.

is fitted with a stronger input shaft and an auxiliary oil pump. The pump provides for sufficient line pressure in the transmission when the engine is restarted during the stop-start sequence.

Batteries Energy for ISAD is stored in three 14-volt valve-regulated lead-acid or AGM batteries. These batteries are used to power the electro-hydraulic power steering system and the starter/generator. A conventional 12-volt under-hood battery supplies the power for all other electrical items, such as lighting, wipers, sound systems, and so on.

Control Module The starter/generator control module controls the flow of electricity in and out of the starter/generator. In doing so, it controls the operation of the hybrid system and the power module. The power module is responsible for all electrical conversion and inversion processes: 42-volt DC is converted to AC for starting and the AC is converted to 42-volt DC for recharging. The module also converts 42-volt DC power to 14-volt DC to charge the under-hood battery; 14-volt DC power is converted to 42-volt AC for jump-starting; and 42-volt AC power is converted to 120 volts AC for use at the electrical outlets.

Accessories To maintain the operation of accessories and auxiliary equipment when the engine is shut off during stop-start, many accessories are powered by the 42-volt battery. When the engine is off during stop-start, an electric pump continues to circulate hot water through the heating system during cold weather; in warm weather, cold, dry air is moved through the vehicle's ventilation system. Power steering is provided by an electrically driven hydraulic pump. The **electro-hydraulic power steering (EHPS)** system operates whether the engine is running or not, and provides fluid under pressure for the Hydroboost power brake system.

Belt Alternator Starter Hybrid System

A **Belt Alternator Starter (BAS)** system (Figure 7-4) replaces the traditional starter and generator in a conventional vehicle. The unit is located where the generator would normally be, and is connected to the engine's crankshaft by a drive belt. This unit, an electric motor, serves as the starting motor and generator. When the engine is running, a drive belt spins the rotor of the motor and the motor acts as a generator to charge the batteries. To start the engine, the motor's rotor spins and moves the drive belt, which in turn cranks the engine. These systems have the capability of providing stop-start, regenerative braking, and high-voltage generation.

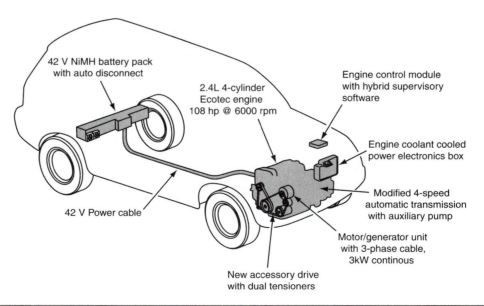

Figure 7-4 GM's BAS hybrid system and key components.

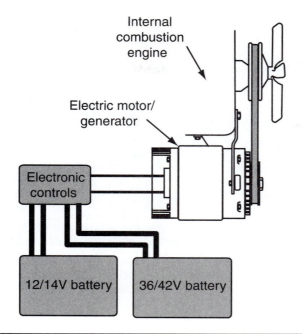

Figure 7-5 The basic architecture for a BAS hybrid system.

A typical BAS includes the motor/generator, electronic controls, and a 42-volt battery. The ability to start the engine quietly and quickly is important to the operation of the stop-start feature, therefore the system uses a rather robust motor. The motor can be either a permanent magnet or induction motor. The required electronic controls for the system depend on the type of motor used. Some systems are also equipped with a conventional starting motor. These are used during extremely cold temperatures. Once the engine is warm, the BAS takes over.

General Motors

GM is scheduled to use the BAS hybrid system in 2006 or 2007 in their Saturn VUE and Chevrolet Malibu. The stop-start feature should result in an 8-10% fuel economy improvement in city traffic, and lower emissions.

The system is based on a dual-voltage architecture of 42V/14V. A 42-volt NiMH battery pack provides power for the motor and some accessories (Figure 7-5), whereas the traditional 12-volt battery powers typical auxiliary devices, such as lights. The system's electronic circuitry monitors many vehicle operating conditions, and controls the operation of the motor/generator and the engine. The electronics must synchronize the activity of the motor/generator with engine systems, such as the fuel injection system. Without precise control, early fuel shutoff during deceleration and quick restarts would not be possible. Nor could there be regenerative braking.

When working as a generator, the electric motor/generator provides more than twice the output of a typical generator. It is capable of providing 3000 watts of continuous power. The generator's output of 42 volts of AC is converted to 42 volts of DC, and is used to charge the battery pack. A DC-DC converter is used to convert the 42-volt output to 14 volts to charge the conventional 12-volt battery and power most of the vehicle's electrical accessories. An inverter takes the 42 volts of DC from the batteries and converts it to 42 volts of AC to power the motor.

Assist Hybrids

Assist hybrids use an engine as the primary source of power for propulsion. An electric motor is used to add torque to the engine's output, when extra power is needed. When the motor is not assisting the engine, it works as a generator to charge the batteries and power some of the

Figure 7-6 Dodge Ram HEV Concept Vehicle Electric Power Access. *Image is Courtesy of DaimlerChrysler Corporation*

electronic systems. The motor also provides for stop-start. The motor never powers the vehicle on its own; therefore, these hybrids are fitted with lower voltage systems than full hybrids.

Honda's hybrids are the most common assist hybrids on the road today. Many different manufacturers are developing assist hybrids. Nearly all of them are using either a belt-driven motor/generator or an integrated motor/generator positioned between engine and transmission. The ISAD system is very similar to Honda's Integrated Motor Assist (IMA) system. Honda introduced the basic concept of placing an electric motor between the engine and transmission, and many variations of their design have been developed.

DaimlerChrysler

DaimlerChrysler has made available to fleet customers a diesel hybrid Dodge Ram pickup called the "contractor special." This hybrid is built on a conventional heavy-duty chassis, and uses a diesel engine fitted with the ISAD system. The ISAD system provides stop-start and can assist the engine during acceleration and other times of heavy loads. The ISAD unit also serves as a generator when the engine is running and when the vehicle is coasting or braking. The electric motor can assist the engine, but cannot propel the vehicle on its own.

The basic system is similar to that in the GM hybrid trucks, except for the use of a diesel engine. The Ram HEV, like GM's hybrid pickups, offers the capacity of serving as a generator to power tools and appliances through 110/220-volt AC outlets (Figure 7-6). In addition, the basic electrical architecture of the truck is a dual-voltage system using 42 volts for the ISAD system and some accessories, and a separate 12-volt battery for other accessories and devices.

Because fuel economy is a major reason why consumers buy hybrid vehicles, the Ram hybrid with a diesel engine should prove to be popular. A diesel engine tends to be more fuel efficient than a comparable gasoline engine. DaimlerChrysler estimates that the Ram hybrid will use 15% less fuel than the non-hybrid diesel Ram truck. The engine used in these HEVs is a Cummins 325 horsepower (242.5 kW), 5.9-liter turbo-diesel engine with 600 lb/ft (813.4 Nm) of torque. Therefore, there should be no shortage of power from this HEV.

Honda's Hybrids

The public acceptance of hybrid technology can be largely credited to Honda. With the introduction of the Insight in December 1999, Honda became the first manufacturer to offer hybrid vehicles in the North America. The Insight, with a very different look and two seats, received the

Figure 7-7 A Honda Insight. *Courtesy of American Honda Motor Co. Inc.*

attention of the public (Figure 7-7). They especially noticed the high fuel economy ratings. The Insight has been America's most fuel-efficient car every year since its introduction. The Insight has an electric motor positioned between the engine and transmission.

In 2002, Honda applied its hybrid technology to the Civic. This car was not noticeably a hybrid. In fact, it looks much the same as a conventional Civic. Sales of the hybrids were, and continue to be, good. Reduced emissions and great fuel economy contributed to their success. The Civic hybrid is based on the same technology as used in the Insight and has been recently modified to make it more efficient. Today, the Civic Hybrid is listed in the top five most fuel-efficient cars as rated by the Environmental Protection Agency (EPA). The hybrid is also certified as a ULEV (Ultra-Low Emission Vehicle).

In 2005, Honda introduced the Accord hybrid. Following the same formula as the Civic, the Accord looks like a normal car and applies the technology used in the Insight to achieve lower emissions and good fuel economy.

Honda Insight

The Honda Insight is equipped with the first generation of Honda's **Integrated Motor Assist (IMA)** hybrid system. It has a 1.0-liter, three-cylinder engine and a 144-volt electric motor positioned between the engine and the transmission. The Insight was designed to be extremely fuel efficient and has a driving range of 700 miles. It is very light and aerodynamic. The Insight has an aluminum body and reinforced frame. Because the car is light, a small ICE can be used to propel it. These two factors greatly contribute to the high fuel economy ratings of the Insight. The electric assist makes up for any power deficiencies of the engine. The Insight meets CARB's ULEV standard.

The shape of the car also contributes to its efficiency. When a car is moving, it is pushing through air. This is called aerodynamic drag, and horsepower is required to push a car through the air. To minimize air drag, the Insight has a teardrop shape with the back of the car being narrower than the front. The rear wheel openings are partially covered by skirts, preventing air from moving into the wheel wells. The underside of the car is flat, and panels are used to close off some of the openings under the car. Air can easily move under, over, and around the car. The Insight is one of the most aerodynamic cars on the road and has a drag coefficient of only 0.25.

IMA The IMA combines the engine with an ultra-thin (only 2.3 inches [58 mm] thick) permanent magnet electric motor (Figure 7-8). The brushless motor receives its power from a 144-volt NiMH battery pack, through the **PCU (Power Control Unit)**. The engine is the primary power source for driving. The motor provides additional power only when needed. The motor

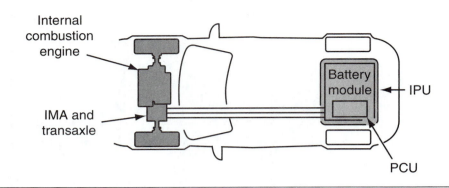

Figure 7-8 The basic layout of the IMA system in an Insight.

produces 13 horsepower and 36 lb/ft of torque. The motor also cranks the engine at 1000 RPM for quick starting. Like other hybrids, the motor also works as a generator to recharge the battery pack. The maximum output of the generator is 10 kilowatts (69.4 amps at 144 volts).

The synchronous AC motor has a three-phase stator and a permanent magnet rotor that is directly connected to the engine crankshaft. There are three commutation sensors mounted inside the motor/generator that give the Motor Control Module information about the rotor's position.

Electronic Controls The gasoline engine is controlled by the Powertrain Control Module (PCM). This module is much like one used in a non-hybrid but has been programmed to interact with the IMA system. The IMA system is controlled by the components housed in a single unit called the **Intelligent Power Unit (IPU)**, which is connected to the motor/generator by high-voltage cables (Figure 7-9). The IPU contains the following:

❑ Battery Module (BM)
❑ Battery Condition Monitor (BCM)

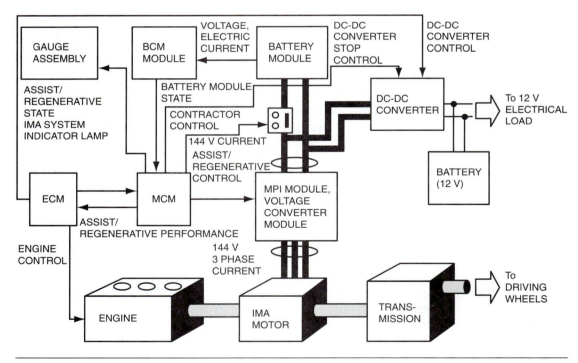

Figure 7-9 A look at the electronics that allow the IMA system to work. *Courtesy of American Honda Motor Co., Inc.*

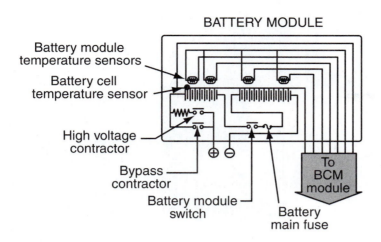

BATTERY MODULE

Battery module temperature sensors

Battery cell temperature sensor

High voltage contractor

Bypass contractor

Battery module switch

Battery main fuse

To BCM module

Figure 7-10 The battery module in an Insight.

- ❏ Motor Control Module (MCM)
- ❏ Power Control Module (PCU)
- ❏ Motor Drive Module (MDM)
- ❏ Voltage Converter Module
- ❏ DC / DC Converter

The battery module (Figure 7-10) is a 144-volt nickel-metal hydride battery pack weighing only 48 pounds. The battery pack has 20 modules connected in series. Within each module are six 1.2-V cells, each about the size of a conventional D battery. The maximum capacity of the battery pack is 6.5 amp-hours (Ah).

The high-voltage contactor and bypass contactor are connected at the positive (+) output terminal of the battery module. These contactors are controlled by the motor control module, connecting the battery to the high-voltage circuits. Current flows through the bypass contactor and bypass resistor when the capacitors in the power control module are being charged.

The battery condition monitor measures battery voltage, battery input/output current, and battery temperature. The module has four thermistor-type temperature sensors, ten voltage sensors, and a current sensor. The condition of the battery is sent to the motor control module.

Heat is generated during charging and discharging, and this heat can have a negative effect on the battery. To control the temperature, the battery condition monitor controls a two-speed cooling fan that prevents the battery module from overheating. If the monitor detects a fault in the battery module, it sends a signal to the **Motor Control Module (MCM)**, which then turns on the IMA system indicator in the instrument panel.

The motor control module controls the motor/generator through the **motor power inverter (MPI)** and the voltage converter unit in the motor drive module. Through communication with the PCM and **Battery Condition Monitor (BCM)**, the MCM directs the motor drive module to control the motor assist feature, as well the action of the regenerative braking system. The operating conditions, such as engine load, are used to determine the appropriate action of the motor/generator. By communicating with the BCM, the motor control module limits the activity of the motor/generator to prevent excessive battery drain and overcharging. On some vehicles, the battery condition monitor is built into the MCM.

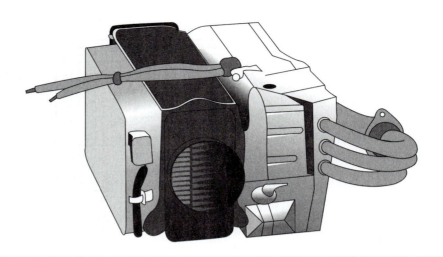

Figure 7-11 The PCU (Power Control Unit) used in an Insight.

The motor power inverter and voltage converter change DC power into three-phase AC power for the motor during assist and converts AC power into DC to charge the batteries during regeneration. The MCM always needs to know the position of the motor/generator's rotor within the stator. Three Hall effect sensors, called the Motor Commutation Sensors, monitor the rotor.

The MCM also sends signals to control the IMA display in the instrument cluster. This display informs the driver of the operating modes and condition of the system. The MCM also stores fault (DTC) codes and operates the IMA warning lamp. Scan tools are used to retrieve the DTCs through a 16-pin data link connector.

The power control unit (Figure 7-11) controls the distribution of electricity throughout the IMA system. This unit holds the motor drive module and the DC-DC converter. Both of these are mounted in a heat sink to prevent them from overheating. A fan moves air across the heat sink to keep it and the modules cool. The MDM is comprised of the motor power inverter module, voltage converter module, capacitor, and the U/V/W phase motor current sensors. The latter sensors measure the current in and out of the three stator windings of the motor/generator. The voltage converter and inverter are responsible for the AC to DC and DC to AC changes.

The DC-DC converter reduces some of 144-volt DC from the battery module to 12-volt DC needed to charge a separate 12-volt battery. It also provides the power to operate the 12-volt electrical system. The converter is mounted to a finned aluminum heat sink inside a magnesium housing. This is necessary because during the reduction of the voltage, much heat is generated. If the temperature of the converter becomes too high, the converter's temperature monitoring system will send a signal to the MCM, which in turn will turn off the converter.

Operation When the ignition switch is turned to the start position, the motor/generator instantaneously spins the engine at 1000 RPM to start the engine (Figure 7-12). Once the engine has started, it runs at all times except during stop-start. The car is fitted with an auxiliary starter motor that is only used to start the engine when the battery module's voltage is very low, when the outside temperature is extremely low, or if there is a problem with the IMA system.

When the driver depresses the accelerator, the PCM sends a signal to the motor control module. The MCM, in turn, sends a signal to the motor drive module. The MDM then sends 3-phase AC power to the motor/generator to trigger motor assist (Figure 7-13).

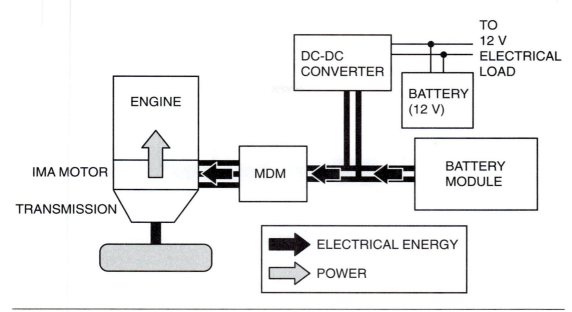

Figure 7-12 The IMA motor, attached directly to the engine's crankshaft, starts the engine under normal conditions. When outside temperature is extremely low, when the battery state of charge is low, or if there is a problem with the IMA system, the conventional starter starts the engine. *Courtesy of American Honda Motor Co., Inc.*

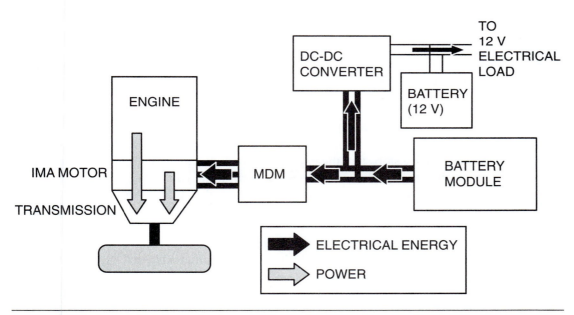

Figure 7-13 When the driver depresses the accelerator, the output of the motor supplements the output of the engine to accelerate the car. *Courtesy of American Honda Motor Co., Inc.*

As the engine overcomes the load of acceleration, the motor is turned off and the car is powered only by the engine. While the car is cruising at a steady speed, the motor/generator can work as a generator to charge the battery module and power the 12-volt system. The engine will not drive the generator unless there is need for charging. This feature minimizes the work of the engine and therefore maximizes fuel efficiency.

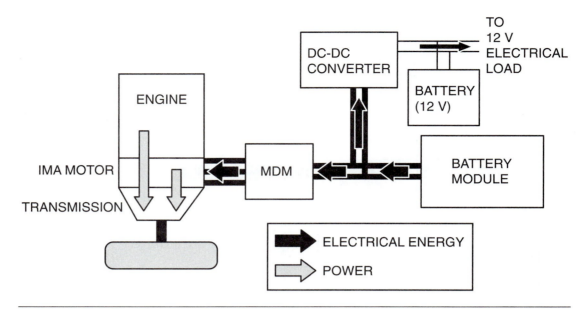

Figure 7-14 During deceleration, the system automatically switches the motor/generator into a generator for regenerative braking. *Courtesy of American Honda Motor Co., Inc.*

During deceleration, the system automatically switches into the Charge Mode. The PCM sends a signal to the MCM, which then directs the MDM to turn the motor/generator into a generator. Because the generator is now being driven by the wheels of the car, the car's kinetic energy is changed to electrical energy (Figure 7-14). Regenerative braking produces 3-phase AC power that is sent to the MDM where it is converted to DC to charge the battery. The DC output of the MDM is also sent to the DC-DC converter and used to charge the 12-volt battery. When the brakes are applied, regeneration is increased and the batteries are charged at a higher rate. It is important to know that if the driver puts the car into neutral while coming to a stop, there will be no regenerative braking because the wheels can no longer drive the generator.

The stop-start (Honda refers to this feature as Idle Stop) mode is initiated when the car is at a stop, the shift lever is placed in neutral, the clutch pedal is released, and/or when the brake pedal is held down. During stop-start, the engine is turned off and the green "Auto Stop" light in the instrument panel illuminates. Stop-stat does not occur if the voltage of the battery module is low. Also, if the transmission remains in a gear, the engine will not be turned off. This allows the engine to be responsive in heavy or stop-and-go traffic. The engine restarts immediately when the driver releases the brake pedal, a gear is selected, and/or the clutch pedal is depressed.

Electronic Instrument Displays The Insight is equipped with analog and digital instruments. The instruments display the typical conditions and fuel level for a gasoline engine, and they display the operation of the IMA system and the car's fuel efficiency. On cars equipped with a manual transmission, the instrument panel also includes upshift and downshift lights that are triggered by the PCM to inform the driver when it is most economic to shift gears.

The instrument panel is divided into three separate sections. The displays for the gasoline engine are located on the left side of the panel. Included in the engine displays are a tachometer, coolant temperature gauge, and warning lights for oil pressure, PCM-related problems, required maintenance, and problems in the 12-volt system.

The center part of the cluster is a combination Odometer/Fuel Economy Meter. This includes a digital speedometer, odometer, and lifetime fuel economy readout. The display can be reconfigured by the driver to show current fuel consumption, the fuel economy for two different trips, and other fuel efficiency indicators.

The right side of the instrument cluster displays the status of the battery and IMA system (Figure 7-15). A charge/assist indicator shows when the system's electric motor is assisting the

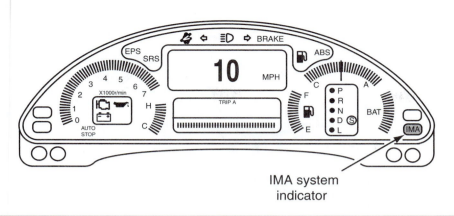

IMA system
indicator

Figure 7-15 The instrument panel in an Insight.

engine. The amount of assist is indicated by amber-colored bars. The number of bars illuminated indicates how much assist is being provided. This same display shows the amount of charge going to the batteries. The amount of recharge is indicated by green-colored bars. When more bars are illuminated, the batteries are being recharged at a higher rate. Also on this side of the cluster is a state-of-charge indicator for the battery module, a fuel gauge, and the shift indicator.

The entire cluster is designed to help the driver achieve maximum fuel economy by minimizing the amount of time there is heavy electrical drain on the battery module. When the engine must drive the generator, additional fuel is used to do this.

Engine The engine in an Insight was specially designed to provide great fuel economy and reduced exhaust emissions. The 1.0-liter, three-cylinder engine incorporates many different technologies to achieve these goals. The engine is very light, weighing only 124 pounds (56 kg) and produces 67 horsepower at 5,700 rpm. The engine is claimed to be smallest, cleanest, and lightest 1.0-liter production automobile engine in the world. Engine weight was reduced by using aluminum, magnesium, and plastic. The engine also uses Honda's **Variable Valve Timing and Lift Control for Economy (VTEC-E)** lean-burn technology, along with several friction-reduction techniques to minimize power losses.

Much of the efficiency of the Insight's engine results from the use of the VTEC-E cylinder head and valve train. The VTEC system used in the Insight has an expanded stratified charge area within the combustion chamber, advanced fuel-injection mapping, and a **Linear Air-Fuel (LAF) sensor**. Like other Honda vehicles, the fuel injection system is based on Honda's sequential programmed fuel injection and direct ignition with individual ignition coils for each cylinder and iridium-tipped spark plugs.

In a VTEC engine, there are two intake valves and two exhaust valves in each cylinder. When the engine is running at low speed, only one intake and one exhaust valve are used. When the engine reaches a particular speed, all of the valves are used. This technology allows the engine to run at very lean mixtures when there is light load on the engine. Lean mixtures provide excellent fuel economy with lower HC and CO emissions.

A VTEC-E engine can burn very lean mixtures at low engine speeds because as the air-fuel mixture enters into the combustion chamber through one of the two intake valves, a strong air-fuel swirl is created (Figure 7-16). This swirl or vortex of the mixture creates a "stratified" charge. This means the mixture close to the spark plug is richer than the mixture in the rest of the combustion chamber. This richer mixture ignites first and the heat from that area moves quickly to ignite the rest of the mixture. Very lean mixtures are not easily ignited by the firing of a spark plug. This is why the stratified charge is an important feature of the VTEC engine: the heat created by the ignition of the rich area in the combustion chamber is much higher than the heat generated by a spark plug. As engine speed builds, the other intake valves open to provide more air. In addition, when more power from the engine is needed, the engine, through PCM control, runs at the stoichiometric (14.7:1) ratio or richer.

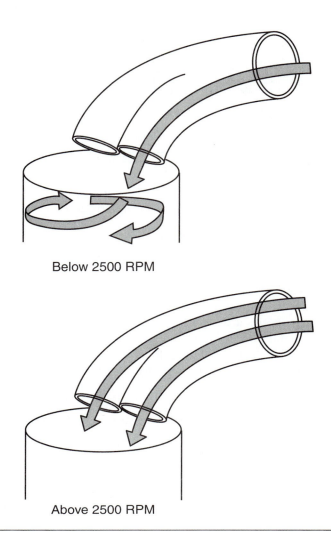

Below 2500 RPM

Above 2500 RPM

Figure 7-16 A VTEC-E engine can burn very lean mixtures at low engine speeds because as the air-fuel mixture enters into the combustion chamber through one of the two intake valves, a strong air-fuel swirl is created that provides for a "stratified" charge.

One of the problems with running an engine lean is that a normal oxygen sensor will not work. Oxygen sensors are used to monitor the air-fuel mixture so the PCM can control the activity of the fuel injectors to ensure the desired mixture ratio is maintained. Most normal engines use a zirconia oxygen sensor that can only measure the exhaust's oxygen content when the mixture is very close to stoichiometric. A zirconia oxygen sensor has a piece of zirconia that has one end exposed to the atmosphere and the other end exposed to the exhaust stream. A voltage is generated when the amount of oxygen surrounding the two ends is different. Because the amount of oxygen in the atmosphere is somewhat constant, the voltage generated by the sensor actually indicates the amount of oxygen in the exhaust. The voltage from an oxygen sensor fluctuates between zero and one volt, depending on the oxygen level.

When the engine is running in the lean-burn mode, the oxygen content of the exhaust is higher than a normal oxygen sensor is capable of measuring. Therefore, to maintain the desired air-fuel ratio, a special oxygen sensor must be used. The Insight uses a LAF sensor that can detect the oxygen level in the exhaust during very lean conditions (25:1). Using this type of sensor, the PCM is able to precisely control the air-fuel mixture, regardless of the operational mode.

The linear air-fuel sensor is located ahead of the three-way catalyst and measures the oxygen in the exhaust. A LAF sensor has two zirconia elements that share a diffusion chamber

Figure 7-17 A LAF sensor can be identified by its five-wire configuration: 1) Heater positive, 2) Heater ground, 3) Sensor element positive, 4) Control element positive, and 5) Common ground for sensor and control elements.

(Figure 7-17). Within the sensor, there is an exhaust flow chamber, atmosphere reference chamber, and a diffusion chamber, plus a heater circuit that allows the sensor to work when the engine is cold. Unlike a non-lean burn sensor, the voltage can be positive or negative. Positive voltage indicates a lean mixture and negative voltage indicates a rich mixture (Figure 7-18). The

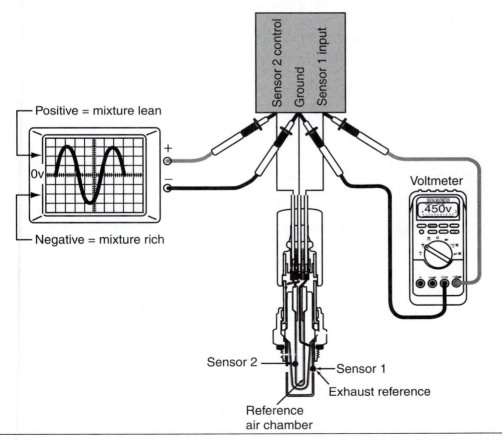

Figure 7-18 Unlike a conventional zirconia oxygen sensor, the voltage from a LAF can be positive or negative. A positive voltage indicates a lean mixture and negative voltage indicates a rich mixture.

normal operating voltage range is about 1.5 volts. The zirconia element that is exposed to the exhaust is the sensor element. The diffusion chamber is the space between the two zirconia elements. By applying varying voltages to the control element, the PCM can control the amount of oxygen in the diffusion chamber. Because the diffusion chamber is the reference chamber for the sensor element, this action changes the output of the sensor element. The PCM monitors the output of the sensor element as the oxygen content of the exhaust changes, and it applies voltage to the element to try to maintain the sensor output at 0.45 volts. It then monitors the control voltage to determine the actual air-fuel ratio. The oxygen sensors located to the rear of the catalytic converters are the conventional zirconia type.

The engine in an Insight is equipped with **a nitrogen-oxide-adsorptive catalytic converter** (Figure 7-19), in addition to a conventional three-way catalytic converter (Figure 7-20). The nitrogen-oxide-adsorptive converter is necessary to keep NOx emission levels low. Lean-burn engines have high combustion temperatures and an excessive amount of oxygen in the combustion chamber. Both of these contribute to the formation of NOx. Conventional three-way catalysts are not very effective in converting NOx into nitrogen when excess oxygen is present. Lean mixtures require the use of a special catalytic converter. The Insight's nitrogen oxide absorptive catalytic converter uses two NOx catalyst beds to trap and convert the oxides of nitrogen in the exhaust. The converter attracts NOx molecules to the surface of the catalyst metals during lean-burn operation. When the engine is running at stoichiometric or richer ratios, the converter combines these NOx molecules with the hydrocarbons and CO in the exhaust to form water vapor, carbon dioxide, and nitrogen.

The efficiency of the engine in an Insight is further enhanced by the use of many friction-reducing features, such as roller-type rocker arms mounted to a single shaft. The engine also has offset cylinder bores. This feature places the crankshaft slightly away from the center of the cylinders. By offsetting the cylinders, the piston and connecting rod move at a more efficient angle during the power stroke, and friction from piston side loads is reduced. The pistons have a small skirt area and the surface of the skirt has been shot-peened, which improves their ability to hold lubricant that, in turn, reduces the friction created by the piston moving in the cylinder.

The weight of the engine is kept low using a magnesium-alloy oil pan. The oil pan also serves as the engine oil-filter bracket and AC-compressor mount. Further, to reduce weight, plastics are used to make the intake manifold, water pump pulley, and valve cover.

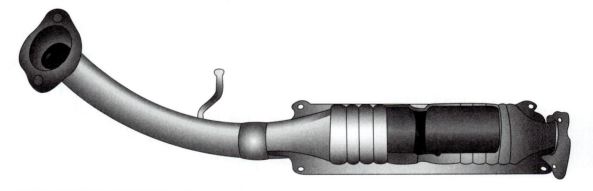

Figure 7-19 A nitrogen-oxide-adsorptive catalytic converter.

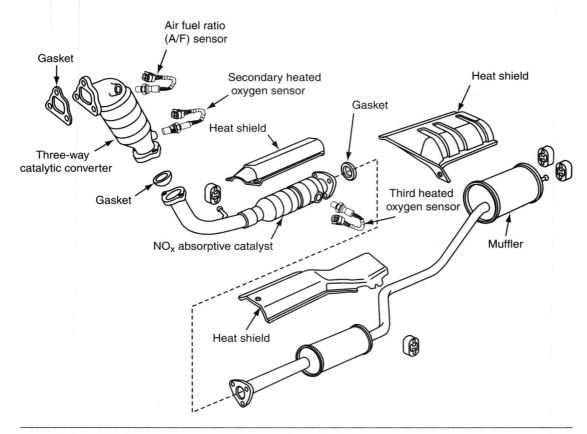

Figure 7-20 The complete exhaust system for an Insight.

The IMA system is also designed to smooth the operation of the engine. Three-cylinder engines are inherently unbalanced and prone to vibrations. To overcome this problem, the IMA system uses its electric motor to dampen the vibrations. The motor applies a reverse torque to the engine's crankshaft. These reverse torque pulses are precisely in phase but in the opposite direction of the torque fluctuations of the engine (Figure 7-21). The reverse torque cancels the vibrations of the engine. This feature requires precise control of the motor/generator as the function of the motor/generator is quickly changed to match the movement of the engine's crankshaft.

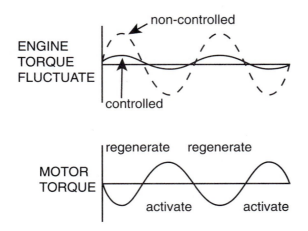

Figure 7-21 The IMA system is equipped with an idle vibration reduction control that minimizes fluctuations in the engine's crankshaft. The motor applies opposite phase torque to the engine when the crankshaft is rotating. *Courtesy of American Honda Motor Co., Inc.*

During a power stroke, the IMA system momentarily switches the motor/generator to the generator mode to help absorb the power pulse. Immediately after the power stroke, the system momentarily switches to the motor mode to speed up the slowing crankshaft.

Transmission Insights are available with either a manual transmission or a **continuously variable transmission (CVT)**. The 5-speed manual transaxle weights just 91 pounds (41.3 kg) and is designed to reduce power loss through friction and to make shifting easy. The CVT (Honda Multimatic) uses computer controlled drive and driven pulleys and a steel "push" belt. It provides two driving modes: *SPORT,* which maximizes power but reduces fuel economy (for acceleration), and *DRIVE,* which reduces power but improves fuel economy (for cruising). The transmission is also equipped with an "anti-rollback brake assist" system that prevents the car from rolling backwards when it is moving from a stop on a hill.

A CVT (Figure 7-22) provides varying drive ratios by varying the position of a high-strength steel belt between two metal pulleys. The sides of the pulleys are controlled by hydraulic pressure. One pulley is connected to the output of the engine and the other is connected to the

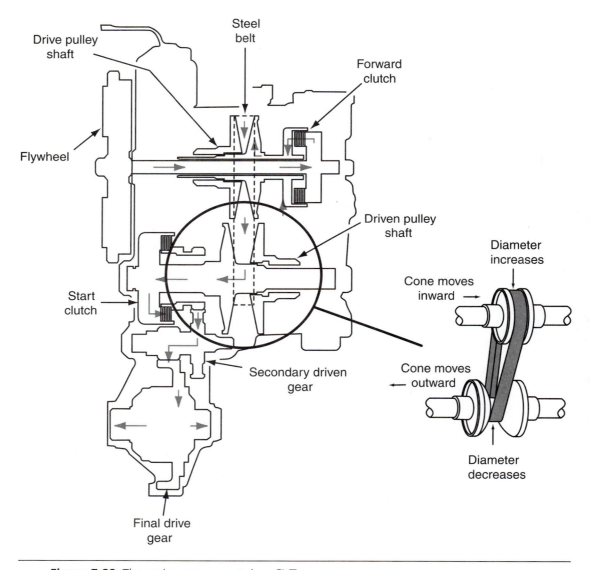

Figure 7-22 The major components in a CVT.

power output side of the transmission. Through various guides within the transmission, the diameter of the pulleys can change, thereby changing the drive to driven ratio of the two pulleys. Because the length of the belt never changes, the effective drive ratio changes. A CVT provides the most suitable gear ratio for any vehicle speed and throttle input.

Honda Civic Assist Hybrid

The 2003 through 2005 Honda Civic Hybrids use the second generation of Honda's IMA system. Compared to the Insight, the Civic is fitted with a larger engine, more powerful electric motor, and refined electronics. Also, many electrical components have been combined, lightened, and reduced in size, to keep the car very fuel efficient. The Civic Hybrid seats five and looks very similar to a conventional Civic; however, through minor changes the hybrid model has a lower drag coefficient than a regular Civic. The EPA mileage estimates for a Civic Hybrid with a five-speed manual transmission are 46 mpg in the city and 51 mpg during highway driving. Models equipped with a CVT are rated at 48 mpg for city driving and 47 mpg on the highway.

This assist-type hybrid relies primarily on power from the engine. The engine is a 1.3-liter, 4-cylinder engine. It uses Honda's VTEC and i-DSI technologies. The **i-DSI** or **intelligent - Dual & Sequential Ignition** system uses two spark plugs per cylinder to provide for more complete combustion and a lean-burn mode (Figure 7-23). The IMA system is an updated version of the system used in the Insight. Although the power output from the electric assist motor is the same 10 kW; 13 hp), the motor provides more torque.

The operation of the IMA system and its controls are the same as those used in the Insight. Only the components and systems that are unique to the early Civic Hybrids will be discussed in the following.

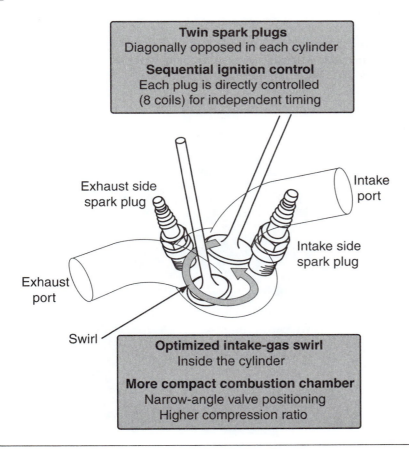

Twin spark plugs
Diagonally opposed in each cylinder

Sequential ignition control
Each plug is directly controlled
(8 coils) for independent timing

Exhaust side spark plug

Intake port

Intake side spark plug

Exhaust port

Swirl

Optimized intake-gas swirl
Inside the cylinder

More compact combustion chamber
Narrow-angle valve positioning
Higher compression ratio

Figure 7-23 The Intelligent Dual Sequential Ignition (i-DSI) system has eight ignition coils, eight spark plugs that can fired sequentially or simultaneously, and is totally controlled by the PCM.

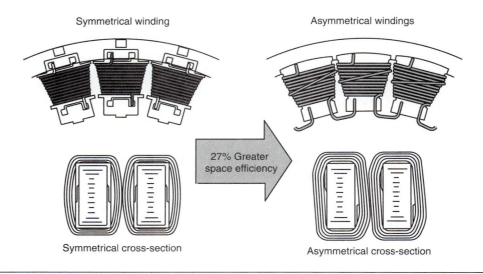

Symmetrical winding

Asymmetrical windings

27% Greater space efficiency

Symmetrical cross-section

Asymmetrical cross-section

Figure 7-24 The stator windings of the IMA motor in the Civic were wound differently to increase the torque output from the motor.

IMA System Several items in the IMA system were redesigned to make them more compact and increase their efficiency. The motor/generator's stator windings were wound differently to increase the density of the magnetic flux lines. This change resulted in an increase of 30% more torque from the motor (Figure 7-24). The torque ratings for the motor (Figure 7-25) vary with the transmission. In a manual transmission-equipped car, the motor's maximum torque is 46 ft/lb torque at 1000 rpm. When used with a CVT, the torque rating is 36 ft/lb torque at 1000 rpm. Each transmission has acceleration characteristics, and the control system provides the appropriate amount of assist to match the transmission. Rewinding the stator also increased the ability of the generator to produce electricity during regenerative braking.

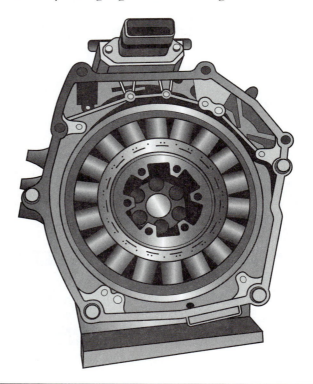

Figure 7-25 The Civic's IMA motor.

Electronic Controls For the Civic Hybrid, Honda made significant changes to the IMA system. These changes resulted in a smaller and lighter unit. In fact, when compared to similar equipment in an Insight, the Civic's IMA system takes up nearly 40% less space and weighs about 30% less. The function of the components and the system remain the same.

The Intelligent Processing Unit (IPU) controls the power of the IMA system. The IPU in a Civic was made smaller so that valuable trunk space was not taken up by the unit. The IPU houses the Power Control Unit (PCU), control unit for the motor, motor power inverter, battery module, and a cooling system.

The PCU controls the flow of electricity between the IMA motor and battery pack. The Motor Control Module (MCM) controls the IMA motor, through the motor power inverter. The MCM monitors the state-of-charge of the battery pack and controls the IPU module fan. The MCM uses inputs of the batteries' voltage, temperature, and input and output current readings to determine the batteries' state of charge. In the Civic, one fan is used to cool the IPU and the batteries. By using one, the Civic's cooling system uses less energy and weighs less. The cooling air is drawn into the battery module from the top of the tray behind the rear seat and passes over the heat sinks or the inverter and DC-DC converter before it is exhausted into the trunk (Figure 7-26). The motor power inverter was also redesigned to make it lighter and smaller.

The battery module is still comprised of twenty 7.2-volt modules and has a nominal output of 144 volts and a capacity of 6.0 amp-hours. However, the battery box has been reduced in size and placed into the IPU. The reduction in size also resulted in a decrease in weight.

Operation The IMA system functions in the same way as it did in the Insight. The only differences are linked to the stop-start feature, the car's transmission type, and the automatic climate control system. The Civic Hybrid automatically turns off the engine during complete stops under most circumstances. The stop-start (Idle Stop) feature does not work during the first few minutes of engine warm-up or if the automatic climate control system is being used in the "air conditioning" mode with the "Economy Mode" not selected.

On cars with a CVT, if the car had been moving at more than 9 mph (15 km/h) and the brake pedal is depressed, the engine will shut off. During this time, the auto stop indicator in the instrument panel blinks. If the driver's door is opened during stop-start, the idle stop indicator blinks and the warning buzzer sounds to remind the driver that stop-start is in operation. The engine restarts when the brake pedal is released and the accelerator is pressed; when the gear selector is placed in the P, R, or L position; and when the control unit senses there is low vacuum in the brake booster.

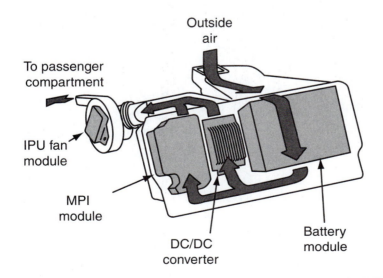

Figure 7-26 The cooling fan assembly for the IPU in a Civic Hybrid.

With a manual transmission, the engine will shut down when the clutch pedal is depressed with the car's speed at 19 mph (30 km/h) or less, when the car is at a stop and the shifter is in neutral with the clutch pedal depressed, and when the brake pedal is depressed and the engine's speed is at 1,000 rpm or less. When the automatic idle stop system is operating, the auto stop indicator in the instrument panel is illuminated. The engine will restart when the gear shifter is moved into any position other than neutral with the clutch pedal pressed, when the accelerator pedal is pressed while the transmission is in neutral or the clutch pedal is pressed, when the vacuum in the brake booster is low while the transmission is in neutral or the clutch pedal is pressed, or when the IMA battery is low and the transmission is in neutral or the clutch pedal is pressed.

Engine One of the biggest differences between the power train of the Insight and the Civic Hybrid is the engine. Both engines were designed to burn clean and fuel efficient. Because the Civic is larger and is designed to carry a heavier load, the engine is larger. Larger, however, does not mean the engine lacks for advanced systems.

The Civic Hybrid uses a single overhead cam, eight valve, 1.3-liter VTEC engine with i-DSI that produces 85 hp at 5700 rpm and 87 lb/ft torque at 3300 rpm. This aluminum-alloy engine has a compression ratio of 10.8:1. The ignition and fuel injection systems provide for a lean-burn mode, and the engine is fitted with a nitrogen oxide absorptive catalytic converter. The engine is designed to provide high amounts of torque at low and mid engine speeds. To reduce fuel consumption, the engine is equipped with i-DSI, a VTEC Controlled Cylinder Idling System, and has many fuel-saving technologies. Some of these were also used in the Insight, such as offset-cylinder bores, low-friction pistons, and roller rocker arms (Figure 7-27).

The aluminum engine uses thin sleeves for the cylinder walls. The walls' surface is plateau honed. Plateau honing creates a very smooth surface for the piston rings to move on within the cylinder walls, which lowers the friction created by that movement. To achieve this smooth finish, the walls are machined twice rather than once, as in a conventional engine. Plateau honing also creates a surface that holds lubricant; this also decreases friction. The pistons are fitted with low friction piston rings. The engine is also fitted with several lightweight, plastic parts, such as the intake manifold and idler pulley.

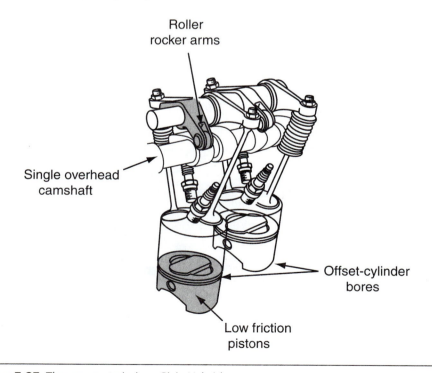

Figure 7-27 The power train in a Civic Hybrid.

The engine uses a newly designed intelligent dual-point sequential ignition (i-DSI) that uses two ignition coils firing two spark plugs per cylinder to ensure complete combustion for economy and power. The eight ignition coils are independently controlled by the PCM to respond to engine speed and load, and can change the timing of the firing of the individual spark plugs, which are located close to the intake port and the exhaust port within the cylinder head. The front spark plugs can be advanced, rear plugs can be retarded, or both plugs can be fired at the same time. These actions depend on the load on the engine and the speed of the engine. Here is a summary of how it works:

❑ When the throttle is opened and engine speed is low, the ignition operates in its sequential mode. The spark plug at the intake side has advanced ignition and fires first. Then the plug located near the exhaust port ignites, forcing the flame to grow throughout the mixture in the combustion chamber.

❑ When the throttle is opened and engine speed is in its mid-range, the ignition operates in its simultaneous mode. During this mode, both spark plugs fire at the same time.

❑ When the throttle is fully opened and the engine's speed is low, the ignition operates in a sequential mode with the spark plug on the intake side being advanced and the one on the exhaust side delayed or retarded. This combination results in maximum torque from the engine.

❑ When the throttle is fully opened and the engine's speed is high, the ignition operates in its simultaneous mode. During this mode, both spark plugs fire at the same time to produce maximum horsepower from the engine.

To improve the regenerative qualities of the IMA system, Honda modified its VTEC system, enabling it to close the intake and exhaust valves on up to three cylinders during deceleration. This system is called the **cylinder idling system** and increases the amount of energy captured during regenerative braking.

With the valves closed, the pistons in those cylinders move quite freely. This, in turn, reduces the amount of engine braking or resistance that takes place during deceleration. By reducing the amount of mechanical resistance of the engine during deceleration, the motor/generator can provide maximum resistance by producing more electricity.

The cylinder idling system actually reverses the role of the standard VTEC system. In conventional VTEC engines, high engine speed and engine oil pressure are used to lock a pair of rocker arms at each cylinder together to increase horsepower. The system relies on the movement of a hydraulic piston that makes a connection between the two rocker arms, each riding on a differently shaped camshaft lobe and each working a separate valve. When oil pressure is high, both valves in each cylinder open to improve engine breathing. In the cylinder idling system (Figure 7-28), low engine speeds and engine oil pressure deactivate the rocker arms. The system also has two rocker arms per valve: a valve lift rocker arm that follows the lobe of the camshaft and a cylinder idle rocker arm that actually opens the valve. The two rocker arms are connected by a hydraulic piston (Figure 7-29). When a cylinder is deactivated or idled, hydraulic pressure moves the piston, and the connection between the two rocker arms is gone. This allows the valve lift rocker arm to move along the camshaft lobe without opening the valve.

Transmissions Civic hybrids are available with either a five-speed manual transmission or a CVT. The manual transmission is basically the same one used in the Insight except for a few modifications. The primary changes are a larger clutch assembly, different gear ratios, the use of a double-cone synchronizer in both first and second gears, and a redesigned lubrication system that reduces internal friction.

The CVT transmission is also based on the unit used in the Insight. It too was modified for the Civic. Most of these changes were necessary to handle the increased power from the drive train and the weight of the vehicle. One notable change is the addition of a creeping aid system that minimizes the amount the car will roll backwards during a stop on hills. Other changes

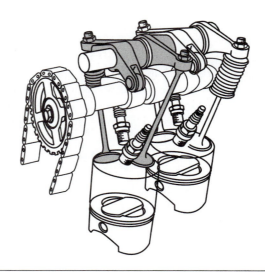

Figure 7-28 The cylinder idling system.

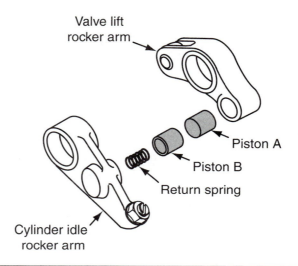

Figure 7-29 A rocker arm assembly for the cylinder idling feature.

include improved hydraulic controls, the use of carbon fiber as the lining material for the starting clutch, and stronger materials used throughout without adding much weight to the unit.

Honda Accord Hybrid

In 2005, Honda released the Accord Hybrid. This car features the third generation of the IMA system. It also incorporates other technologies that result in reduced fuel consumption and emission levels. The changes made to the IMA system include increased motor output, improved battery performance, and greater total system efficiency. The system is used in concert with a 3.0L i-VTEC V-6 engine equipped with **Variable Cylinder Management (VCM)**. VCM shuts down three of the engine's six cylinders during certain times, such as cruising and deceleration.

Normally, performance is sacrificed for economy, but the application of hybrid technology in the Accord did not result in this. The combination of the gasoline engine and the IMA system resulted in performance gains as well as in fuel economy. The total available output from the combination of power plants is 255 horsepower (non-hybrid Accord V-6 Sedans have 240 hp) and 232 lb/ft of torque. When compared to a conventional Accord equipped with a 3.0-liter V-6 engine, the acceleration times of the hybrid are quicker by more than 5%. At the same time, the fuel economy rating for city driving increased 38%, and by 23% for highway driving. The hybrid also is categorized as a ULEV.

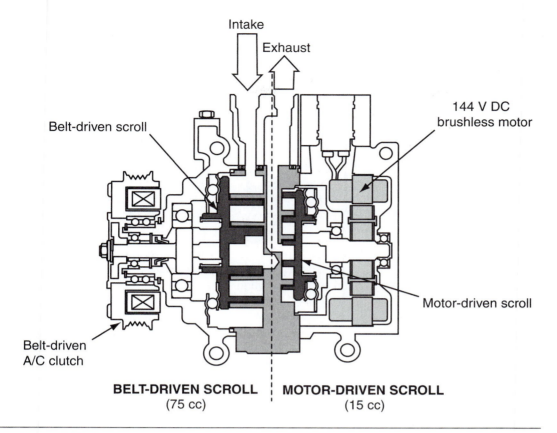

Figure 7-30 The dual scroll "hybrid" air conditioning compressor used in the Accord Hybrid.

To increase efficiency, Honda equipped the Accord with VCM and the redesigned IMA system. They also used other technologies to reduce power losses and vehicle weight. Air drag was subtlety reduced through the addition of a rear spoiler and lightweight aluminum alloy wheels designed to control airflow. Although the hybrid looks like most Accords, these changes result in a drag coefficient of only 0.29. To reduce the weight, the Accord Hybrid is built with a lightweight aluminum hood, aluminum front and rear bumper beams, aluminum rear suspension knuckles, a magnesium engine head cover, and a magnesium intake manifold.

The Accord Hybrid is also equipped with EPS and a dual scroll "hybrid" air conditioning compressor (Figure 7-30). The air-conditioning system uses two compressors built into one housing; one compressor is driven by the engine and the other driven by an electric motor powered by the high-voltage battery. When full cooling is needed, the A/C unit relies on both power sources to provide maximum cooling. During normal cooling, the A/C is powered by either the belt-driven compressor or the electric-motor-driven compressor. This allows the A/C system to work at all times, even when the stop-start feature is activated.

IMA System Like other Honda IMA systems, the third generation is comprised of a primary power unit, the gasoline engine, and the IMA motor. The IMA system was modified to work with a more powerful gasoline engine and its overall efficiency has been increased. The system uses a 144-volt battery pack and an AC synchronous motor (Figure 7-31). The PCM controls the output of the engine and the electric motor depending on the needs of the driving conditions, transmission, cruise control, traction control system, and anti-lock brake system.

▲ **WARNING:**
Because the electric motor inside the compressor case is in contact with the oil in the compressor, only the specified oil (Sanden SE-10Y) should be used in the compressor. This oil has electrical insulating qualities that protect you from dangerous electrical shocks. Also, if you use the wrong oil, the air-conditioning unit will be contaminated and this may result in a need to replace the compressor, condenser, evaporator, and/or all of the refrigerant lines.

● **NOTE:**
The following discussion includes only those features or equipment that have been changed with new generation of the IMA system. Components and systems that are not mentioned are similar to those used in Insight and Civic hybrids.

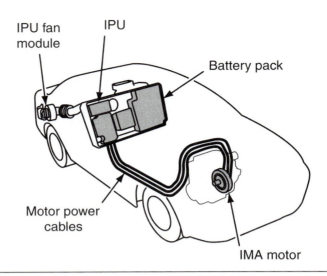

Figure 7-31 The IMA system in an Accord Hybrid.

The PCU, located between the passenger compartment and the trunk, controls the power flow to and from the IMA motor/generator. The motor used in the Accord Hybrid has a redesigned rotor that increases output density and makes the motor more efficient than previous IMA motors. Dimensionally, the new motor is only 0.12 inches (3 mm) thicker than the motor used in the Civic. However, its torque output has been increased by 26% during startup, and during driving it can provide more than twice the output of the motor used in the Civic. The total output of the motor is 16.1 hp (12 kW) and 100.4 lb/ft (136 Newton-Meters) of torque.

The redesigned motor is also able to generate more electricity. During regenerative braking, the motor/generator produces 12% more power (14 kW total). This reduces the amount of time the engine must drive the generator to keep the batteries charged.

The third-generation IMA system has a redesigned IPU that can respond to conditions quicker than previous systems. The IPU also contains a new inverter and DC-DC converter. Although the voltage and capacity ratings for the battery are unchanged, the output density of the battery has been increased by 45%.

The IPU is equipped with an integrated cooling system mounted directly on the battery pack's outer box (Figure 7-32). Air is pulled, by the cooling fan, into the battery module through the top of the tray behind the rear seats. The air passes over the heat sinks of the inverter, DC-DC converter, and A/C compressor driver before it is exhausted to the outside.

Engine The engine in the Accord hybrid is a 3.0-liter i-VTEC V-6. It has the same displacement as other Accord V-6s but has been modified to achieve better fuel economy, without sacrificing performance. Without the power assist from the IMA system, the engine provides 240 hp and 217 lb/ft of torque. With the IMA system, the rated peak power is 255 hp at 6000 rpm and 232 lb/ft. at 5000 rpm. This combination results in quicker acceleration times than the nonhybrid models.

Engine modifications include a dual-stage intake manifold (Figure 7-33) and a new airflow sensor. The i-VTEC system provides lean-burn and a valve pause system that reduces engine-pumping loss and increases the effectiveness of regenerative braking. The engine is also fitted with a two-axis belt drive with an automatic tensioner designed to prevent power losses due to friction. To keep down the weight of the engine, the engine has an aluminum alloy block with cast-iron cylinder liners, and other parts are made of magnesium.

A major contributor to fuel economy is the VCM system. The system monitors throttle position, vehicle speed, engine speed, transmission gear selection, and other factors to determine if full engine power is required. If full power is not required, the system shuts down the rear bank of three cylinders. This is accomplished by closing the valves and disabling the fuel injectors at

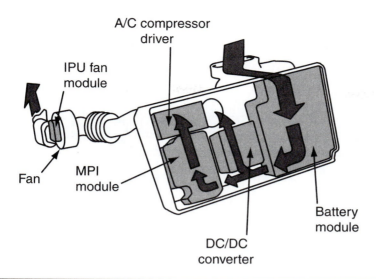

Figure 7-32 The IPU is equipped with a fan. Air is drawn into the battery module from the top of the rear tray, then it is exhausted into the trunk compartment and outside of vehicle through the MPI module heat sink, the DC-DC converter and A/C compressor driver heat sink.

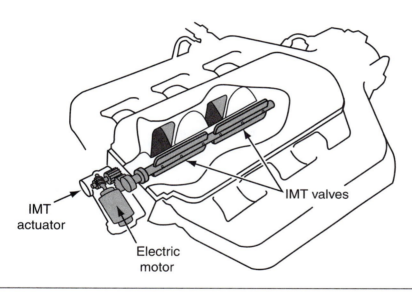

Figure 7-33 Engine power is enhanced by closing and opening the intake manifold tuning (IMT) valve, which changes the effective length of the intake runners. When the valve is closed, there is high torque at low engine speed. When the valve is open, there is high torque at high engine speed.

those cylinders. The deactivation of the cylinders normally takes place during cruising and deceleration. The iridium-tipped spark plugs continue to fire when the cylinders are deactivated to minimize spark plug temperature loss and prevent fouling caused by incomplete combustion.

When full engine power is needed, the system quickly gets the three idling cylinders back into action. While the cylinders are idling, the VCM system controls the ignition timing and cycles the torque converter lock-up clutch to suppress any torque-induced jolting caused by the switch from six- to three-cylinder operation.

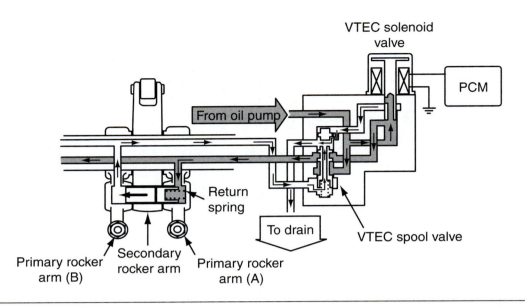

Figure 7-34 During the six-cylinder mode, on the intake side, the spool valve switches oil pressure so that the primary rocker arm moves the switching piston. This causes the piston to slide into the primary rocker arm, locking the rocker arms together. The primary rocker arms are then actuated by the secondary rocker arm.

Like the cylinder idle system in the Civic hybrid, the VCM uses a hydraulic circuit with two systems, each capable of providing the hydraulic pressure required to push the synchronizing or locking piston in the required direction (Figure 7-34). To deactivate the cylinders, the piston moves to separate the valve-lift rocker arms from the deactivated cylinders' valve rocker arms. This action keeps the valves of those cylinders closed. When full engine power is needed, a solenoid-operated hydraulic valve opens and applies pressure to the piston. The piston moves to lock the two rocker arms together. At the same time, the system reinitiates the fuel injection system and the three cylinders work in harmony with the others.

To reduce the vibrations that result from the deactivation of the three cylinders, the engine is equipped with an **Active Control Engine Mount (ACM)** system (Figure 7-35). This system uses sensors monitored by the ECU. The ECU, in turn, directs the engine mounts to move with the vibrations. This stops the vibrations from moving into the passenger compartment. Also, to reduce the noise level of the engine while the cylinders are deactivated, an **Active Noise Control (ANC)** system works in concert with the ACM. The ANC uses microphones in the front and rear of the passenger cabin to monitor the low-frequency noise created by engine when it is running on only three cylinders. The system creates an equal but opposite noise, which cancels out the noise from the engine. The created noise is sent out through the car's speaker system.

Operation The engine is normally started with the IMA motor, however if the charge of the IMA batteries is low, an auxiliary 12-volt starter is used. When the driver initially presses on the accelerator, the power from the engine is used to move the car. If hard acceleration is needed, the IMA motor assists the engine while it is running on all six cylinders. The motor also assists the engine when the car is moving up a sharp incline or is moving a heavy load (Figure 7-36). When assist is provided, the blue indicator bars in the IMA display will light. If the state of charge of the IMA batteries is below a specified level, the PCM and MCM will not allow assist, regardless of the operating conditions. Also, assist is not available when the battery pack is very cold or hot.

When the car is driven at high speeds but without the accelerator fully open, the engine runs on six cylinders but is not assisted by the motor. Once the car is cruising at a steady speed and with a light load, the VCM deactivates the rear bank of cylinders. During this time, if slightly

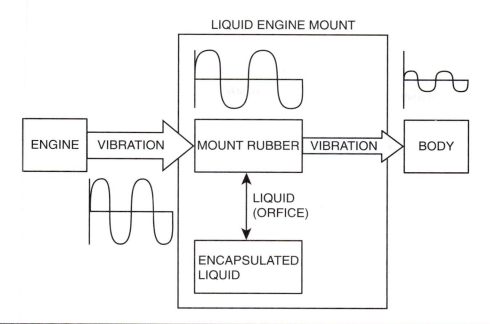

Figure 7-35 The ACM system decreases engine-to-chassis vibration at low rpm and when the engine is in cylinder pause mode. The system includes conventional, liquid-filled engine mounts that absorb vibration and an actuator that cancels engine vibration by producing a counter, or reverse vibration. *Courtesy of American Honda Motor Co., Inc.*

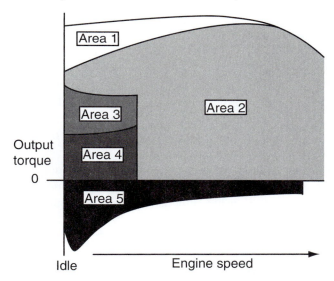

Figure 7-36 Explanation: **Area 1** - When maximum output torque is required, the engine is driven by six cylinders with wide open throttle, and the IMA motor assists the engine to increase torque. **Area 2** - At high engine speed except wide open throttle and deceleration, the engine is driven by six cylinders without assist from the IMA motor. **Area 3** - When partial output torque is required at low engine speed, the engine is driven by three cylinders and the IMA motor assists the engine to increase output torque. This control method saves fuel consumption during acceleration. **Area 4** - When low output torque is required at low engine speed, the engine is driven by three cylinders without assist from the IMA motor. This control method saves fuel consumption during cruise and light-load conditions. **Area 5** - During deceleration, the PCM stops the fuel injection to all cylinders, and the IMA motor functions as a generator to charge the IMA battery. The intake and exhaust valves of the rear bank are deactivated by the VCM to reduce mechanical friction and increase the energy for charging.

more power is needed, the IMA motor will assist the engine as the engine continues to run on three cylinders. When more power is needed, the VCM system will reactivate the cylinders, and the motor assist will stop unless maximum power is needed.

When the driver releases the accelerator to slow the car, the IMA motor, driven by the car's wheels, works as a generator to charge the batteries. The intake and exhaust valves of the rear bank of cylinders are deactivated by the VCM to reduce mechanical friction and to increase the effectiveness of the regenerative braking. During this time, the green bars in the IMA display illuminate to show the amount of electricity being generated. When the battery is fully charged, regeneration is stopped to prevent overcharging of the battery. When the driver presses on the brake pedal to decelerate more rapidly, regenerative braking continues and the VCM system shuts off the fuel injectors in the front bank of the engine.

Like other hybrids, the Accord hybrid has the stop-start (idle-stop) feature to reduce fuel consumption and emissions. The IMA system temporarily turns off the gasoline engine when the vehicle comes to a stop from speeds over 10 mph. Unlike the Insight and Civic Hybrid, the idle-stop system in an Accord Hybrid will continue to operate even while the car's automatic climate control system is in use. This is because the Accord is fitted with the dual source A/C compressor. When idle-stop is operating, the auto stop indicator in the instrument panel blinks (Figure 7-37). If the driver's door is opened during idle-stop, the indicator blinks and a warning buzzer sounds to remind the driver that auto-stop is in operation. Idle-stop will not operate when the car is first started on an extremely hot day and maximum cooling is required.

Instrumentation The biggest difference between a conventional Accord and a hybrid Accord is the instrument panel. In hybrid models, the panel displays the level of IMA motor charge or assist, the state of charge of the IMA battery, an "ECO" light to indicate when the VCM is operating in three-cylinder mode, a light to indicate when the Idle Stop mode is active, and trip and lifetime fuel economy readouts.

Transmission The Accord Hybrid is equipped with a five-speed automatic transmission. The transmission is much the same as used in the conventional Accord but has been modified to work with the IMA system and to provide better performance and fuel economy. The transmission is fitted with different gear ratios, a redesigned lock-up torque converter, and an electric oil pump. The electric oil pump maintains consistent pressure within the transmission, whether the engine is running or not. In a conventional transmission, the engine drives the transmission's oil pump and there is only pressure in the unit when the engine is running. The electric oil pump maintains hydraulic pressure when the engine is not running during idle-stop. The pressure allows the transmission to be responsive immediately after the engine is restarted.

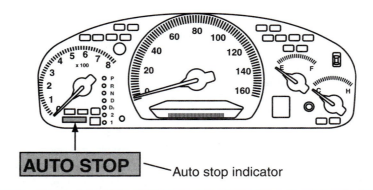

Figure 7-37 When idle stop is operating, the indicator blinks. If the driver's door is opened during auto idle stop, the auto stop indicator blinks and the warning buzzer sounds to remind the driver that auto stop is in operation.

Honda Hybrid Safety Issues

These hybrid systems rely on very high voltages. All of the high-voltage cables are covered in orange sleeves for easy identification and you should follow the procedures for disarming the high-voltage system before performing any service on or near the high-voltage circuits. There is a main switch on the IPU that is used to disconnect the battery module from the rest of the car. There are three large capacitors in the MDM that will take at least 5 minutes to discharge after the switch is turned off. To disconnect the high-voltage system:

1. Remove the switch cover from the IPU cover (Figure 7-38).
2. Turn the switch OFF (Figure 7-39).
3. Wait at least 5 minutes.
4. Remove the IPU cover.
5. Measure the voltage at the output terminals (Figure 7-40). Make sure the voltage is low enough for safe operation before any service is done to the car.

The high-voltage circuits are isolated from car's chassis ground. The car has a ground-fault detection system that can warn the driver if there is a short between the high-voltage circuit and the chassis.

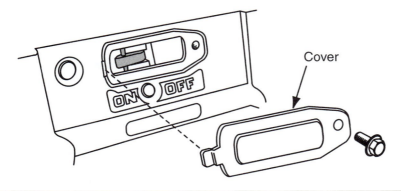

Figure 7-38 To shut off the high-voltage circuit in a Honda Hybrid, turn the ignition switch OFF, remove the rear seat back, and remove the battery module switch lid (A) from the battery module.

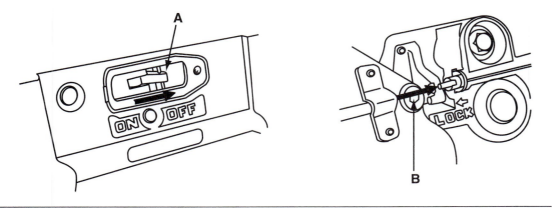

Figure 7-39 Continue disabling the high-voltage system by turning the battery module switch (A) OFF. Make sure the bolt (B) is showing. Then, wait at least 5 minutes to allow the MDM capacitors to discharge before removing the IPU cover.

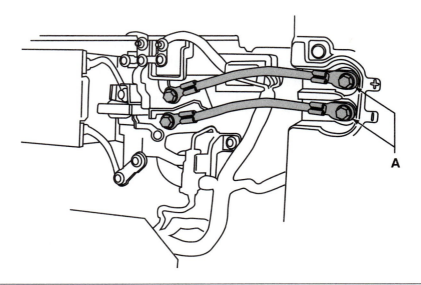

Figure 7-40 Before doing any service with or near the high-voltage system. measure the voltage at the battery module terminals (A). There should be 30 V or less. If more than 30 V is present, there is a problem in the circuit; do the DTC troubleshooting first.

● **NOTE:** Throughout Honda's literature, service information, and press releases, the IMA motor is described as a DC brushless type or an AC synchronous type. It cannot be both and it is not. In fact, it is an AC synchronous motor and many of the controls and sensors indicate this. An explanation for the apparent conflicting information comes from a source within Honda. I include this to prevent confusion and the possibility of a reader deeming the contents in this chapter as incorrect. Honda's position is "The terminology for 'DC brushless' and 'AC synchronous' are both correct when viewed in the proper context. 1) DC brushless refers to the Honda IMA system from a total system standpoint, best described in two parts. A) The initial power source from the battery pack for the electric motor is DC (which is converted into AC via an inverter to power the electric motor). Simply put, DC is used to clearly indicate that the initial power source is DC. The brushless part is used to best describe the long lifecycle advantage of the motor's design. 2) AC Synchronous refers to the electric motor specifically. It is indeed an AC Synchronous design. This terminology is used to describe the electric motor when referencing it on an individual component basis. The best way to summarize this situation is that many terminologies exist for electric motor technology. The dual descriptions provided by Honda are attempting to best explain the technology from a total systems standpoint and from an individual component standpoint, as appropriate." After many discussions with Honda, I was asked how they should describe the motor. I suggested they describe the motor as a brushless AC synchronous motor. That description is accurate and tells the public what Honda wants them to know and it says what the motor really is.

Review Questions

1. As described in this chapter, what is the difference between a mild hybrid and an assist hybrid?
2. What are the basic components of a belt alternator starter hybrid system?
3. What are the main reasons that a mild hybrid consumes less fuel than a conventional vehicle?
4. Describe the location of the main parts of the motor used in Honda's IMA system.
5. In Honda hybrids, what is the purpose of the DC-DC converter?
6. What are the basic steps that should be followed to disable the high-voltage system in a Honda Hybrid?
7. Which of the following statements about the high-voltage circuits in a Honda Hybrid is NOT true?
   A. All of the high-voltage cables are covered in orange sleeves for easy identification.
   B. You should always follow the disarming of the high-voltage systems before performing any service work on or near the high-voltage circuits.
   C. There is a main switch on the instrument panel that is used to disconnect the battery module from the rest of the car.
   D. There are three large capacitors in the MDM that will take at least 5 minutes to discharge after the switch is turned off.

8. Describe the different ways voltage is changed in a GM mild hybrid pickup.
9. Which of the following statements about a linear air-fuel (LAF) sensor is NOT true?
   A. The linear air-fuel sensor is located at the outlet of the three-way catalyst.
   B. Within the sensor, there is an exhaust flow chamber, atmosphere reference chamber, and a diffusion chamber, plus a heater circuit that allows the sensor to work when the engine is cold.
   C. Positive voltage indicates a lean mixture and negative voltage indicates a rich mixture.
   D. The normal operating voltage range is about 1.5 volts.
10. In an Accord hybrid, which of the following systems relies on microphones to function properly?
    A. VTEC
    B. ACM
    C. ANC
    D. VCM

FULL HYBRIDS

After reading and studying this chapter, you should be able to:

❏ Describe the difference between a mild and a full hybrid.

❏ List and explain the purpose of the basic components used in Toyota's hybrid system.

❏ Explain why hybrids work well with Atkinson cycle ICEs.

❏ Describe the basic operation of the two electric motors used in Toyota's hybrids.

❏ Explain why Toyota uses very high voltage to power the electric motors.

❏ Describe the purpose of an inverter.

❏ Describe the basic changes Toyota made to the ICEs used in its hybrid vehicles.

❏ Explain how Toyota provides four-wheel drive in its hybrid SUVs.

❏ Describe how Toyota modified the power-split unit for the high-voltage motors used in its SUVs.

❏ Describe the basic operation of the hybrid system used by Ford Motor Company.

❏ Explain how Ford provides four-wheel drive in its hybrid SUVs.

❏ Describe the changes Honda made to change their assist-type hybrid to full hybrids.

❏ Describe the operation of the two-mode hybrid system.

Key Terms

Atkinson cycle
Battery smart unit
Brake System Control
    Module (BSCM)
Electric motor-
    assisted power
    steering (EMPS)
HC Adsorber
    and Catalyst
    System (HCAC)
Hybrid Vehicle
    Control Unit
    (HV ECU)
Planetary carrier
Planetary pinions
Power-split device
Resolver
Ring gear
Sun gear
System Main
    Relay (SMR)
Transmission Control
    Module (TCM)
Two-mode
    hybrid system
Variable Cylinder
    Management (VCM)

Introduction

Full hybrids (referred to as "strong hybrids" by some) are vehicles that can be powered by their engine, one or motor electric motors, or a combination of these. Some full hybrids are described as having a series-parallel design because the engine is used to drive a generator, as well as power the vehicle (Figure 8-1).

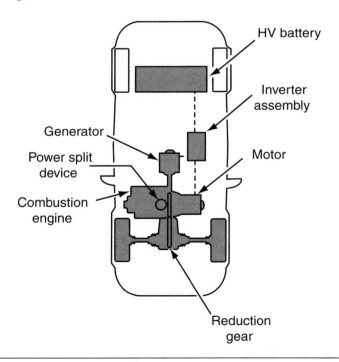

Figure 8-1 The basic setup for a series-parallel hybrid system.

Common full hybrids include the Toyota Prius, Ford Escape, and late-model Honda Civics. Toyota has also introduced four-wheel-drive hybrid SUVs and, by the time this book is released, there will be full hybrids available from General Motors, Daimler Chrysler, and other automobile manufacturers.

To function as a full hybrid, the vehicle must be fitted with a powerful electric motor and a high-voltage battery pack. These are necessary to provide the power required to move the vehicle on battery power alone. Full hybrids also require electronic and mechanical devices that allow for seamless switching between the power sources. Full hybrids also have regenerative braking, stop-start capabilities, and the motor is used to assist the engine.

This chapter will look at the most common full hybrids on the road today as well as those that will be released shortly.

Toyota's Hybrids

Toyota introduced the first mass-produced hybrid vehicle in 1997. The hybrid, the Prius, was only available in Japan until the 2000 model year, when it is was brought to North America (Figure 8-2). Sales of the first-generation Prius were good, but increased drastically with the second generation in 2004. Toyota also offers hybrid cars and SUVs, and Lexus (Toyota's premium brand) has introduced several hybrid models.

The original Prius did not look like a regular automobile. Its shape dictated by aerodynamics made it noticeable. Unlike the Honda Insight, which was the only other hybrid at that time, the Prius was a family car that could carry up to five passengers. The second-generation Prius grew up and now provides enough interior space to be classified as a midsized vehicle. With the size increase came an increase in power, making it more usable for everyday driving. The fuel economy ratings for the Prius have a combined EPA mileage estimate of 55 mpg. The Prius has also been certified as SULEV, or Super Ultra Low Emission Vehicle. The Prius is the world's best-selling hybrid. In fact, there is a waiting list of potential buyers.

Toyota's approach to hybridization is based on both the series and parallel designs. The engine can power the vehicle or a motor/generator in the series configuration. Another motor/generator is used to assist the engine, power the vehicle by itself, or charge the batteries. The control of the motor/generators and the engine is one of the keys to Toyota's system. Complex electronics are required to monitor operating and driving conditions and to control the flow to and from the electric motor/generators. Toyota also relies on a power-splitting device

Figure 8-2 A 2001 Toyota Prius. *Courtesy of Dewhurst Photography and Toyota Motor Sales, U.S.A., Inc.*

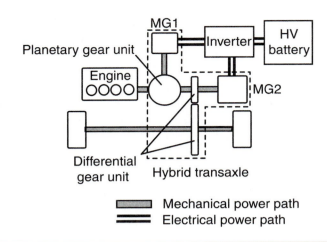

Mechanical power path
Electrical power path

Figure 8-3 The THS uses a combination of two types of motive forces: an engine and a motor. The THS can be divided into two systems: the series hybrid system, and the parallel hybrid system.

that mechanically blends the output from the motors and the engine. This system (Figure 8-3) was called the **Toyota Hybrid System (THS)** when it was released, and the newer designs of the THS are called the **Hybrid Synergy Drive (HSD)** system.

Toyota has applied this technology to their midsized sport utility vehicle (SUV), the Highlander, and Lexus has a hybrid version of its RX 330 luxury utility vehicle, the RX 400H. Both use a version of HSD with electric motors and a V-6 engine. Four-wheel drive models of these SUVs are also available. These vehicles are also based on the HSD system and have front and rear electric motors in addition to the engine.

Toyota Prius

The first-generation Prius had a fuel-efficient 1.5-liter, four-cylinder gasoline engine rated at 70 horsepower (52 kW) and a 45-horsepower (33 kW) electric traction motor. A separate motor/generator was used to charge the battery pack and to start the engine; this motor does not contribute power to move the car. During deceleration and braking, the traction motor worked as a generator to provide for regenerative braking.

Power for the second-generation Prius is provided by the same 1.5-liter, four-cylinder gasoline engine, now rated at 76-horsepower (57 kW), and a 68-horsepower (50 kW) electric motor. This model also has a second motor/generator connected to the engine. With the second generation came a larger car (Figure 8-4) and a redesigned version of Toyota's hybrid system. The additional size and more powerful engine did not increase fuel consumption; rather, fuel economy improved. The improvements made to the hybrid system allow the motor to power the vehicle by itself for longer periods and the amount of energy captured during regenerative braking was increased. Both of these reduce the amount of time the engine must drive the generator to keep the battery pack charged. Other changes to the Prius include a reduction in electrical loads, friction-reducing techniques applied to the engine and transaxle, and an improved air conditioning system.

Figure 8-4 A 2006 Toyota Prius. *Courtesy of Dewhurst Photography and Toyota Motor Sales, U.S.A., Inc.*

The revised HSD system uses a smaller battery pack with a slightly lower nominal voltage. However, the system can boost the voltage to the motor up to 500 volts (Figure 8-5). This additional voltage reduces the amount of current required to power the motor and decreases the amount of electrical energy lost as the motor is powered. If the motor's power (output watts) is held constant, the amount of current drawn by the motor will inversely increase or decrease with

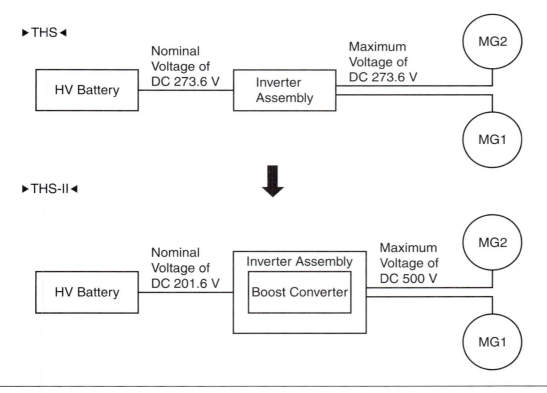

Figure 8-5 A comparison of the voltages used by the first and second generations of the THS. *Courtesy of Toyota Motor Company*

an decrease or increase in voltage. Therefore, if the voltage is doubled, the current will be reduced by half. Also, if the current to the motor is held constant and the voltage is increased, the motor's power will be increased.

As current flows through a wire, some of the electrical energy is changed to heat as the electrons move through the wire. The heat is a result of the resistance in the wire. When current is low, less heat is produced, and therefore less energy is lost. According to Joule's Law, if the current is reduced by half and circuit resistance remains the same, the amount of energy lost is reduced by 75%. This decrease in energy loss increases the efficiency of a hybrid vehicle.

The engine in a Prius drives the wheels and/or a generator only when needed. The primary goal of the HSD is to use electric power as often as possible. The electronics of the system responds to operating conditions, driver inputs, and driving conditions and attempts to keep the car operating in its most efficient mode at all times. The engine only runs when extra drive power is needed or when the battery pack needs recharging. The engine is connected to the drive wheels and a generator through the transmission. The transmission houses a single planetary gear assembly, the power split device. This device controls where the engine's output is directed: to the drive wheels, the generator, or both.

Operation The basic operation of the first and second generations of the hybrid system are the same. The biggest difference between the two is the role of the electric traction motor. In the second generation, this motor has the ability to power the car more often and with greater power. Both versions of the THS rely on an engine, a motor/generator that serves as the starter motor and a generator (referred to as MG1, Motor Generator 1), and a traction motor and generator (called the MG2, Motor Generator 2).

The start procedure also varies between the two generations. A 2001 through 2003 Prius uses an ignition key to operate the ignition switch. This switch activates the system's power mode and effectively starts the system. With the brake pedal depressed, the ignition switch is moved and held in the START position until the READY lamp on the instrument panel is lit and a beep is heard. At that time, the car is ready to move.

On 2004 and later Prius models, a push button is used to get the system ready after the ignition key is inserted into the key slot or after the system recognizes the code of the key. The latter is part of the "smart entry and start" system. With the ignition key in position and the brake pedal depressed, the system is started or placed in the READY mode by depressing the POWER button. Once the system is ready, the READY lamp will illuminate and a buzzer will sound. There is a power mode switch that can be used to select various states of readiness: OFF, ACC, IG-ON, or READY. An indicator on the switch shows the selected mode.

In the second-generation system, the battery's state of charge and temperature, engine coolant temperature, and the electrical load determine whether the engine starts when the POWER button is depressed. During conditions that do not meet minimum requirements, the engine is started to drive MG1 to charge the battery.

When drive is selected and the driver releases the brake pedal, the traction motor (MG2) moves the vehicle forward (Figure 8-6). The motor is able to supply propulsion power until the battery pack's voltage drops or when the driver calls for rapid acceleration. When MG2 is powering the vehicle, the engine is off and MG1 rotates freely and does not function as a generator. If battery recharging is needed, the engine starts and drives MG1. When reverse gear is selected, MG2 rotates in a reverse direction. The engine remains off, as does MG1.

During normal operating conditions, as vehicle speed increases the engine starts. The power from the engine is split according to the needs of the system. Part of the engine's power drives MG1 to supply energy to MG2, which is working to assist the engine. The remaining power from the engine is used to propel the vehicle. The amount of engine power sent to the

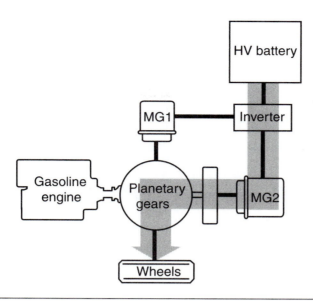

Figure 8-6 During initial acceleration, electrical power from the battery pack to MG2 provides the energy to drive the wheels.

generator and drive wheels is controlled by the system. When the vehicle is at a cruising speed, both the engine and MG2 power the vehicle (Figure 8-7). If engine power is not needed to maintain the speed, the system will shut down the engine and the vehicle is powered by electricity alone. If the battery's state of charge gets low, the engine is restarted to drive MG1.

To overcome a heavy load, such as during acceleration, climbing a hill, or passing another vehicle on the highway, both the engine and MG2 power the car. MG2 receives energy from MG1 and the battery pack. This enables the traction motor to work under full power. The engine

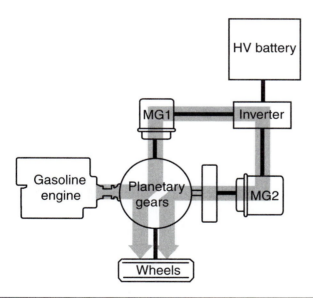

Figure 8-7 While the wheels are being driven by the engine through the planetary gears, MG1 is rotated by the engine to supply voltage to MG2.

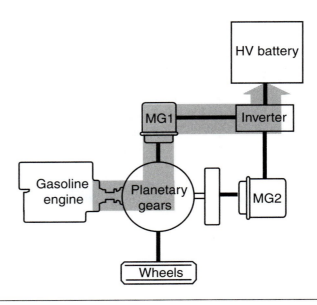

Figure 8-8 To charge the battery pack, MG1 is rotated by the engine through the planetary gears.

also works under full power to ensure smooth acceleration and to drive MG1 for maximum electrical generation. Once the car returns to a normal cruising speed, battery power for MG2 is shut off and battery recharging is provided by MG1, driven by the engine (Figure 8-8).

During deceleration or braking, MG2 acts as a high-output generator, driven by the car's wheels (Figure 8-9). As soon as the driver lets off the accelerator pedal, the engine is shut down, MG2 becomes a generator, and regenerative braking begins. When the driver presses the brake pedal, most of the initial braking force is actually the force required to turn MG2. The hydraulic brake system supplies the rest of the braking force and brings the car to a halt.

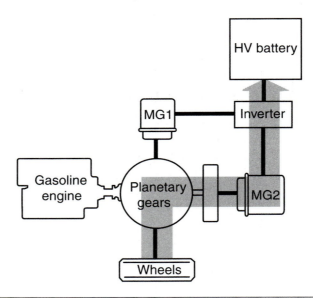

Figure 8-9 When the vehicle is decelerating, the vehicle's kinetic energy is captured by MG2 and used to charge the battery pack.

The engine remains off when the car is stopped, unless the battery pack needs recharging. In this case, the engine automatically restarts to drive MG1. However, if the air conditioning system on a first-generation Prius is set on MAX A/C, the engine will not shut down because the air conditioning compressor is driven by the engine. Second-generation models have an electrically operated compressor and the engine can be shut down without affecting A/C operation. Stop-start in a Prius operates differently than the systems used in mild and assist hybrids. Because the traction motor can power the car by itself, the engine does not restart as soon as the driver releases the brake pedal or depresses the accelerator. The engine in a Prius is only restarted when it is needed to power the vehicle or to drive the generator.

After the car has been driven and the driver stops it and puts the gear selector in the PARK position, the engine may run and drive MG1 to recharge the batteries. This activity is determined by the control unit.

Motor/Generators The hybrid system in both generations of the Prius is basically the same. However, improvements were made from one generation to the next. As the major parts are discussed, the difference between the generations will be noted. Current Prius models feature more advanced engine controls, more efficient motor/generators, a revised inverter, and an advanced battery pack with a lower nominal voltage.

The main components of the hybrid system include the engine, MG1, MG2, a planetary gear set (**power-split device**), the control unit, an inverter, and the battery pack. The system is capable of instantaneously switching from one power source to another or combining the two. Electronic controls and the power-split device (Figure 8-10) channel the outputs of these power sources.

Both the MG1 and the MG2 are permanent magnet AC synchronous motors that can also function as generators. MG1 starts the engine when engine power is needed. It also is driven as a generator, by the engine, whenever electrical energy is needed to run MG2 or to recharge the battery pack. In addition, when acting as a generator, its speed varies with the charging needs of the system and, therefore, MG1 effectively controls the power-split device. This motor/generator was modified in the second-generation THS to allow it to operate at higher speeds, which in turn increased its charging capacity. This was done by changing the size and magnetic density of the rotor.

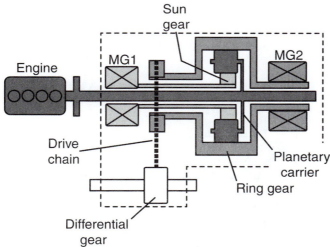

Figure 8-10 The transaxle assembly with MG1, MG2, and the power-split device.

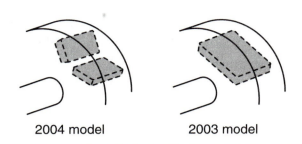

2004 model   2003 model

Figure 8-11 On the 2004 and later models of Prius, the permanent magnets in the rotor of MG2 are in a V-shape to improve the power output of the motor.

MG2 is used to move the car by itself, assist the engine, or serve as a generator. In THS I, MG2 has a rated voltage of 273.6 with a maximum output of 33 kW. MG2 in the THS II design is rated at 201.6 volts and has a maximum output of 50 kW. This increase in power output came from a redesigned rotor. The new rotor features neodymium magnets shaped in V structure (Figure 8-11). The new motor is also capable of providing higher torque at a broader speed range (295 lb/ft [400 Nm] at 0 to 1200 rpm compared to 259 lb/ft [350 Nm] at 0 to 400 rpm). This increased output also results in greater generation of electricity during regenerative braking.

AC synchronous motors require a sensor to monitor the position of the rotor within the stator. It is necessary to time, or phase, the three-phase AC so it attracts the rotor's magnets and keeps it rotating and producing torque. AC creates a rotating magnetic field in the stator and the rotor chases that field. The control system monitors the position and speed of the rotor and controls the frequency of the stator's voltage, which controls the torque and speed of the motor.

Toyota uses a sensor (Figure 8-12), called a **resolver**, in MG1 and MG2 to monitor the position of the rotor. The sensor has three individual coil windings: one excitation coil and two detection coils. The detection coils are electrically staggered at 90 degrees. The flow of alternating current into an excitation coil results in a constant frequency through the coil.

This frequency is sent to the detection coils and is altered by the presence of the rotating magnetic field. Because the rotor has an oval shape, the gap between the stator and rotor varies as the rotor rotates. A strengthening and weakening of the magnetic field occurs with the changing gap and alters the frequency at the detection coils. The MG ECU bases the exact position of the rotor on the difference between the frequency values of the detection coils. The MG ECU also calculates the rotational speed of the rotor based on the speed at which the position changes.

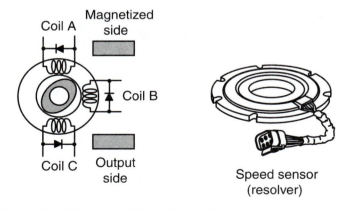

Figure 8-12 A resolver located in each motor precisely detects the magnetic pole position of the rotor and allows the control unit to monitor and control the activity of MG1 and MG2.

Electronic Controls The ultimate control of the hybrid system is the responsibility of the **Hybrid Vehicle Control Unit (HV ECU)**. This microprocessor receives information from sensors and other processors and in turn sends commands to a variety of actuators and controllers (Figure 8-13). In 2004 and newer Prius models, CAN communication is used to link the various microprocessors together. In addition, these late-model cars have a 32-bit CPU rather than a 16-bit unit as used in earlier models. This change increased the processing speed of the control unit.

The control system:

❏ Controls MG1, MG2, and the engine based on torque demand, vehicle speed, regenerative brake control, and the battery pack's state of charge (SOC). The primary inputs for this calculation are from speed sensors, the shift position, accelerator pedal position, and the skid control ECU that controls regenerative braking
❏ Controls the operation of the hybrid transaxle.
❏ Monitors the SOC and temperature of the battery, MG1, and MG2.
❏ Monitors the operation of the hybrid system and runs continuous self-diagnostic routines. If a malfunction is detected, the unit stores a diagnostic code and controls the system according to data stored in its memory rather than current conditions (fail-safe) or it may shut down the entire system, depending on the malfunction.
❏ Sends information and commands to the Engine Control Module (ECM), inverter assembly, battery ECU, and skid control ECU to control the system.

The ECM and skid control ECUs were also updated to 32 bits on late models. The ECM coordinates the engine's activity with the hybrid system. It starts and stops the engine as needed, as well as controls the operation of the engine. The skid control ECU or brake ECU calculates the total amount of braking force needed to stop or slow down the car based on the pressure

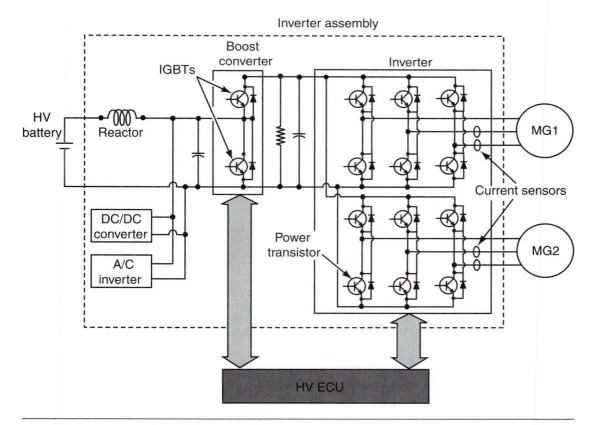

Figure 8-13 An electrical schematic for the hybrid control system.

exerted on the brake master cylinder. This in turn determines how much regenerative braking should take place and how much pressure should be sent to the brakes through the hydraulic system. This information is sent to the HV ECU, which controls the regenerative braking of MG2. The brake ECU also controls the hydraulic brake actuator solenoids and generates pressure at the individual wheel cylinders. The total amount of force applied to the hydraulic brake system is the total required brake force minus the force supplied through regenerative braking. The skid control ECU also controls the operation of the anti-lock brake system.

Battery All models of the Prius rely on Nickel Metal Hydride (NiMH) batteries for hybrid operation. They also have an auxiliary battery that serves as the power source for the ECM, lights, and other 12-volt systems. The battery module (Figure 8-14) contains the hybrid (HV) battery pack, battery ECU, and the **System Main Relay (SMR)**. The module is positioned behind the rear seat, in the trunk.

The hybrid battery pack in the early Prius models had 38 modules of six cells, connected in series. Each cell provided 1.2 volts, which meant the battery pack had a total voltage of 273.6 volts of DC. Later models have 168 cells and a total voltage of 201.6 volts. Later models also have two internal connections between the modules instead of one; this reduces the internal resistance of the battery pack. Reducing the resistance makes the battery pack more efficient by reducing internal power losses. The new battery is 15% smaller, 25% lighter, and has 35% more specific power than the first generation.

The battery ECU receives information about the HV battery's current SOC, temperature, and voltage from various sensors. This information is then sent to the HV ECU, which controls MG1 to keep the battery pack at the proper charge. The battery ECU also calculates the charging and discharging amperage required to allow MG2 to power the car. This information is also sent to the HV ECU, which sends commands to the ECM to control the engine's output. This continuous loop of information is done to maintain at least a 60% SOC at the battery. On late models, the battery ECU is a 32-bit microprocessor.

The battery ECU also monitors the temperature of the batteries during the charge and discharge cycles. The battery ECU also estimates the temperature change that will result from the cycling. Based on this information, it can adjust the battery's cooling fan or, if a malfunction is present, it can slow down or stop charging and discharging to protect the battery. To monitor and calculate battery cooling needs, the battery ECU relies on three temperature sensors housed in the battery module and a temperature sensor in the air intake for the module. On early models, the cooling fan could be set at three separate speeds by the battery ECU. In late models, the cooling fan has no fixed speeds and is duty cycle controlled.

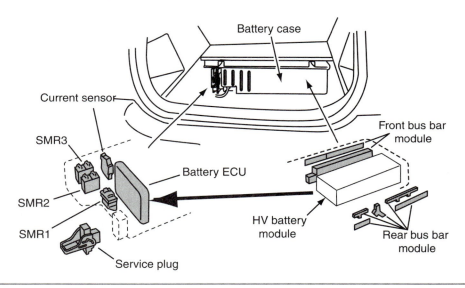

Figure 8-14 The battery pack assembly in a late-model Prius.

If the battery ECU determines there is a problem with the battery pack, it will illuminate a warning light on the instrument panel and store the appropriate DTC in its memory.

The SMR connects and disconnects the high-voltage circuit based on commands from the HV ECU. This relay is actually composed of three separate relays and a resistor. Two of the relays (SMR1 and SMR2) are connected to the positive side of the battery pack and the other (SMR3) is in the negative circuit. When the high-voltage circuit is initially turned on, SMR1 and SMR3 are energized by the HV ECU. The resistor is in series with SMR1 and serves to control the amount of current that flows through the relay to eliminate a current surge when the circuit is turned on. Once the circuit is turned on, SMR1 turns off and SMR2 is turned on. This allows full current flow out of the battery pack. When the circuit is shut down, SMR2 and SMR3 are deenergized. If the HV ECU receives a deployment signal from the airbag sensor or an actuation signal from the inverter's circuit breaker sensor, the HV ECU will open the high-voltage circuit by deenergizing the SMR.

A service plug and main high-voltage fuse is also inserted in the high-voltage circuit. The fuse protects the circuit by opening if there is excessive current in the circuit. The service plug is used to disconnect or isolate the high-voltage circuit so that service can be performed on the circuit. The service plug is positioned in the middle of the battery modules. When removed, the circuit is open.

Individual high-voltage cables are used to connect the battery pack to the inverter, the inverter to MG1, the inverter to MG2, and the inverter to the air conditioning compressor on 2004 and newer Prius'. The cables run from the rear of the car, through the floor panel, to the engine compartment. The high-voltage cables are shielded to reduce electromagnetic interference and are orange in color for easy identification.

Inverter The inverter controls current flow to and from the motor/generators and the batteries. A single unit, in 2004 and newer Prius models, contains the inverter, DC-DC converter, boost converter, and air conditioning inverter (Figure 8-15). On earlier models, the inverter assembly housed the inverter and DC-DC converter.

An inverter changes the battery's high-voltage DC into high-voltage AC to power MG1 and MG2. It also rectifies the AC generated by MG1 and MG2 to recharge the battery pack. The inverter is basically a three-phase parallel circuit with each leg containing two power transistors

■ CAUTION:
Always turn OFF the vehicle before removing the service plug. Wear insulating gloves when removing the service plug and when working on high-voltage systems. Keep the removed service plug in your pocket to prevent others from installing it while you are servicing the vehicle. After removing the service plug, do not touch the high-voltage connectors and terminals for at least five minutes. Also, do not operate the power switch with the service plug removed. Doing so may damage the HV ECU.

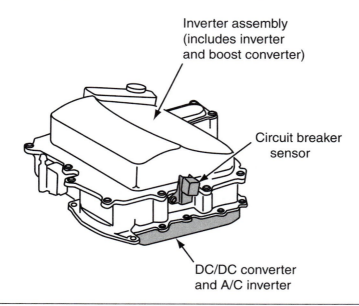

Inverter assembly
(includes inverter
and boost converter)

Circuit breaker
sensor

DC/DC converter
and A/C inverter

Figure 8-15 The inverter assembly is located under the hood with the engine.

connected in series. The HV ECU controls the power transistors to ensure proper phasing of the AC to the motors. This circuit is connected between the motor/generator and the battery pack. The inverter assembly contains two of these circuits, one for each of the motor/generators.

The DC-DC converter changes the high DC voltage from MG1 and MG2 to 12 volts and sends it to the auxiliary battery. The converter is composed of four transistors, four diodes (to maintain DC), a transformer, and two more diodes. The four transistors and diodes are used to control the high voltage at one side of the transformer. The other diodes rectify the voltage on the output side of the transformer. The voltage of the auxiliary battery is monitored by the converter, which attempts to keep it at a constant level. The voltage of the battery does not, as in conventional vehicles, change with engine speed or electrical load. If a problem occurs in the hybrid system, the HV ECU will stop the operation of the converter.

The late-model Prius has a boost converter that provides high voltage (up to 500 volts) to MG2. This increased voltage increases the power output of the motor. This converter is composed of the boost Integrated Power Module (IPM) that contains two Insulated Gate Bipolar Transistors (IGBT), a reactor to store the energy, and a signal processor (Figure 8-16).

The inverter assembly in the late-model Prius also contains an air conditioning inverter. This inverter changes the DC voltage from the battery to AC voltage to power the A/C compressor.

The inverter unit, MG1, and MG2 are kept within a desired temperature range by a separate water pump and radiator. By controlling the water pump based on inputs from the battery ECU, the HV ECU controls the cooling system.

Engine The engine used both generations of the Prius is a 1.5-liter, four-cylinder engine. The horsepower ratings of the engine increased over time. These increases were largely due to changes that also reduced exhaust emissions. The changes include improved engine controls and the inclusion of a coolant heat storage system (Figure 8-17). This system takes hot coolant from the engine and stores it in an insulated tank where it can stay hot for up to three days. When the engine is restarted, an electric water pump circulates the hot coolant through the engine to preheat it. This reduces the amount of HC emissions normally associated with cold starting.

The engine is based on the Atkinson cycle and is equipped with the **Variable Valve Timing-intelligent (VVT-i)** system and **Electric Throttle Control System-intelligent (ETCS-i)**. The Prius operates with low displacement (Atkinson cycle) or normal displacement (1.5 liters). In an **Atkinson cycle** engine, the intake valve is kept open well into the compression stroke. The timing of the opening and closing of the intake valves is controlled by the VVT-i system. While the

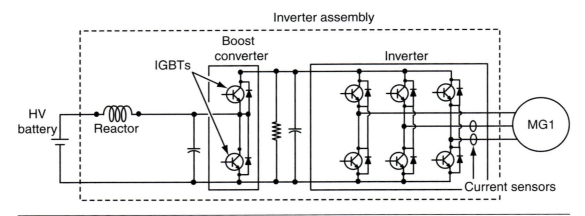

Figure 8-16 A schematic of the boost converter and inverter assembly.

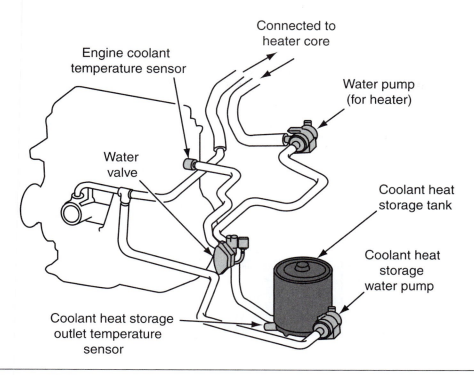

Figure 8-17 The coolant heat storage system.

valve is open during the compression stroke, some of the air/fuel mixture is pushed back into the intake manifold. This reduces the volume in the cylinder and effectively reduces the displacement of the engine. When the displacement is low, fuel consumption is minimized, as are exhaust emissions. The engine runs with normal displacement when the intake valves close earlier. This action provides for more power output. Because the VVT-i system responds to operating conditions, the displacement of the engine changes accordingly.

A "surge tank" is located in the intake manifold to hold the mixture pushed out of the cylinder during the Atkinson compression stroke.

The VVT-i system is controlled by the ECM. With this system, the intake valve can change within a range of 43 degrees. (Note: On late-models cars, the maximum retard closing timing of the intake valve has been decreased from 115 degrees ABDC [After Bottom-Dead-Center] to 105 degrees ABDC.) The ECM adjusts valve timing according to engine speed, intake air volume, throttle position, and water temperature. In response to these inputs, the ECM sends commands to the camshaft timing oil control valve.

The VVT-i controller, which is housed at the end of the camshaft, is a housing driven by the crankshaft. The ECM controls the oil pressure sent to the controller. A change in oil pressure changes the position of the camshaft and the timing of the valves. The camshaft timing oil control valve is duty cycled by the ECM to advance or retard intake valve timing. The various valve timing settings is shown in Figure 8-18.

The camshaft timing oil control valve selects the path for oil pressure to the VVT-i controller according to signals from the ECM to advance, retard, or hold (keep things the same) the timing. The controller rotates the intake camshaft in response to the oil pressure. An advance in timing results from oil pressure being applied to the timing advance side vane chamber (see Figure 8-19).

Operation State	Range	Valve Timing	Objective	Effect
During Idline	1	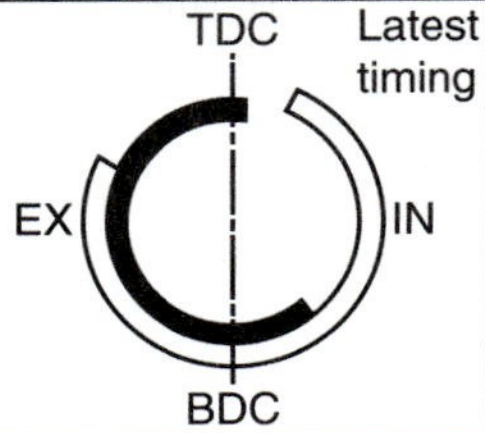 TDC / Latest timing / EX / IN / BDC	Eliminating overlap to reduce blow back to the intake side	Stabilized idling rpm Better fuel economy
At Light Load	2	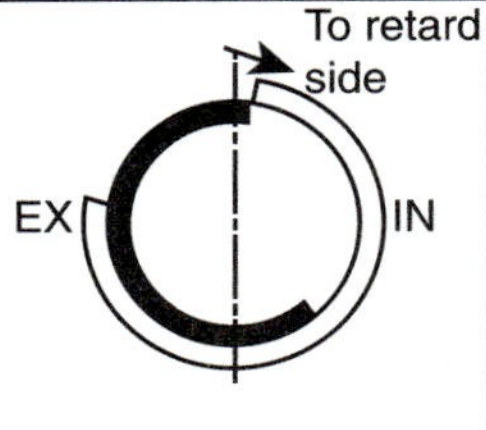 To retard side / EX / IN	Decreasing overlap to eliminate blow back to the intake side	Ensured engine stability
At Medium load	3	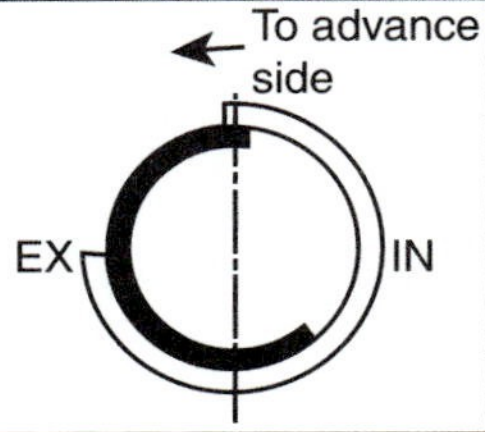 To advance side / EX / IN	Increasing overlap to increase internal EGR for pumping loss elimination	Better fuel economy Improved emission control

Operation State	Range	Valve Timing	Objective	Effect
In Low to Medium Speed Range with Heavy Load	4	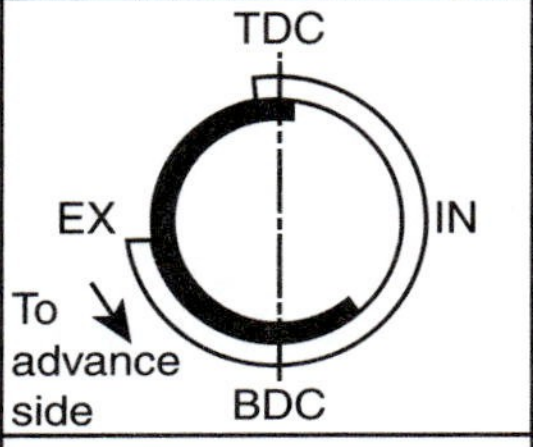 TDC / EX / IN / To advance side / BDC	Advancing the intake valve close timing for volumetric efficiency improvement	Improved torque in low to medium speed range
In High Speed Range with Heavy Load	5	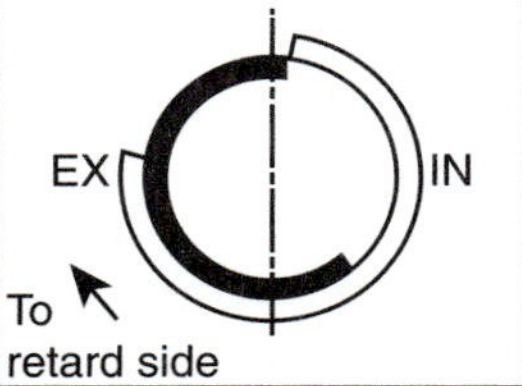 EX / IN / To retard side	Retarding the intake valve close timing for volumetric efficiency improvement	Improved output
At Low Temperatures	—	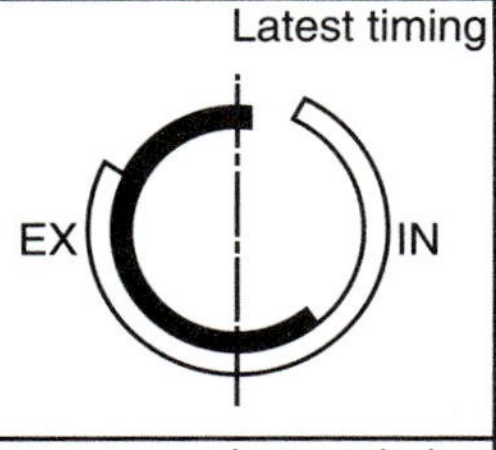 Latest timing / EX / IN	Eliminating overlap to prevent blow back to the intake side for reduction of fuel increase at low temperatures, and stabilizing the idling rpm for decreasing fast idle rotation	Stabilized fast idle rpm Better fuel economy
Upon Starting/ Stopping the Engine	—	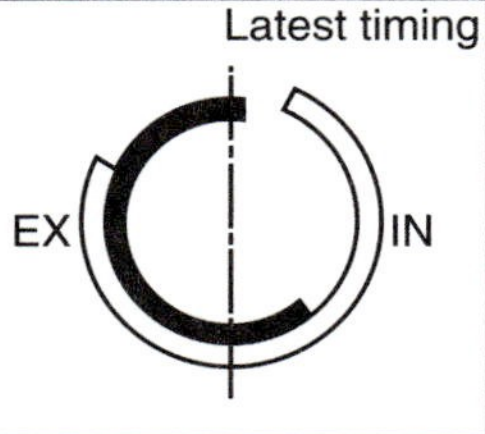 Latest timing / EX / IN	Eliminating overlap to eliminate blow back to the intake side	Improved startability

Figure 8-18 The various valve timing settings for the VVT-i system. *Courtesy of Toyota Motor Company*

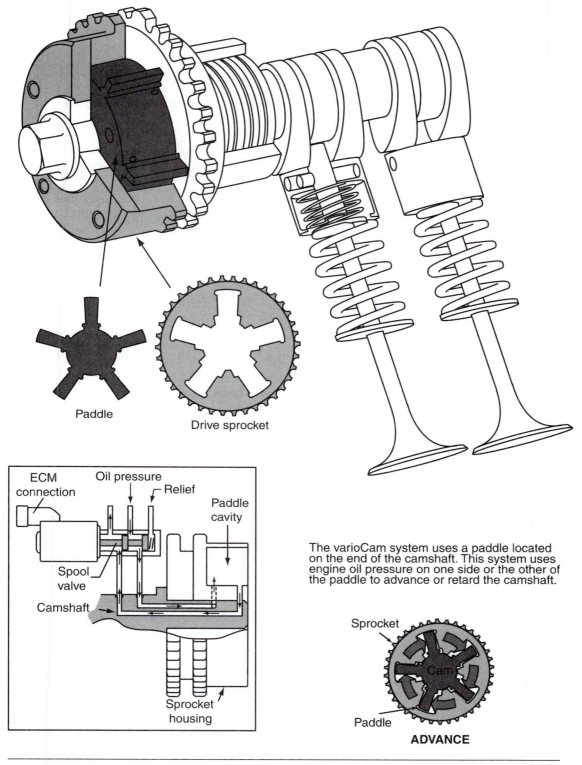

Paddle

Drive sprocket

ECM connection

Oil pressure

Relief

Paddle cavity

Spool valve

Camshaft

Sprocket housing

The varioCam system uses a paddle located on the end of the camshaft. This system uses engine oil pressure on one side or the other of the paddle to advance or retard the camshaft.

Sprocket

Cam

Paddle

ADVANCE

Figure 8-19 The camshaft timing oil control valve is duty cycle controlled by the ECM. When the camshaft timing oil control valve is positioned as shown above, the resultant oil pressure is applied to the timing advance side chamber to advance the camshaft.

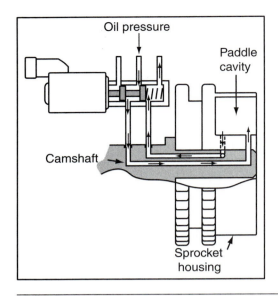
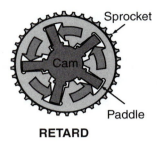

RETARD

Figure 8-20 When the camshaft timing oil control valve is moved by a retard signal from the ECM, oil pressure is applied to the retard side vane chamber to rotate the camshaft in the timing retard direction.

When the oil control valve is moved and the oil pressure is applied to the timing retard side vane chamber (Figure 8-20), the timing is retarded. The ECM constantly calculates the desired or required valve timing based on the conditions of the car and the engine.

In addition to controlling the VVT-I system, the vehicle's ECM also controls the operation of the fuel injection and ignition systems. The input from many sensors is used to govern the engine in order to provide maximum efficiency. On the 2004 and newer Prius, an air/fuel sensor is placed before the catalytic converter rather than a conventional oxygen sensor.

The ETCS-i system controls the position of the throttle plate. There is no mechanical connection between the accelerator pedal and the throttle plate; the connection is made electronically. The ECM calculates the appropriate throttle opening and sends commands to the throttle control motor. The ECM receives inputs from the accelerator pedal position sensor and the HV ECU. The HV ECU monitors current operating conditions and the battery's state of charge to determine the optimal engine speed and sends that information to the ECM.

The 2001 to 2003 Prius was equipped with a **HC Adsorber and Catalyst System (HCAC)**. This system was composed of a HC adsorber, three-way catalytic converter (TWC), an ECM-controlled actuator, and a bypass valve. The system captured hydrocarbons in the exhaust when the TWC was cold and basically ineffective. Once the engine and TWC warmed up, the hydrocarbons were released through the warm TWC.

Transaxle The power-split device is also called the hybrid transaxle assembly (Figure 8-21). The unit functions as a continuously variable transaxle, although it does not use the belts and pulleys normally associated with CVTs. The variability of this transaxle depends on the action of MG1 and the torque supplied by MG2 and/or the engine. The transaxle unit contains:

❏ Differential assembly
❏ Reduction unit
❏ Motor Generator 1 (MG1)
❏ Motor Generator 2 (MG2)
❏ Transaxle damper
❏ Planetary gear set

A conventional differential unit is used to allow for good handling when the car is making a turn. The reduction unit increases the final drive ratio so that ample torque is available to

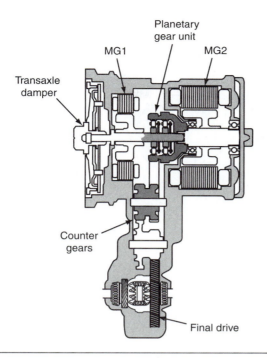

Figure 8-21 The hybrid transaxle in a late-model Prius.

the drive wheels. The reduction unit is a gear set linked by a chain. MG1, which generates energy and serves as the engine starter, is connected to the planetary gear set. This is also true for MG2, which is also connected to the differential unit by the drive chain. This transaxle does not have a torque converter or clutch. Rather a damper is used to cushion engine vibration and the power surges that result from the sudden engagement of power to the transaxle.

The planetary gear set provides continuously variable gear ratios and serves as the power-splitting device. The engine, MG1, and MG2 are mechanically connected at the planetary gear set. The gear set can transfer power between the engine, MG1, MG2, drive wheels, and in nearly any combination of these. The unit splits power from the engine to different paths: to drive MG1, drive the car's wheels, or both. MG2 can drive the wheels or be driven by them.

To understand how this power split device and transaxle works, a look at the operation of a planetary gear set is necessary. A simple planetary gear set (Figure 8-22) has three parts: a sun gear, a carrier with planetary pinions mounted to it, and an internally toothed **ring gear**. The **sun gear** is located in the center of the assembly. It meshes with the teeth of the planetary pinion gears. Planetary pinion gears are small gears fitted into a framework called the **planetary carrier**. The planetary carrier is designed with a shaft for each of the planetary pinion gears (planetary pinion gears are typically called **planetary pinions**). The carrier and pinions are considered one unit—the midsized gear member. The planetary pinions surround the sun gear and are surrounded by the ring gear, which is the largest part of the gear set. The ring gear acts like a band to hold the gear set together and provide great strength to the unit.

Each member of a planetary gear set can spin (revolve) or be held at rest. Power transfer through a planetary gear set is only possible when one of the members is held at rest, or if two of the members are locked together. Any one of the three members can be used as the driving or input member. Another member might be kept from rotating and thus becomes the held or stationary member. The third member then becomes the driven or output member. Depending on which member is the driver, which is held, and which is driven, either a torque increase or a speed increase is produced by the planetary gear set. Output direction can also be reversed through various combinations. The amount of torque and speed change depends on the size of the gears serving as the input and output. Also, remember that when a combination of gears

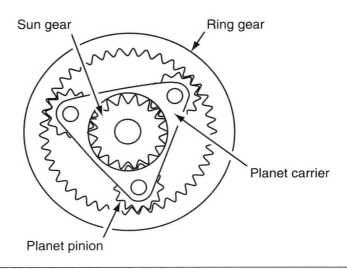

Sun gear

Ring gear

Planet carrier

Planet pinion

Figure 8-22 The components of a simple planetary gear set.

results in an increase in speed, there will be a corresponding decrease in output torque. Likewise, a decrease in output speed results in an increase in torque.

The following summarizes the results of various gear combinations in a simple planetary gear set (see Figure 8-23).

- ❏ When the planetary carrier is the drive (input) member, the gear set produces an over-drive condition. Speed increases; torque decreases.
- ❏ When the planetary carrier is the driven (output) member, the gear set produces a forward underdrive direction. Speed decreases; torque increases.
- ❏ When the planetary carrier is stationary (held), the gear set produces an output in the opposite direction as the input.
- ❏ When two members of the gear set are locked together, direct drive results.
- ❏ When no member is held, a neutral condition exists.

In the planetary gear set used in the power-split device, the sun gear is attached to MG1. The ring gear is connected to MG2 and the final drive unit in the transaxle. The planetary carrier is connected to the engine's output shaft. The key to understanding how this system splits power

SUN	CARRIER	RING	SPEED	TORQUE	DIRECTION
Input	Output	Held	Maximum reduction	Maximum increase	Same as input
Held	Output	Input	Minimum reduction	Minimum increase	Same as input
Output	Input	Held	Maximum increase	Maximum reduction	Same as input
Held	Input	Output	Minimum increase	Minimum reduction	Same as input
Input	Held	Output	Reduction	Increase	Opposite of input
Output	Held	Input	Increase	Reduction	Opposite of input

Figure 8-23 The basic laws of reduction in a simple planetary gear set.

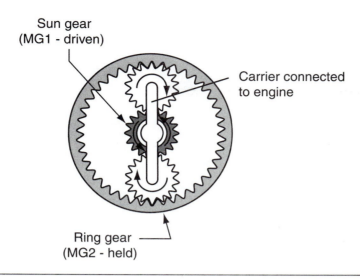

Figure 8-24 Planetary gear action during engine startup.

is to realize that when there are two sources of input power, they rotate in the same direction but not at the same speed. Therefore, one can assist the rotation of the other, slow down the rotation of the other, or work together. Also, keep in mind the rotational speed of MG2 largely depends on the power generated by MG1. Therefore, MG1 basically controls the continuously variable transmission function of the transaxle. Here is a summary of the action of the planetary gear set during different operating conditions.

❏ To start the engine, MG1 is energized and the sun gear becomes the drive member of the gear set (Figure 8-24). Current is sent to MG2 to lock or hold the ring gear. The carrier is driven by the sun gear and walks around the inside of the ring gear to crank the engine at a speed higher than that of the sun gear.

❏ After the engine is started, MG1 becomes a generator. The ring gear remains locked by MG2 and the carrier now drives the sun gear, which spins MG1.

❏ When the car is driven solely by MG2 (Figure 8-25), the carrier is held because the engine is not running. The ring gear rotates by the power of MG2 and drives the sun gear in an opposite direction. This causes MG1 to slowly spin in the opposite direction without generating electricity.

❏ If more drive torque is needed while running with MG2 only, MG1 is activated to start the engine. There are now two inputs to the gear set, the ring gear (MG2), and the sun gear

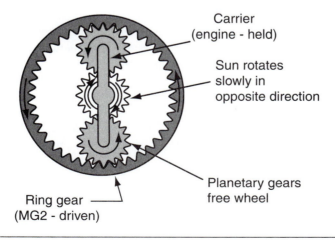

Figure 8-25 Planetary gear action when MG2 is propelling the vehicle.

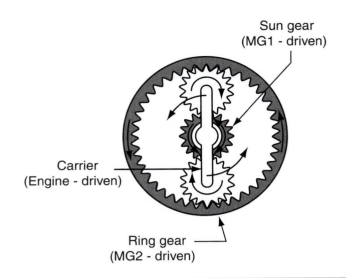

Sun gear
(MG1 - driven)

Carrier
(Engine - driven)

Ring gear
(MG2 - driven)

Figure 8-26 Planetary gear action when the engine and MG2 are propelling the vehicle.

(MG1). The carrier is driven by the sun gear and walks around the inside of the rotating ring gear. This cranks the engine at a faster speed than when the ring gear is held.

❑ After the engine is started, MG2 continues to rotate the ring gear and the engine rotates the carrier to drive the sun gear and MG1, which is now a generator.

❑ When the car is operating under light acceleration and the engine is running, some engine power is used to drive the sun gear and MG1, and the rest is rotating the ring gear to move the car (Figure 8-26). The energy produced by MG1 is fed to MG2. MG2 is also causing the ring gear to rotate and the power of the engine and MG1 combine to move the vehicle.

❑ This condition continues until the load on the engine or the condition of the battery changes. When the load decreases, such as during low-speed cruising, the HV ECU increases the generation ability of MG1, which now supplies more energy to MG2. The increased power at the ring gear allows the engine to do less work while driving the car's wheels and do more work driving the sun gear and MG1.

❑ During full throttle acceleration, battery power is sent to MG2, in addition to the power generated by MG1. This additional electrical energy allows MG2 to produce more torque. This torque is added to the high output of the engine at the carrier.

❑ When the car is decelerating and the transmission is in DRIVE, the engine is shut off, which effectively holds the carrier (Figure 8-27). MG2 is now driven by the wheels and acts as a generator to charge the battery pack. The sun rotates slowly in the opposite direction and MG1 does not generate electricity. If the car is decelerating from a high speed, the engine is kept running to prevent damage to the gear set. The engine, however, merely keeps the carrier rotating within the ring gear.

❑ When the car is decelerating and the transmission is moved into the B range, MG2 acts as a generator to charge the battery pack and to supply energy to MG1. MG1 rotates the engine, which is not running at this time, to offer some engine braking.

❑ During normal deceleration with the brake pedal depressed, the engine is off and the skid control ECU calculates the required amount of regenerative brake force and sends a signal to the HV ECU. The HV ECU, in turn, controls the generative action of MG2 to provide a load on the ring gear. This load helps to slow down and stop the car. The hydraulic brake system does the rest of the braking.

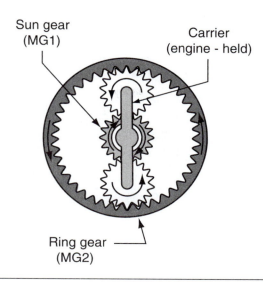

Sun gear
(MG1)

Carrier
(engine - held)

Ring gear
(MG2)

Figure 8-27 Planetary gear action during deceleration and MG2 is driven by the vehicle's wheels.

❏ When reverse gear is selected, only MG2 powers the car. MG2 and the ring gear rotate in the reverse direction. Because the engine is not running, the carrier is effectively being held. The sun gear is rotating in its normal rotational directional, but slowly, and MG1 is not acting as a generator. Therefore, the only load on MG2 is the drive wheels.

❏ It is important to remember that at any time the car is powered only by MG2, the engine may be started to correct an unsatisfactory condition, such as low-battery SOC, high battery temperature, and heavy electrical loads.

Electronic Displays A Prius has a multi-information display located on the center cluster panel. The display, a 7.0-inch LCD (Liquid Crystal Display), with a pressure sensitive touch panel serves many functions. Many of these are typical, but some are unique to hybrid technologies. One of the unique features is the fuel consumption screen, which shows average fuel consumption, current fuel consumption, and the current amount of recovered energy.

Another unique display is the energy monitor screen (Figure 8-28), which shows the direction and path of energy flow through the system, in real time. By observing this display, drivers can alter their driving to achieve the most efficient operation for the current conditions.

Like other vehicles, the Prius is equipped with a variety of warning lamps and indicators. Here are some of the indicators that are unique to the Prius hybrid.

❏ READY light—This lamp turns on when the ignition switch is turned to START to indicate that the car is ready to drive.

❏ Output control warning light—This lamp turns on when the temperature of the HV battery is too high or low. When this lamp is lit, the system's power output is limited.

❏ HV battery warning light—This lamp is illuminated when the charge of the HV battery is too low.

❏ Hybrid system warning light—When the HV ECU detects a problem with MG1, MG2, the inverter assembly, the battery pack, or the ECU itself, this lamp will be lit.

❏ Malfunction indicator light—This lamp is tied into the engine control system and will be lit when the ECM detects a fault within that control system.

❏ Discharge warning light—This lamp is tied to the 12-volt system and DC-DC converter. It will illuminate when there is problem in that circuit.

Electric Motor-assisted Power Steering (EMPS) System To reduce fuel consumption and provide power steering when the engine is not running, the Prius uses an **electric motor-assisted power steering (EMPS)** system. The EMPS is a variable ratio, speed-sensitive power

● NOTE: When the gear selector is placed in the N position, all power to the drive wheels is shut off. In this position, all power transistors in the inverter are turned off. This action shuts down MG1 and MG2. This means that no generation can take place. This is true whether the engine is running or not, or if the wheels are driving MG2 during deceleration.

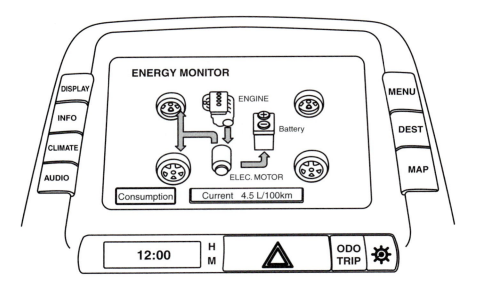

Figure 8-28 The multi-information display on the instrument panel shows average fuel consumption, current fuel consumption, and the current amount of recovered energy. It also has an energy monitor screen, which shows the direction and path of energy flow through the system.

steering system. Besides providing for appropriate steering ratios according to vehicle speed, this system eliminates the drag on the engine that results from rotating a hydraulic power steering pump. The EMPS provides steering assist when the engine is on and when it is off.

The system relies on a DC motor to add torque to the driver's steering effort, rather than hydraulic pressure. The system is based on a conventional rack-and-pinion steering gear. When the steering wheel is turned, torque is transmitted through the steering column to a torsion bar that links the column to the pinion of the steering gear. The torsion bar twists until the torque and the reaction force equalize. A torque sensor detects the twist of the torsion bar and converts it to an electrical signal that is sent to the EPMS ECU. The ECU then calculates the required amount of assist and controls the electric assist motor accordingly. With this system, the DC motor consumes electrical energy only when power assist is required.

Racing a Prius This is included just for fun and to get an idea of the potential of hybrid vehicles. In 2004, a specially equipped Prius ran on the Bonneville Salt Flats. This run was the first by a hybrid vehicle. Therefore, it obviously set the top speed record for its class. That record was 130.794 miles per hour.

The Prius was modified for this run, but used a standard hybrid synergy drive system.

❏ The output of the engine was increased slightly to 96 hp at 5,800 rpm.
❏ The transmission final drive gear ratio was changed from the production 4.23:1 to 3.2.
❏ The inverter's maximum voltage was increased to 550 volts from 500 volts.
❏ A transmission cooling system was added to decrease the temperature of the inverter and electric motors and maximize efficiency.
❏ Ice was constantly added to the MG1, MG2, and inverter cooling system to prevent overheating. This was due to the high temperatures present during the run.
❏ The interior of the car was removed to lower the car's weight and a roll cage installed to protect the driver.
❏ The suspension was lowered 5 inches to improve aerodynamics. The inner fender wells were altered to allow for lowered ride height.
❏ Finally, 26-inch front and 25-inch rear Goodyear Eagle Landspeed Record tires were used to reduce rolling friction and withstand the heat generated by traveling over the hot sand.

Toyota's Hybrid SUVs

Toyota modified its HSD to work with a high-output six-cylinder engine to offer a hybrid option on its midsized SUVs: the Highlander Hybrid and RX 400h. Based on the system used in the Prius, this new hybrid system was developed to reduce emissions and fuel consumption and increase the performance of these SUVs. Also, a four-wheel-drive (4WD) option is available (Figure 8-29).

These SUVs have the engine and transaxle assembly in front to drive the front wheels. The 4WD models have an additional electric motor and transaxle in the rear to drive the rear wheels. The two motor/generators used to drive the front wheels are still referred to as MG1 and MG2. The motor/generator at the rear on 4WD models is referred to as **motor/generator rear (MGR)**. MG1 is used to start the engine and serve as an engine-driven generator. MG2 and MGR provide power for propulsion and serve as generators during regenerative braking. As in the Prius, the amount of voltage generated by MG1 controls the effective gear ratio of the transaxle.

These SUVs have a MG2 with much more power and are capable of higher rotational speeds. The motor works with a 3.3 liter V-6 engine that has been slightly modified. The combined drive system output (electric motor plus engine output) is nearly 268 horsepower (200 kW). The redesigned MG2 produces its peak torque (twice the amount available in a Prius) from zero to 1,500 RPM. The output from the motor, in addition to that of the engine, allows these SUVs to accelerate quicker than their pure ICE-powered cousins. Additionally, these hybrids use less fuel and have a combined EPA fuel economy rating of about 29 miles per gallon (9l/100 km).

Along with the new motor, these vehicles have a higher-voltage battery and a variable voltage system that boosts the voltage when extra power is needed. These vehicles have a 288-volt NiMH battery pack and a revised inverter. Inside the inverter is a boost converter that can increase the operating voltage to a maximum voltage of 650 volts.

Like the Prius, the front transaxle is a continuously variable transmission and it serves as the power-split device. Because of the vehicle's weight, changes have been made to the transaxle. The transaxle still contains the MG1, MG2, planetary gear set (power-split unit), a counter gear set, and the differential unit. However, an additional planetary gear set (the Motor Speed Reduction unit) has been tied to the power-split unit. This gear set reduces the speed of MG2 and, thereby, increases its torque.

The 4WD system is capable of responding to operating conditions by varying the distribution of torque between the front and rear axles and by controlling the front and rear electric motors. MGR is incorporated into a rear transaxle assembly that also houses the rear differential.

The Highlander Hybrid and RX400h are also equipped with many energy-saving, safety, and comfort systems. These include an electrically operated A/C compressor, an electrically operated water pump, electric power steering (EPS), and a Vehicle Dynamics Integrated Management

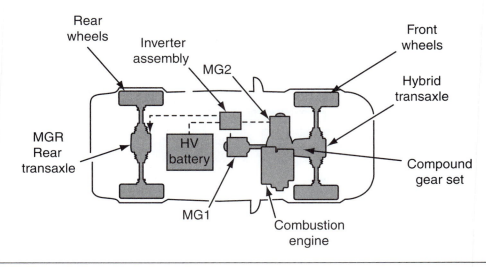

Figure 8-29 The basic layout of the hybrid components in a Toyota four-wheel-drive SUV.

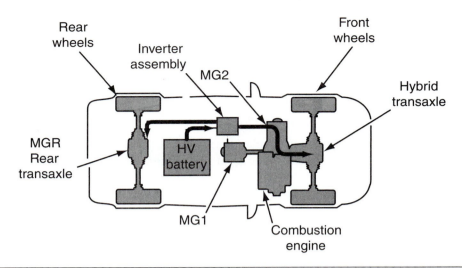

Figure 8-30 When the vehicle accelerates from a stand still, it is powered only by MG2 and MGR. The engine is off and MG1 is not generating electricity.

(VDIM) system that incorporates the electronically controlled brake (ECB) system, anti-lock braking system (ABS), electronic brakeforce distribution (EBD), and brake assist systems.

Operation The operation of this new HSD system is similar to that in the Prius except for 4WD operation and the speed reduction unit in the front transaxle. The latter will be explained in detail with the discussion of the transaxle. The front drive unit is the same assembly for two-wheel and four-wheel-drive vehicles.

In 4WD models, MGR and MG2 are used to move the vehicle during initial acceleration (Figure 8-30). Both motors are powered directly by the battery pack. The engine is not running and MG1 is inactive. However, the engine may start if the control unit senses a need for more drive torque or if the battery's SOC, battery temperature, and/or engine temperature are not within a specified range.

Once the vehicle is operating under a low load and at a constant speed, power to MGR stops and the engine starts to power the vehicle and MG1. The energy from MG1 powers MG2, which works with the engine to keep the vehicle moving.

During hard acceleration, the vehicle is powered by the engine, MG2, and MGR (Figure 8-31). The engine's output drives the wheels and MG1. Electrical power for MG2 and MGR comes from

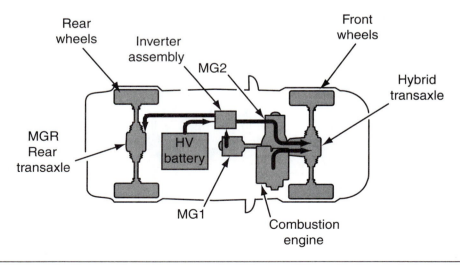

Figure 8-31 During full throttle acceleration, the vehicle is powered by the engine, MG2, and MGR. Electrical power for MG2 and MGR comes from the battery pack and MG1.

the battery pack and MG1. If needed, higher voltages from the boost converter may be sent to MG2 and MGR.

During deceleration, the system calculates the torque distribution at the front and rear axles. Based on this calculation, it controls the regenerative capability of MG2 and MGR. This prevents any "push-pull" that can result if one axle was braking or slowing more than the other.

Torque distribution is constantly being monitored by the skid control ECU. If the system detects any variance in wheel or axle speeds, which normally occur on slippery roads or while cornering at high speeds, adjustments are immediately made to rectify the situation. These adjustments affect the power output of the engine, MG2, and MGR.

When the vehicle is driven in reverse, MG1 and the engine are shut down and the vehicle is powered by battery power to MG2 and MGR.

Motor/Generators The three motors used in the HSD system are AC synchronous motors. The rotors are constructed in the same way as those in the second-generation Prius. The stator windings use high-voltage resistant wire to handle the boosted voltages.

As in the Prius, MG1 is primarily used as a generator. Power from MG1 is used to charge the batteries and supply power to the A/C compressor, power-steering motor, and the water pump. MG1 is also used to start the engine.

The new MG2 has a maximum output rating of 167 hp (123 kW) at 4,500 rpm and a torque rating of 247 lb/ft (333 Nm) at 0-1,500 rpm. MG2 is the main traction motor and drives the front wheels, by itself or with the engine. MG2 also serves as the main generator during deceleration and braking.

The electric 4WD-i system uses the MGR to power the rear wheels of the vehicle. MGR has a maximum output of 68 hp (50 kW) at 4,610-5,120 rpm and a maximum torque rating of 96 lb/ft. (130 Nm) at 0-610 rpm. The final drive ratio in the rear transaxle is very low. This provides a large amount of torque to the rear wheels. Of course, the amount of torque produced by MGR depends on the voltage it receives. The source of its voltage can be the battery pack and/or MG1. During deceleration, MGR acts as a generator.

The operation of the motors is controlled by a motor/generator ECU through commands from the HV ECU (referred to now as the THS ECU). The MG ECU is linked to the Intelligent Power Module (IPM) and switches its six Insulated Gate Bipolar Transistors (IGBTs) to turn the various MGs on or off and to change their operation from a motor to a generator and vice versa. The MG ECU receives input from each of the motors' resolvers to determine the speed of the unit's rotor and its exact position within the stator.

The motors are also fitted with a temperature sensor. The THS ECU monitors the temperature of each and will alter the power to them if there is evidence of overheating.

Electronic Controls The THS ECU is the main control unit for the hybrid system and is very similar to the one used in the Prius. The same basic systems and operating conditions are monitored by the ECU. However, the THS ECU in the SUVs, especially those equipped with 4WD, has additional systems to control (Figure 8-32).

The THS ECU receives information from a variety of sensors and other control units. This information is used to calculate how much power is required from the motors and engine to meet current driving conditions. The primary inputs used for this calculation are from the shift position sensor, motor resolvers, accelerator pedal position sensor, vehicle speed sensor, battery smart unit, skid control ECU, and EPS ECU. Based on the calculations, the THS ECU sends the appropriate commands to the MG ECU, inverter, and skid control ECU to control the powertrain. The THS ECU also sends commands to the ETCS-i, fuel injection, ignition, and VVT-i systems.

The THS ECU constantly monitors the SOC and temperature of the battery pack. When the SOC is below a specific level, the THS ECU orders the engine to rotate MG1 faster. If the engine is not running and the SOC is too low, the engine will start and drive MG1. A temperature sensor in the battery module sends information regarding current battery temperature. Based on this input, the THS ECU controls the operation of the cooling fans in the battery module. If the SOC

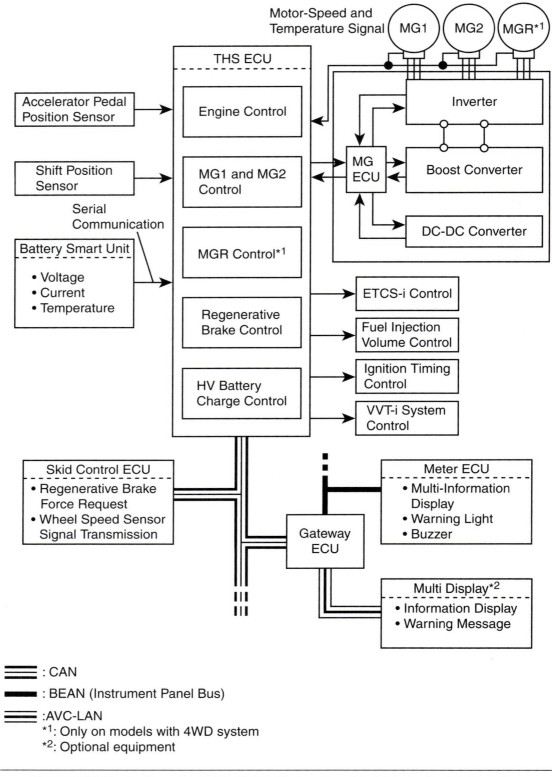

Figure 8-32 The basic configuration of the THS II system with the four-wheel-drive option. *Courtesy of Toyota Motor Company*

is low, or the temperature of the battery, MG1, MG2, or MGR is too high, the THS ECU will decrease the motors' output until the undesirable condition is corrected.

The THS ECU also monitors the speed of each wheel to determine if any slippage is occurring. Slippage is evident by any sudden speed change. The ECU calculates the amount of slippage and controls the operation of MG2 to reduce it. If one axle is turning faster than the other, the rotational speed of MG2 will increase. It may increase enough to damage to the planetary gear set. Because the drive axles are connected directly to MG2, which is directly connected to the sun gear in the planetary gear set, slippage can cause the sun gear to attempt to rotate too fast for the rest of the gear set. Also, this over-speed condition may cause MG1 to generate an excessive amount of electricity. To prevent these from occurring, if the THS ECU detects slippage it will immediately apply a brake force to the spinning axle.

If one of the drive wheels is slipping, the THS ECU will send a command to the skid control ECU to apply the brake at the wheel that is slipping.

The THS ECU monitors the entire system and memorizes all conditions and operating perimeters that are outside a specified range. Depending on the type and severity of the problem, the THS ECU will illuminate or blink the MIL, master warning light, or the HV battery warning light. The THS ECU will assign a DTC to the problem and keep it in memory for diagnosis. If the THS ECU detects a non-working sensor or control unit, it will ignore the inputs from that source and operate on data stored in its memory. This is the fail-safe mode of operation.

The **battery smart unit** constantly monitors the voltage, current flow, and temperature of the batteries. It also has a leak-detection circuit that watches for excessive current drain. Serial communication is used to transfer the digital signals from the battery smart unit to the THS ECU.

The **skid control ECU** determines how much regenerative force, during deceleration, should be applied and sends this information to the THS ECU. This calculation is based on the pressure applied to brake master cylinder and the brake pedal. The ECU also calculates the amount of brake force that should be applied by the hydraulic brake system. On 4WD systems, the skid control ECU also monitors the torque distribution between the front and rear drive axles.

The hybrid system is also linked to the cruise control ECU. Information from the cruise control switch and vehicle speed sensor is sent to the cruise control ECU, which in turn, sends information to the THS ECU. The THS ECU regulates the motor/generators and engine to provide for the selected speed with a minimum use of fuel.

Battery The battery pack (Figure 8-33) is composed of 30 modules of eight 1.2-volt NiMH cells connected in series by a bus bar. The nominal voltage is 288 volts. The battery pack supplies power for MG1, MG2, and MGR according to the commands of the THS ECU. It is also recharged by MG1 and MGR through the commands of the same control unit.

Changes to the previous battery pack include a rubber damper at the mounting point of the SMRs to reduce vibrations, an electrical connection is made between every 10 cells to minimize internal resistance, and there is a more efficient cooling system. The battery module is divided into three separate units and relies on a cooling fan for each unit to keep them within the desired temperature range. The fans pull in air from the passenger compartment. There are eight temperature sensors located within the battery module, when the THS ECU detects an increase in battery temperature it duty cycles the appropriate cooling fan to bring the temperature down.

Inverter The inverter in the hybrid SUVs is substantially different than the one used in the Prius. The inverter includes a boost converter that can deliver up to 650 volts to the motor/generators (Figure 8-34). The inverter assembly also contains the MG ECU, inverter, and the DC-DC converter. The inverter converts the direct current from the battery pack into an alternating current for MG1, MG2, and MGR. When the motor/generators are operating as generators, the inverter converts the generated AC voltage to DC before sending it to the battery pack. The inverter also directly sends the AC generated by MG1 to MG2 and MGR.

The MG ECU ultimately controls the inverter, boost converter, and DC-DC converter through commands received from the THS ECU. These commands control the operation of the inverter's power transistors. The transistors, in turn, control the phasing of the motors and their output. If the THS ECU detects a problem in the high-voltage circuit or if the transmission is placed in neutral, the inverter is turned off to stop the operation of the motor/generators.

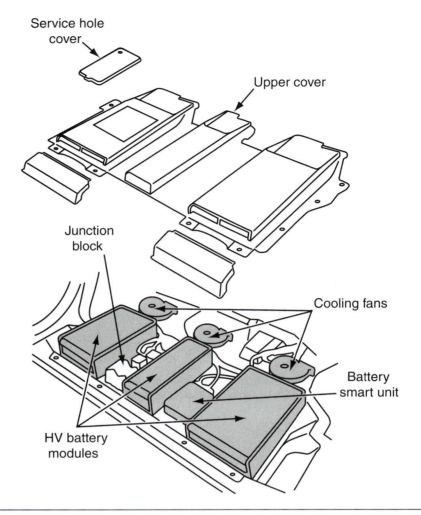

Figure 8-33 The high-voltage battery pack.

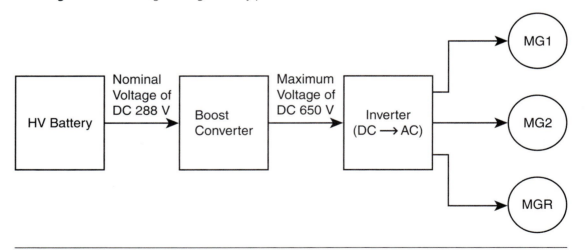

Figure 8-34 The control system uses a boost converter to allow for a variable voltage to and from the motor/generators. *Courtesy of Toyota Motor Company*

The boost converter is also controlled by the THS ECU through the MG ECU. It can boost the nominal voltage of the battery (288 DC volts) to 650 DC volts. After the voltage has been increased, it passes through the inverter and is changed to AC voltage for the motors. The opposite is true when the motor/generators are acting as a generator. During this time, the generators can generate as much as 650 AC volts. This output is passed through the inverter and changed to DC. The DC output from the inverter is dropped to 288 volts by the boost converter and this voltage is sent to the battery pack.

The boost converter contains a boost Integrated Power Module (IPM) that uses IGBTs to control the circuit and a reactor to store the higher voltage. The action of the boost converter is similar to that of an ignition coil. Composed of two parallel coil windings, one side is energized by battery voltage and the other winding is the output. When the magnetic field on the input side collapses, a higher voltage is induced in the output coil. As in an ignition coil, the amount of time current flows through the battery's winding before the field collapses determines the amount of voltage that is induced in the output winding. The longer that current flows through the input winding, the more voltage is induced in the output coil. The output coil is connected to a capacitor, called the reactor, that stores the higher voltage until it is sent to the motor/generators. The IGBTs are duty cycled and control the current to the boost converter.

The DC-DC converter changes a portion of the high voltage to a much lower voltage to charge the auxiliary battery. This voltage is used to power some accessories and safety items.

The inverter assembly, MG1, and MG2 are kept within a specified temperature by a cooling system. In the SUVs, the radiator for the inverter and motors is part of the engine's radiator, but is totally isolated from it.

Engine The engine in these hybrids is basically the same as used in non-hybrid models. Some modifications have been made to make it work with the hybrid system and to make it more efficient. Most of the modifications affect the operation of the ETC-I and VVT-I systems.

The engine is a 24-valve, dual overhead cam, V-6 engine that produces a maximum of 208 hp (155 kW) at 5,600 rpm and a maximum torque of 212 ft/lb (288 Nm) at 4,400 rpm. It is equipped with many electrically operated accessories that are normally belt driven to reduce power loss and to allow these items to operate when the engine is not running.

The ECM is incorporated in the THS ECU and diagnostic communications has been changed from serial communication (ISO9141) to CAN communication.

Transaxle The front transaxle assembly now has a speed reduction unit (Figure 8-35). This unit is a planetary gear set coupled to the power-split planetary gear set. This compound gear set has a common or shared ring gear that drives the vehicle's wheels. The sun gear of the power-split unit is driven by MG1 and the carrier is driven by the engine. In the reduction gear set, the carrier is held and the sun gear is driven by MG2. Because the sun gear is driving a larger gear, the ring gear, its output speed is reduced and its torque output is increased proportionally. High torque is available because MG2 can rotate at very high speeds. As in the Prius, the rotational speed of MG1 essentially controls the overall gear ratio of the transaxle.

The torque of the engine and MG2 moves from their designated gears to the common ring gear to the final drive gear and differential unit.

The transaxle has three distinct shafts: a main shaft that turns with MG1, MG2, and the compound gear unit; a shaft for the counter driven gear and final gear; and a third shaft for the differential. Because a clutch or torque converter is not used, a coil spring damper is used to absorb torque shocks from the engine and the initiation of MG2 to the driveline.

 MGR uses its own transaxle assembly to rotate the drive wheels. Unlike conventional 4WD vehicles, there is no physical connection between the front and rear axles. The aluminum housing of the rear transaxle contains the MGR, a counter drive gear, counter driven gear, and a differential. The unit has three shafts: MGR and the counter drive gear are located on the main shaft (MGR drives the counter drive gear), the counter driven gear and the differential drive pinion gear are located on the second shaft, and the third shaft holds the differential.

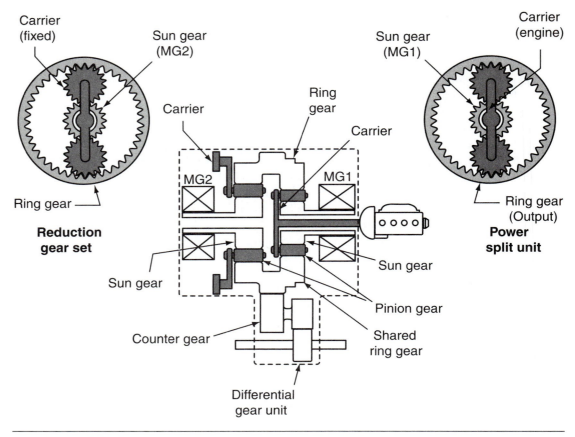

Figure 8-35 The new hybrid transaxle with a compound planetary gear set.

Electronic Displays The hybrid SUVs are equipped with instrument displays that are similar to those found in the Prius. However, the displays show more information and include lamps for the 4WD option. The latter is monitored by a 4WD warning lamp that notifies the driver of any detected fault within the MGR and rear transaxle. When this lamp is lit, the THS ECU will also cause the master warning lamp to light and a warning buzzer to sound. The remaining new segments of the display are typical for near-luxury and luxury vehicles.

GS 450h The Lexus GS450h is a hybrid luxury sedan with a front-mounted engine and rear-wheel drive. This HSD system combines a 3.5-liter V-6 engine with a high-output, permanent magnet AC synchronous motor. The total output of the system is 339 horsepower and the car accelerates from zero to 60 mph (100 km/hour) in a little more than 5 seconds. The transmission is different than previous models and includes a two-stage torque multiplication unit. This unit provides a final drive gear change for the CVT. This model is designed to be a sports or performance luxury sedan. It still achieves good fuel economy and is designated as a SULEV. The drive line is unique for hybrids as the power flow from the engine and transmission passes through a drive shaft to a rear axle, much like a traditional rear-wheel-drive vehicle.

Ford Escape

In 2004, Ford released the Escape hybrid. This was the first hybrid SUV and the first hybrid vehicle built in North America. Sales of the hybrids went very well, although they were limited simply because of a shortage of batteries for the vehicles. The Ford Escape SUV received many

Figure 8-36 One of the only clues that a Ford Escape may be a hybrid.

accolades and looked nearly identical to the non-hybrid Escape. The major differences in appearance are badges or external markings that say it is a hybrid (Figure 8-36) and a barely noticeable vent near the rear quarter window on the driver's side of the vehicle. This vent is for the battery's temperature-management system. The standard Escape Hybrid is front-wheel drive and an Intelligent 4WD system is optional. This option made the Escape the first 4WD hybrid.

In 2006, Ford released a hybrid version of the Escape's cousin, the Mercury Mariner. The Mariner Hybrid is a Ford Escape Hybrid with Mercury luxury items, badges, and styling.

These SUVs are full hybrids and feature a CVT transmission and stop-start technology, as well as the ability to be powered solely by battery power. They have a towing capability of up to 1,000 pounds (454 kg), the same tow rating as a regular 4-cylinder Escape or Mariner. The hybrid package is condensed to minimize its required space, so these small SUVs offer much practicality, in addition to good fuel economy and low emissions.

The hybrid system is based on a four-cylinder engine and two electric motors. The combined power output from the engine and the traction motor is the equivalent of 155 horsepower, which provides comparable acceleration times with conventional V-6 powered Escapes and Mariners. The main components and systems of these hybrids are (Figure 8-37):

- ❏ **Gasoline engine**: The 2.3-liter, DOHC four-cylinder engine uses the Atkinson cycle and advanced engine control systems.
- ❏ **Electric motors**: Two separate electric motor/generators are used. One of these, a 28 kW (equivalent to 38 hp) motor is primarily used as a generator but also serves as the starter motor for the engine and controls the activity of the transaxle. The other motor/generator has a peak power of 70 kW (equivalent to 94 hp). This motor is used to propel the vehicle during low-speed and low-load conditions and to assist the engine during hard acceleration, heavy loads, and/or during high-speed driving.
- ❏ **Battery pack**: The sealed NiMH battery pack is rated at 330 volts and has 250 D-size cells. The battery pack supplies the power for the two motors.
- ❏ **Regenerative braking**: Through intelligent controls, deceleration and braking is accomplished through regenerative braking and a conventional hydraulic brake system.
- ❏ **Electronically Controlled Continuously Variable Transmission (eCVT):** Based on a single planetary gear set, the transaxle controls the direction of power from the engine and motors.
- ❏ **Vehicle System Controller (VSC)**: The vehicle system controller is the primary electronic control unit and based on information from several other control units and inputs, it controls the charging, drive assist, and engine starting functions of the system according to current conditions. The VSC is part of the Powertrain Control Module (PCM).

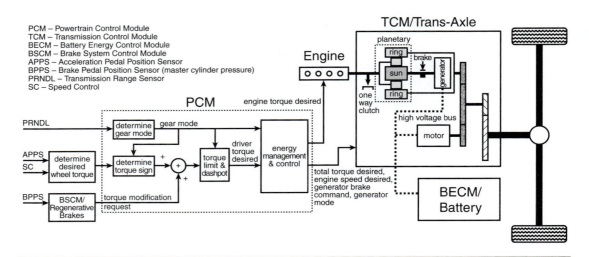

Figure 8-37 Layout of the Ford Escape hybrid power train.

These vehicles are also equipped with an electro-hydraulic brake-by-wire system and an electric power steering system. They also offer an optional 110-volt, 150-watt AC auxiliary outlet. They are equipped with standard instrumentation but have an additional gauge that monitors the economics of the current operating mode and driving style.

These hybrids, when equipped with two-wheel drive, are rated at about 33 to 36 mpg in city traffic and 29 to 31 mpg on the highway. The total driving range per tank of gasoline is 400 to 500 miles (650 to 800 km) in-city driving. The all-wheel-drive model is rated at 33 mpg in the city and 29 mpg on the highway. Each is certified as a SULEV and an AT-PZEV. They emit less than one-half the CO_2 emissions of a conventional vehicle.

Ford also claims the maintenance costs will be lower because the engine runs less often and more efficiently. This means longer oil life and fewer oil changes, increased brake life due to regenerative braking, and less engine wear because the engine idles less than a conventional one.

Currently, Ford Motor Company plans to release another SUV hybrid, the Mazda Tribute, and a new generation of its hybrid system in 2008 with hybrid versions of the Ford Fusion and Mercury Milan.

Operation

The basic components and operation of Ford's hybrid system are very similar to what is found in Toyota hybrids. This has led many to conclude that Ford is simply buying the system from Toyota. This is not true. This is unfortunate because Ford's system is its own and Ford is not getting the credit deserved for its engineering. The concept of using two electric motors to create a transaxle suitable for hybrid technology was first exposed by an American engineer 40 years ago. This concept is actually an adaptation of the planetary gear set Henry Ford used in the original Model T.

Going forth with the basic idea of using a planetary gear set as a power-split device, Toyota entered into a partnership with a Japanese transmission design and manufacturing company, Aisin, in 1986 to develop a hybrid transmission with two electric motors. These companies continued to work together for about six years (when their contract expired), and then Toyota decided to continue development of the transaxle on its own. The result was the transaxle it introduced in the Prius. In the meantime, Volvo joined forces with Aisin and continued to develop the transaxle. Shortly after that, Ford purchased Volvo and the technology. Because of the basic roots, the Toyota and Ford power-split units are quite similar. Due to the similarities and to avoid legal problems, Ford licensed some of the technology from Toyota and

Figure 8-38 The hybrid drive system integrated with an inverter.

Toyota licensed some technology from Ford. Toyota holds over 150 patents on the technology and Ford has received more than 100 patents on its design.

The bottom line is simply, Aisin supplies the transmission used in the Ford hybrids (Figure 8-38) and Toyota makes its own. Toyota does not supply hybrid components to Ford. Both Ford and Toyota state that Ford received no technical assistance from Toyota during the development of the hybrid system.

These are series-parallel hybrid vehicles. Ford divides the operation of the hybrid system into three different modes: positive split, negative split, and electric modes. During the positive split (series) mode, the engine is running and driving the generator, to recharge the battery or directly power the traction motor. The system is in this mode whenever the battery needs to be charged or when the vehicle is operating under moderate loads and at low speeds.

During the negative split (parallel) mode, the engine is running, as is the traction motor. The output of the traction motor tends to reduce the speed of the engine through the action of the planetary gear set. The engine's output is, however, supplemented by the power from the traction motor. During this mode, the traction motor can function as a motor or a generator, depending on the current operating conditions and the demands of the driver.

In the electric mode, the engine is off and the vehicle is propelled solely by battery power. This is the mode of operation during slow acceleration and low speeds, as well as when reverse gear is selected by the driver.

Ford's hybrid technology includes stop-start and relies on a non-traction motor to start the engine, to control the action of the planetary gear set, and to serve as a generator when the engine is running. This motor is claimed to be able to start the engine within 400 milliseconds (0.4 seconds). The engine is shut down when the vehicle is stopping or slowing down, unless the air conditioning system's demands are high or the battery needs to be recharged. Electric assist steering allows for power-assisted steering even when the engine is off. When the vehicle is slowing down or stopping, the system checks the battery's SOC and makes sure there is enough power available for the restart. If there is low power, the engine continues to run in order to drive the generator and charge the battery pack. If the battery has a decent charge, the engine stops running and the electric traction motor (powered solely by the battery) moves the vehicle for initial acceleration after the stop.

The engine can restart immediately after the driver depresses the accelerator or when the electric motor cannot continue to efficiently move the vehicle. The driver has no control over this; the system simply decides what is best for the situation. When operating at low load, low speed, the propulsion power comes from the electric traction motor. This means there are zero exhaust emissions and no fuel is consumed. When the load or vehicle speed increases, the engine starts and more power is available to accelerate or overtake the load at the drive wheels.

When the vehicles are accelerated from a stop to a speed greater than 25 mph, the engine is started and joins in with the electric traction motor to power the vehicle. The amount of engine power and motor power sent to the drive wheels is controlled by the control system and the planetary gear set. When these hybrids are cruising on the highway, the engine provides most of the power. When the driver needs more power for passing or acceleration, the electric motor assists the engine.

Like other hybrids, these vehicles have regenerative braking. They are equipped with an electro-hydraulic brake system with disc brakes at all four wheels and an anti-lock brake system. When the driver releases pressure on the accelerator pedal, the electronic controls change the electrical path to the traction motor so that it operates as a generator. When the brake pedal is depressed, the control unit calculates the required braking force based on the pressure on the brake pedal and the vehicle's speed. The control unit then controls the amount of braking supplied by the motor/generator and the hydraulic brake system. The motor/generator can provide most of the braking force during less-than-panic stops.

Electronic Components

The individual cells in the battery pack (Figure 8-39) are contained in a stainless-steel case and are welded together in groups of five to form a total of 50 separate modules. Each cell has a voltage of about 1.3 volts; therefore, the battery pack has a nominal voltage of 330 volts. The battery pack has a rated capacity of 5.5 amp-hours. To control the temperature of the batteries, the pack is equipped with a thermal management system that operates an electric heater and a forced-air cooling system to keep the battery pack's temperature within a specified range.

The Escape and Mariner hybrids are also equipped with a lead-acid 12-volt battery, located under the hood, to provide power for the various 12-volt systems of the vehicle. This battery is recharged by the DC-DC converter, also located under the hood.

Two inertia-type switches, one in the front and the other in the rear, can disconnect the high-voltage system if the vehicle is in an accident. There is also a high-voltage service disconnect switch (Figure 8-40) that should be moved to the "service" position whenever the work is done to the high-voltage system. This switch is located, in plain view, at the top of the battery housing. It is made of orange molded plastic.

Also, all high-voltage wires, harnesses, and connectors are wrapped in orange-colored insulation or are colored orange. The high-voltage cables are routed under the vehicles from the battery pack to the motors in the front transaxle and from the transaxle to the DC-DC converter. All high-voltage components have a warning label.

Figure 8-39 The battery pack is located behind the rear seat.

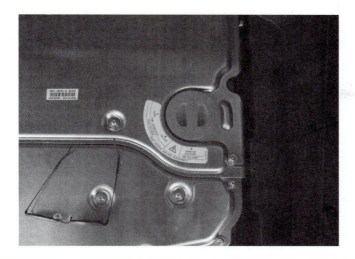

Figure 8-40 The high-voltage service disconnect switch that must be moved to the "service" position whenever the system requires service.

Motor/Generators The operation of the two motors is ultimately controlled by the master control unit (the VSC), through inputs concerning speed and rotor position. The non-traction motor is powered by the battery pack, while the traction motor (Figure 8-41) can be powered by the battery pack and/or the other motor/generator.

Electronic Controls The control system is composed of several different modules, which control the operation of the system. These modules use CAN communications and have diagnostic capabilities. The PCM monitors the activity of the system and has direct control of the engine's operation. The VSC (Figure 8-42) communicates with the other modules and receives inputs from the gear selector (PRNDL) sensor, the accelerator pedal position sensor, the brake pedal position sensor, and many other inputs. Plus, it receives as many as 50,000 signals each second regarding the temperature and SOC of the battery pack. Based on this information, the VSC manages the charging of the battery pack, controls the stop-start function, and controls the operation of the traction motor/generator.

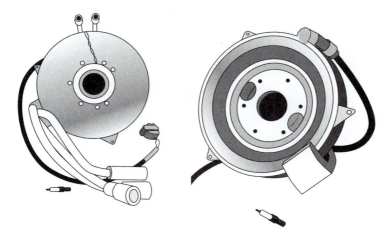

Figure 8-41 The rotor and stator assemblies of the traction motor.

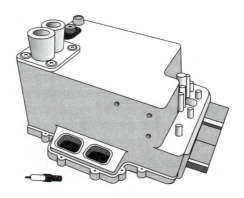

Figure 8-42 The vehicle system controller.

Through commands from the VSC, the **Transmission Control Module (TCM)** directly controls the operation of the motor/generators, and therefore controls the operation of the transaxle. This module is housed inside the transmission case.

The **Battery Energy Control Module (BECM)** is housed in the battery pack and controls the activity of the battery pack. It receives commands from the VSC and sends feedback to the VSC to verify that the hybrid components are operating within the parameters set for the current condition of the battery. The battery pack is divided into two units, one with 26 modules and the other with 24 modules. The voltage of each unit of modules is constantly monitored, as is the current flow to and from them. There are eight temperature senses in each unit to help the BECM keep the battery pack within a specified temperature range. If the temperature is outside that range or if the voltage and current flow is outside their range, the BECM will order the PCM to set a fault code and the system will move to a default setting or shut down.

The **Brake System Control Module (BSCM)** calculates the required brake force to slow down or stop the vehicle. Based on this calculation, it controls the regenerative braking and the hydraulic brake system.

The instrument panel is equipped with gauges that allow the driver to monitor the activity of the hybrid system (Figure 8-43), as well the current SOC of the battery pack.

Figure 8-43 The instrument cluster in a Ford Escape Hybrid.

Figure 8-44 The inverter assembly hides the view of the gasoline engine.

The electric-assist power-steering control module relies on a variety of inputs, one of which measures the amount the steering wheel is turned and how much effort is exerted on the wheel. Based on these inputs, the control module calculates the amount of assist required and controls a small motor in the steering linkage to provide the desired amount of assist.

Engine

These hybrid SUVs are equipped with a 2.3-liter, aluminum, four-cylinder, dual overhead cam engine (DOHC), 16-valve, Atkinson cycle engine (Figure 8-44). The engine is also equipped with electronic throttle control and sequential multi-port electronic fuel injection. The rated output from the engine is 133 hp at 6000 rpm and 129 lb/ft at 4500 rpm. This is less than the output of Ford's conventional 2.3-liter engine because the hybrid utilizes the Atkinson cycle. The decrease in power output is negated by the assist from the electric motor. The combined output of the gasoline engine and the electric traction motor is 155 horsepower.

The engine uses the Atkinson cycle because it is more fuel efficient than a conventional four-stroke cycle engine. Atkinson cycle engines are often described as five-cycle engines. The five cycles are typically called intake, back-flow, compression, power (expansion), and exhaust. During the back-flow cycle, the mixture moves back into the intake manifold to reduce pumping losses. Pumping losses represent the power lost when the engine rotates during periods of high vacuum. Because the intake valve closes after the piston begins its compression stroke, some of the incoming air/fuel mixture is pushed back into the intake manifold, which reduces the effective volume or displacement of the cylinder and provides an instant charge of mixture for the next cylinder in the firing order.

These engines have a high static compression ratio but operate at a low compression ratio. This is due to the loss of compressed gases into the intake manifold. These engines also perform best with low-octane fuel, something that is not characteristic of high compression engines. The Atkinson cycle allows the engine to operate more efficiently but sacrifices are made in the total output of the engine. This is not a problem for hybrids because the motors supplement the engine's output. The end result is satisfactory performance from an engine that barely meets the needs of the vehicle, but achieves good fuel economy and low overall emissions.

Transaxle

Ford's hybrids are equipped with an electronically controlled, continuously variable transmission (eCVT). Based on a simple planetary gear set, like the Toyota's, the overall gear ratios are determined by the motor/generator. Ford's transaxle is different in construction from that found in the Prius. In a Ford transaxle, the traction motor is not directly connected to the ring gear of the gear

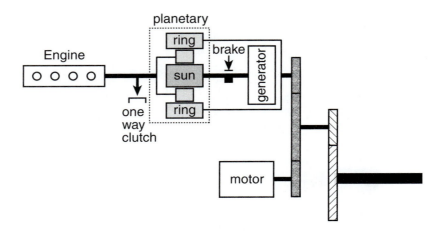

Figure 8-45 In a Ford hybrid transaxle, the traction motor is connected to the transfer gear assembly.

set. Rather it is connected to the transfer gear assembly (Figure 8-45). The transfer gear assembly is composed of three gears, one connected to the ring gear of the planetary set, a counter gear, and the drive gear of the traction motor.

The effective gear ratios are determined by the speed of the members in the planetary gear set. This means the speed or the motor/generator, engine, and traction motor determines the torque that moves to the final drive unit in the transaxle. All three of these power plants are controlled by the VSC through the TCM. Based on commands from the VSC and information from a variety of inputs, the TCM calculates the amount of torque required for the current operating conditions. A motor/generator control unit then sends commands to the inverter. The inverter, in turn, sends phased AC to the stator of the motors. The timing of the phased AC is critical to the operation of the motors, as is the amount of voltage applied to each stator winding.

Angle sensors (resolvers) at the motors' stator track the position of the rotor within the stator. The signals from the resolvers also are used for the calculation of rotor speed. These calculations are shared with other control modules through CAN communications. The TCM, through these sensors, monitors the activity of the inverter and constantly checks for open circuit, excessive current, and out-of-phase cycling. The TCM also monitors the temperature of the inverter and transaxle fluid.

Four-Wheel Drive (4WD)

Unlike the Toyotas with 4WD, the Escape and Mariner do not have a separate motor to drive the rear wheels. Rather, these wheels are driven in a conventional way with a transfer case, rear driveshaft, and a rear axle assembly. This 4WD system is fully automatic and has a computer-controlled clutch that engages the rear axle when traction and power at the rear are needed. The system relies on inputs from sensors located at each wheel and the accelerator pedal, the system calculates how much torque should be sent to the rear wheels. By monitoring these inputs, the control unit can predict and react to wheel slippage. It can also make adjustments to torque distribution when the vehicle is making a tight turn; this eliminates any driveline shutter that can occur when a 4WD vehicle is making a turn.

Honda Civic Full Hybrid

The 2006 Civic has a more powerful electric traction motor (see Chapter 7) and is classified as a full hybrid. The engine used in this hybrid is the same as used in previous non-hybrid models except it has been modified to include the i-VTEC system and other efficiency-increasing technologies. Also, this hybrid is only available with a continuously variable transmission (Figure 8-46).

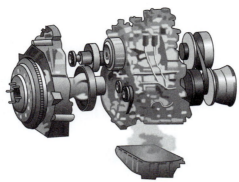

A Continuously Variable
Transmission is standard on every
Civic Hybrid

Figure 8-46 All new Civic Hybrids are equipped with a CVT.

The engine also is capable of full (all) cylinder deactivation, whereas the previous model could only shut down three cylinders. The output from the engine has been increased and is now 93 hp at 6000 rpm compared to 85 hp at 5700 rpm and torque output has increased from 87 lb/ft at 3300 rpm to 89 lb/ft at 4500 rpm (from 93 hp at 5700 rpm and maximum torque of 105 lb/ft at 3000 rpm). The engine and electric motor have a combined maximum output rating of 110 horsepower at 6000 rpm and 123 lb/ft of torque at 2500 rpm. Besides the increase in output, note that the system now provides its peak torque at a lower engine speed.

The IMA motor for the system remains between the engine and the transaxle. However, its power output has been increased to 20 horsepower (15 kW) at 2000 rpm from 13 at 4000 rpm and can provide up to 76 lb/ft at 0-1160 rpm of additional torque. This new motor (Figure 8-47) enables the Civic to be powered solely by the electric motor. Although the motor is less powerful than other hybrid traction motors, it helps the Civic obtain an EPA estimated fuel economy rating of 49/51 mpg city/highway (from 47/48) and the vehicle is rated as a AT-PVEV.

The new motor is based on the same thin motor used in previous models, but has a new rotor whose permanent magnets have been rearranged to make the motor more efficient. The motor also uses flat wire construction to increase the density of the magnetic field around the wires. Honda claims the new motor can now convert 96% of the energy it receives into usable energy. Previous models converted approximately 95%. The generative capabilities have also been increased with this redesigned motor.

Figure 8-47 The new IMA motor produces 20 horsepower.

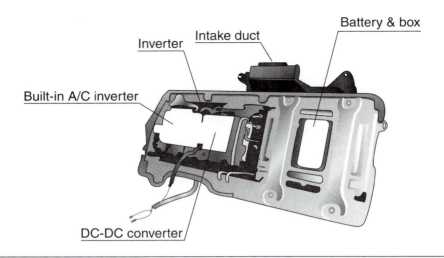

Figure 8-48 The IPU is the control center for the IMA system.

Electronic Controls

This Civic also has a redesigned Intelligent Power Unit (IPU). The new IPU (Figure 8-48), which is the control center for the IMA system, is smaller and weighs less than the previous designs. The voltage of the battery pack has also been increased; it is now 158 volts. The battery pack is composed of 132 NiMH cells rated at 1.2 volts each. The cells are assembled into modules of 12; therefore, there are a total of 11 modules in the pack. The nominal voltage of the battery pack is 158 volts and has a maximum capacity rating of 5.5 amp-hours.

The construction of the battery module is like that in earlier Honda hybrids. The module is fitted with temperature sensors and a cooling system that relies on outside air and a cooling fan. The airflow also cools the inverter, motor control module, DC-DC converter, and the heat sink for the air conditioning compressor driver. A junction board, which consists of the contactors, a bypass resistor, and battery current sensor, is mounted on the battery module. The junction board distributes the high voltage for system operation.

The PCM, which is part of the IPU, continuously receives information from the Battery Condition Monitor (BCM). The BCM monitors the voltage, temperature, and the current in and out of the battery to calculate the battery's SOC. The SOC is then sent to the PCM, which in turn sends commands to the inverter within the motor control module. If the battery's SOC is lower than a specified amount, the PCM will prevent the electric motor from working.

The motor control module also converts the motor to a generator and vice versa. This new Honda hybrid has a revised regenerative braking system. In this system, a control module calculates the amount of brake force required for safe slowing down and stopping. The control module attempts to maximize the amount of power generated by regenerative braking by minimizing the amount of brake force supplied by the hydraulic brake system.

Engine

The engine used in this latest model of the Civic Hybrid is much the same as used in previous models. It is a single overhead cam, eight-valve, 1.3-liter i-VTEC engine with i-DSI. This aluminum alloy engine is equipped with ignition and fuel injection systems that provide for a lean-burn mode and a nitrogen oxide absorptive catalytic converter. The engine is also manufactured with many features to reduce friction and, therefore, fuel usage, such as plateau-honed cylinder walls, offset-cylinder bores, low friction pistons, and roller rocker arms. The engine also uses a very lightweight oil (0W-20) to reduce friction.

The i-DSI system uses two ignition coils firing two spark plugs per cylinder to ensure complete combustion for economy and power. The eight ignition coils are independently controlled by the PCM in response to engine speed and load and can change individual spark plug firing times.

The engine is equipped with an electronically controlled throttle plate and a **Variable Cylinder Management (VCM)** system, which is a three-stage i-VTEC system that increases power output by varying cam timing and allows for deactivation of all four cylinders to reduce pumping losses during deceleration. The latter increases the system's ability to generate electricity during deceleration.

This Civic Hybrid is also equipped with the dual scroll hybrid air conditioning system found in the Accord hybrids. These compressors are actually two compressors built into a single unit. One is driven by the engine and the other is driven by an electric motor. The compressors can work independently or together, depending on the temperature inside the passenger cabin and the SOC of the battery pack.

Three-Stage i-VTEC with VCM The three-Stage i-VTEC valve control system (Figure 8-49) provides normal and high-output valve timing, plus cylinder idling at all cylinders. The 2005 Civic Hybrid had a two-stage VTEC system that provided normal valve timing and cylinder idling at three of the four cylinders. With the three-stage system, the lift and opening duration of the valves can change according to engine speed and driver demands. At high speeds, the system extends valve opening time for improved performance. The VCM system allows for deactivation of all systems during deceleration. It is claimed that this deactivation reduces pumping losses by 66% and improves the regenerative ability by 1.7 times because the engine now provides very little engine braking during deceleration. This effect is accomplished by the generator.

The three stages of the VTEC system are accomplished through hydraulic pressure and an advanced rocker arm design. A pin moves inside the rocker arm shaft to engage or disengage a rocker arm. There are three separate oil passages leading to the pin. As the pressure moves through a passage, the pressure moves the pin. The pressure and the passage for the pressure are controlled by a spool valve that is controlled by the PCM.

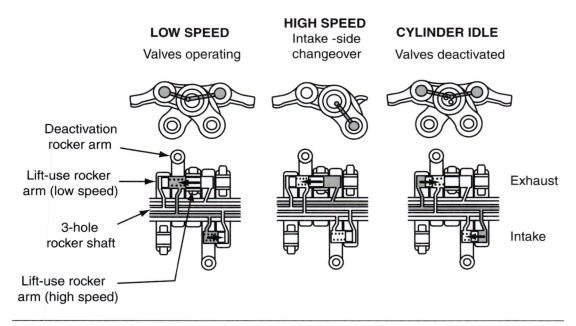

Figure 8-49 The action of the three-stage i-VTEC system.

Operation

The system operates much like previous generations of the IMA hybrid system, except the vehicle can now operate solely by battery power for a short period of time (Figure 8-50). The vehicles have the stop-start function that shuts down the engine during idle times if the battery does not need charging. When the engine is off but the system is turned on, the auto stop indicator blinks. If the driver's door is opened during auto idle stop, the auto stop indicator blinks and the warning buzzer sounds. The engine will not shut off if the engine is cold, the pressure on the brake pedal is low, the A/C system is running, when the battery's SOC is low, or if the vehicle did not reach a speed of over 10 mph during the last drive cycle. The latter allows for battery charging during a traffic jam.

When the brake pedal is released and the accelerator pedal depressed, the vehicle moves by both electric and ICE power. The engine at this time is running in the economy mode with the valves opening by the low lift camshaft profile. During slow acceleration, the motor is turned off and the vehicle moves only by the engine, which is still operating in the economy mode.

When the driver is maintaining a very low cruising speed, the engine shuts off and the motor powers the car by itself. During this time, the engine's rocker arms are not opening the valves. If the battery needs charging during this time, the engine will continue to run.

During acceleration from a low speed, the engine starts and runs in the economy mode. The motor is also running and supplements the power output of the engine. If the battery pack's temperature is outside the specified range, the motor will not run and the vehicle is powered solely by the ICE. Once a cruising speed has been established, the motor turns off and the vehicle is propelled solely by the engine. During heavy acceleration, the engine runs in its high-output mode and the electric motor assists the engine.

During deceleration, the motor begins to work as a generator and the valves in the engine close and remain closed, which allows for maximum regenerative braking and reduces fuel consumption. If the battery is fully charged, the system causes the hydraulic brake system to do all of the slowing down and braking. The activity of the regenerative braking system is totally dependent on the charge needs of the battery pack. Once the vehicle is stopped, the engine shuts down until the brake pedal is released.

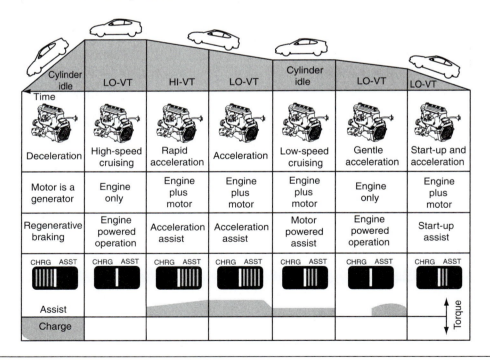

Figure 8-50 The operation of the IMA and i-VTEC systems during different driving conditions.

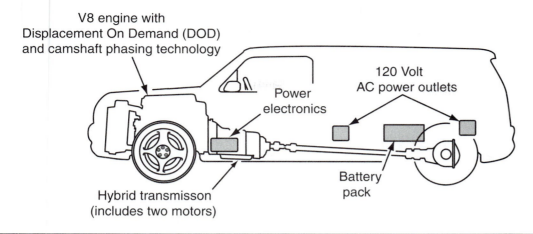

V8 engine with Displacement On Demand (DOD) and camshaft phasing technology

Power electronics

120 Volt AC power outlets

Battery pack

Hybrid transmisson (includes two motors)

Figure 8-51 The basic arrangement of the components in a two-mode hybrid system.

GM/DCX Two-Mode Hybrid System

GM, BMW, and DaimlerChrysler are working together to develop a two-mode full hybrid system that can be used with gasoline or diesel engines. Although no vehicles with this system are available at this time, they will be shortly. The **two-mode hybrid system** relies on advanced hybrid, transmission, and electronic technologies to improve fuel economy and overall vehicle performance. It is claimed that the fuel consumption of a full-sized truck or SUV will be decreased by at least 25% when it is equipped with this parallel hybrid system.

The system fits into a standard transmission housing and is basically two planetary gear sets coupled to two electric motors, which are electronically controlled. This combination results in a continuously variable transmission and motor/generators for hybrid operation (Figure 8-51). The system has two distinct modes of operation. It operates in the first mode during low-speed and low-load conditions and the second mode is used while cruising at highway speeds.

General Motors is expected to have two-mode Chevrolet Tahoe, Cadillac Escalade, and GMC Yukon hybrids available for the 2008 model year. DaimlerChrysler will use this technology in a Dodge Durango, scheduled to be released the same time as GM's hybrids. These vehicles will also feature other fuel savings technologies, such as variable valve timing, Atkinson cycle, and cylinder deactivation (Displacement on Demand).

Operation

Two compact AC synchronous motors (some reports say they are AC induction motors) are connected to the transmission's gear sets (Figure 8-52). The result is a continuously variable transmission that is based totally on planetary gears. The gears work to increase the torque output of the motors. This enables the system to rely on a relatively low voltage, which in turn means the inverter, converter, and controller can be made lighter and smaller. The NiMH battery pack has a nominal voltage of 300 volts and is contained in a housing equipped with a cooling circuit. Like other hybrids, keeping the battery pack within a specified range is a top priority.

The two-mode hybrid system can operate solely on electric or engine power or by a combination of the two. Electronic controls are used to control the output of the motors and the engine. Typically, when one or both of the motors are not providing propulsion power, they are working as generators driven by the engine or by the drive wheels for regenerative braking.

The first mode of operation is called the input split and second is the compound split (Figure 8-53). During the input split mode, the vehicle can be propelled by battery power, engine power or both. This is the normal mode of operation when the vehicle is slowly accelerating from a stop and when it is cruising at slow speeds. When the control unit determines that battery power is sufficient for the current conditions, the engine shuts off. During this time, one motor is working to move the vehicle, while the other may be working as a generator to supply power for the traction motor or to recharge the battery. If the engine is commanded to

GM Two-Mode Hybrid Electric Powertrain

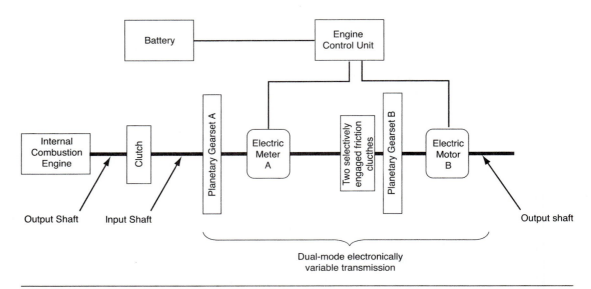

Figure 8-52 The two-mode hybrid system uses two electric motors connected to planetary gear sets to propel the vehicle or assist the engine during propulsion.

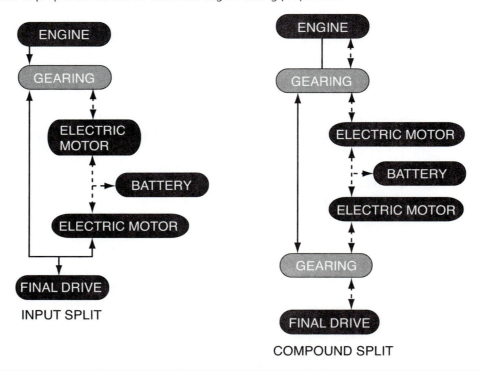

Figure 8-53 The flow of energy during the two modes of operation.

start, the traction motor may shut down but the second motor can continue to operate as a generator if needed.

During normal driving under light loads, the vehicle is powered solely by the engine. Depending on the load and other conditions, the engine may switch off some of its cylinders. When vehicle speed increases or when a heavy load is introduced to the vehicle, such as hard acceleration, climbing a hill, or towing, the system switches to the compound split mode.

In this mode, the control unit can order both motors to supply assistance to the engine or require one of them to operate as a generator. The goal of the control unit is to maximize fuel economy while meeting the needs of the current operating conditions. The control unit also works with engine controls to determine if the other fuel-savings features, such as cylinder deactivation and late intake valve closing (Atkinson cycle), should be initiated. It is important to realize that the deactivation of cylinders and the initiation of the Atkinson cycle reduces the power output of the engine. These fuel-savings features do not hurt the performance of the vehicle because the engine's output is supplemented by the electric motor.

This feature distinguishes a two-mode hybrid from other full hybrids. Typically, electric assist is available only when there is a high demand for power. In the two-mode system, the motors make it possible to reduce the work of the engine, even during light and moderate loads.

Hybrid Buses

The two-mode hybrid technology has been used for a few years in city buses. Although the hybrid components in these buses are different from those used in automobiles, the concept is the same. In 2003, GM equipped city buses in the Seattle area with a parallel two-mode hybrid system. The use of these buses has moved to many other major cities.

The GM/Allison hybrid system uses a 600-volt NiMH battery pack. There are two 100-kW electric motors and an electronically variable transmission. The combination of torque-increasing and speed-reducing gears allows the size of the engine to be decreased while providing the performance of a larger engine. This technology has also resulted in much less fuel consumption; in fact, a GM spokesperson claimed, "If we replaced the 13,000 buses (with the hybrid buses) in the nine largest U.S. cities, we would save 40 million gallons of fuel a year." The reduced engine size, along with a particulate filter, results in a 90% reduction in emissions.

This hybrid is called the "EP System" and combines a hybrid drive and an Allison transmission into one unit. This system provides 430- to 900-volt power to the motors and is able to capture regenerative energy at speeds up to 50 mph. The unit is cooled by a common oil-cooling system for the motors, drive unit, and control unit. When the vehicle accelerates from a stop, it uses the torque of electric motors, and then maintains the desired speed with its engine. During this time, the engine also drives the generator to charge the batteries.

Graphyte Concept

To demonstrate the practicality of the two-mode hybrid system, GM introduced two concept vehicles: the Graphyte SUV and the Opel Astra Diesel Hybrid. The Graphyte combines the two-mode hybrid system with a Vortec 5300 V-8 with displacement on demand. This midsized hybrid uses the same technology as will be used in the Tahoe and Envoy. The Graphyte is a front-engined, all-wheel-drive, five-passenger SUV. It has a 300-volt battery pack under the rear passenger seat and uses a regular four-speed transmission to house the hybrid components. The gasoline-powered engine is a 5.3-liter V-8 with displacement on demand technology. The vehicle is also equipped with electrically operated power steering, power brakes, and A/C system. There is also a LCD screen in the instrument panel that allows drivers to monitor the action of the hybrid system so they can achieve the highest possible fuel economy.

Astra Diesel Hybrid

To demonstrate how a diesel engine will work with a two-mode hybrid system, GM revealed its concept Astra Diesel Hybrid. This is a front-wheel-drive, five-passenger vehicle based on the (Europe only) Opel Astra GTC. It uses a 125-hp (93-kW) 1.7-liter turbocharged diesel engine and two electric motors (rated at 30 kW and 40 kW). The motors are integrated into the vehicle's automatic transmission.

The hybrid system's operation includes stop-start and runs on battery power during slow acceleration from a stop. During braking and coasting, regenerative braking charges the battery pack. The concept vehicle also has a video animation, in the instrument panel that shows the current state of propulsion as well as the current SOC of the battery pack.

Other Planned Full Hybrids

At the time of this writing, only Toyota, Honda, and Ford offer full hybrids. There are many other manufacturers that plan to release full hybrids and the following discussion merely introduces what they may present to the public. General Motors and Chrysler are geared up to offer their two-mode hybrids in the near future. Many other manufacturers have plans to do the same, each with its own system. However, many of the new hybrids that will be released in the near future will be modifications of the systems introduced by others.

Hyundai Motor Co. and Kia Motors Corp.

Hyundai Motor Co., its affiliate, Kia Motors Corp., announced they will release hybrid versions of Hyundai's Accent and Kia's Rio. These hybrids probably will be modeled after the Toyota Prius. Hyundai also is investing heavily to develop fuel cell vehicles.

Mazda Premacy Hydrogen RE Hybrid This "minivan" hybrid design combines the advantages of a rotary engine with hybrid technology. The rotary engine in this concept vehicle can be fueled by hydrogen or gasoline. The bi-fuel engine is connected to an electric motor and the vehicle can be driven by either the motor or engine, or both. Mazda claims a drive range of 124 miles on hydrogen fuel. Along the same design is the RX-8 Hydrogen RE, which is a hydrogen-fueled, rotary-engined RX-8. This also is a dual fuel engine: gasoline or hydrogen.

Major European Manufacturers

Audi, VW, and Porsche are all developing hybrid vehicles. Some are seeking fuel economy whereas others are striving for increased performance. The systems they are working on depends on their goals, but it us fair to say that if overall performance is sought, they will use electric motors to supplement the engine and/or four-wheel-drive system. If economy is sought, they will use electric motors to propel the vehicles when the engine is shut down.

Keep in mind that BMW is one of the partners with GM and DCX in developing the two-mode hybrid system. When BMW releases a hybrid, it will undoubtedly be based on that system (Figure 8-54).

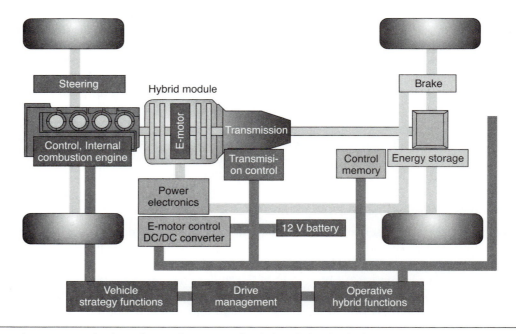

Figure 8-54 The basic layout for a rear-wheel-drive hybrid vehicle.

Review Questions

1. Describe the basic construction of a simple planetary gear set.
2. Which of the following statements about the Atkinson cycle are *not* true?
   A. The Atkinson cycle allows the effective displacement of the engine to change according to need.
   B. In an Atkinson cycle engine, the intake valve is kept open well into the compression stroke.
   C. The delay in intake valve closing reduces the power output of the engine and decreases its emissions.
   D. During the Atkinson stroke, some unburned air/fuel mixture is pushed into the exhaust manifold.
3. General Motors and DaimlerChrysler have developed a two-mode hybrid system. What are the two modes and how is this system constructed?
4. In Honda's three-stage i-VTEC system, what are the three stages?
5. What two major things did Honda do to change the assist hybrid Civic into a full hybrid?
6. Compare the power-split unit found in a Prius to that found in a Ford Escape hybrid.
7. Which of the following units provides the increased voltage in Toyota's latest hybrid system?
   A. HV battery
   B. Inverter
   C Boost converter
   D. Resolver
8. In a Toyota Prius, what members of the planetary gear set are connected to the motor/generators and the engine?
9. The ECM for the Variable Valve Timing-intelligent (VVT-i) system looks at several inputs when calculating the proper valve timing for the conditions. Name three of the more important inputs.
10. Which of the following statements about the motor/generators used in the Toyota Prius is *not* true?
    A. MG1 controls the action of the planetary gear set in the power-split device.
    B. MG1 is used to start the ICE.
    C. Only MG2 is used to supplement the power from the engine.
    D. Both MG1 and MG2 are used during regenerative braking.

HYBRID MAINTENANCE AND SERVICE

After reading and studying this chapter, you should be able to:

❏ Describe the major consideration of servicing a hybrid, compared with performing the same service on a conventional vehicle.

❏ List and describe the commonsense precautions to be adhered to while working around or on a hybrid vehicle.

❏ Describe the procedure for de-powering the high-voltage system in Honda hybrids.

❏ Describe the procedure for de-powering the high-voltage system in Toyota hybrids.

❏ Describe the procedure for de-powering the high-voltage system in Ford hybrids.

❏ Describe the procedure for de-powering the high-voltage system in GM hybrids.

❏ Explain how the manufacturers have designed their hybrid vehicles to ensure the safety of the passengers and technicians.

❏ Explain the vital role the 12-volt battery has in the operation of a hybrid vehicle.

❏ Describe how the 12-volt battery should be serviced.

❏ Describe how the high-voltage battery pack should be serviced.

❏ Describe what preventative maintenance procedures are unique to a hybrid vehicle.

❏ Explain the proper steps to take when diagnosing a problem in a typical hybrid vehicle.

❏ Describe the special diagnostic tools that must be used on a hybrid vehicle.

❏ Describe the special considerations and procedures that must be followed when servicing a hybrid vehicle's engine.

❏ Describe the special considerations and procedures that must be followed when servicing a hybrid vehicle's cooling system.

❏ Describe the special considerations and procedures that must be followed when servicing a hybrid vehicle's transmission.

❏ Describe the special considerations and procedures that must be followed when servicing a hybrid vehicle's brake system.

❏ Describe the special considerations and procedures that must be followed when servicing a hybrid vehicle's steering system.

❏ Describe the special considerations and procedures that must be followed when servicing a hybrid vehicle's air conditioning system.

Key Terms

Auxiliary power
   outlet (APO)
Category 3 (CATIII)
Conductance test
Energy storage
   box (ESB)
Insulation resistance
   tester
Lineman's gloves

Introduction

Hybrid vehicles (Figure 9-1) are maintained and serviced in the same manner as conventional vehicles, except for the hybrid components. The latter includes the high-voltage battery pack and circuits, which must be respected when doing any service on the vehicles. This chapter covers the steps that should be followed to work safely around the high-voltage system and other service procedures unique to hybrid vehicles. Some of the latter are normal everyday services that must be completed in a different way.

For the most part, actual service to the hybrid system is not something that is done by technicians, unless they are certified to do so by the automobile manufacturer. Diagnosing the systems varies with the manufacturer, although certain procedures apply to all. Keep in mind, a hybrid has nearly all of the same basic systems as a conventional vehicle, and these are diagnosed and serviced in the same way. Through an understanding of how the hybrid vehicle operates, you can safely service them. This thought will be explained throughout this chapter.

Figure 9-1 A late-model hybrid vehicle.

Before performing any maintenance, diagnosis, or service on a hybrid vehicle, make sure you understand the system found on the vehicle and try to experience what it is like to drive a normal operating model. These vehicles offer a unique driving experience, and it is difficult to say what is working correctly if you do not have firsthand seat experience.

One of the things to pay attention to, both from the owners' or service manuals and driving experience is the stop-start feature. You need to know when the engine will normally shut down and restart. Without this knowledge, or the knowledge of how to prevent this, the engine may start on its own when you are working under the hood. Needless to say, this can create a safety hazard. Imagine reaching into the engine compartment and the engine starts. There is a possibility that your hands or something else can be trapped in the rotating belts or hit by a cooling fan. Unless the system is totally shut down, the engine may start at any time when its control system senses the battery needs to be recharged.

In addition, there is a possibility that the system will decide to power the vehicle electrically. When it does this, there is no noise, just a sudden movement of the vehicle. This can be alarming and dangerous. To prevent both of these incidents, always remove the key from the ignition. Make sure the "READY" lamp in the instrument cluster is off; this lets you know the system is also off.

Precautions

Hybrid vehicles have high-voltage systems (Figure 9-2). Careless handling of some components can lead to serious injury, including death. Always follow and adhere to the precautions given by the manufacturer. These precautions are clearly labeled in their service manuals. Also, emergency response guides are available from each manufacturer. These allow emergency workers to safely work around their vehicles in case of an accident. All emergency workers and technicians, especially body repair technicians, should be familiar with the contents in these guides. All service procedures should be followed exactly as defined by the manufacturer. Carelessness and/or not following the procedures can cause serious injury and can cause the battery to explode! Below is a list of commonsense items to consider when working on a hybrid vehicle.

❏ Before doing any service on a hybrid vehicle, refer to the service manual for that specific vehicle. All hybrids have similar operation but have different systems and components; this is true for vehicles made by the same manufacturer.

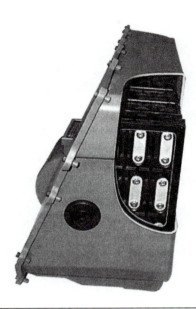

Figure 9-2 A typical high-voltage battery pack assembly.

❏ All high-voltage wires and harnesses are wrapped in orange-colored insulation. Respect the color and stay away from it unless the system is de-powered.
❏ Warning and/or caution labels are attached to all high-voltage parts. Be careful not to touch these cables and parts without the correct protective gear, such as safety gloves.
❏ Make sure the high-voltage system is shut down and isolated from the vehicle before working near or with any high-voltage component.
❏ If the vehicle needs to be towed into the shop for repairs, make sure it is not towed on its drive wheels. Doing this will drive the generator(s) to work, which can over-charge the batteries and cause them to explode. Always tow these vehicles with the drive wheels off the ground or move them on a flat bed.
❏ When working on or near the high-voltage system, even when it is de-powered, always use insulated tools.
❏ Never leave tools or loose parts under the hood or close to the battery pack. These can easily cause a short.
❏ Never wear anything metallic, such as rings, necklaces, watches, and earrings, when working on a hybrid vehicle.
❏ In the case of a fire, use a Class ABC powder type extinguisher or very large quantities of water.

Gloves

Always wear safety gloves during the process of de-powering and powering the system back up again. These gloves must be class "0" rubber insulating gloves (Figure 9-3); rated at 1000 volts (these are commonly called "**lineman's gloves**"). The condition of the gloves must be checked before each use. Make sure there are no tears or signs of wear. Electrons are very small and can enter through the smallest of holes in your gloves. To check the condition of your gloves, blow enough air into each one so they balloon out. Then fold the open end over to seal the air in. Continue to slowly fold that end of the glove toward the fingers. This will compress the air. If the glove continues to balloon as the air is compressed, it has no leaks. If any air leaks out, the glove should be discarded. All gloves, new and old, should be checked before they are used. Keep in mind, these insulating gloves are special gloves and not the thin surgical gloves you may be using for other repairs. Also, to protect the integrity of the insulating gloves, as well as you, while doing a service, wear leather gloves over the insulating gloves.

Figure 9-3 A pair of lineman's gloves.

De-Powering The High-Voltage System

The procedure for properly de-powering and isolating the high-voltage system from the rest of a hybrid vehicle is very important and not very difficult. However, each manufacturer has its own procedure that must be followed in the order presented. Fortunately, the various hybrid models offered by a particular manufacturer have much the same procedure. This does not mean you should not look for the correct procedure for a specific model you are working on. With the correct information and following the procedures, you may safely work on a hybrid vehicle.

The following discussion covers the de-powering procedures for the currently available assist and full hybrids. Other hybrids also have high-voltage systems. These systems should also be de-powered before performing any service.

Honda Hybrids

Honda's procedure for de-powering the high-voltage systems includes the following steps, which must be followed in the order they appear to avoid serious personal injury and damage to the vehicle's electrical system. Always wear undamaged electrically insulated gloves when inspecting and handling any high-voltage cable or component.

1. Turn the ignition switch OFF.
2. On many models, the rear seatback must be removed to gain access to the battery module.
3. Remove the switch lid from the IPU cover on the battery module.
4. Then remove the locking cover.
5. Turn the battery module switch OFF (Figure 9-4), and then reinstall the locking cover to secure the switch in the OFF position.
6. Wait for five or more minutes after turning the battery module switch OFF, to allow the ultra-capacitors to discharge.
7. Then disconnect the negative cable from the 12-volt battery.
8. Locate the terminals for the junction board.
9. Measure the voltage at those terminals. If the voltage is less than 30 volts, the system can be serviced. If the voltage is greater than 30 volts, there is a problem. This problem must be diagnosed before continuing with any service.
10. When the system is ready to be powered again, make sure all circuits have been securely reconnected, then reverse the de-powering sequence.

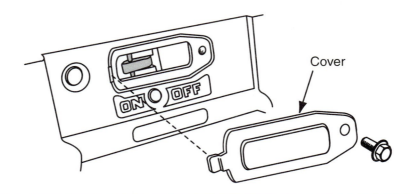

Figure 9-4 The battery module switch in a Honda hybrid.

⚠ **WARNING:**
Always use the tools designed for rotor installation and removal. If you use your hands to install the rotor, the magnetic field may suddenly pull the rotor toward the stator. If your hands are on the back of the rotor, they can be seriously injured by the force.

Honda also recommends that certain precautions be taken when inspecting, diagnosing, and/or working on or around their IMA system:

❏ When the IMA system indicator is on, perform diagnostics on the IMA system before proceeding with any other service or check.
❏ After disconnecting any part of the high-voltage system, wrap all electrical contacts with insulated electricians tape.
❏ Attach a sign to the steering wheel that says *"WORKING ON HIGH VOLTAGE PARTS. DO NOT TOUCH!"*
❏ When handling or servicing parts of the system that are not covered with insulating material, make sure you have insulated gloves on and that you only use insulated tools and equipment.
❏ The rotor assembly is comprised of very strong permanent magnets. Keep the rotor away from magnetically sensitive devices. Individuals with heart pacemakers or other magnetically sensitive medical devices should not handle the rotor.

Toyota Hybrids

All Toyota and Lexus hybrids have the same de-powering sequence, although some use a much higher voltage than others do. These higher voltage systems require only one additional precaution, that being the wait time for the ultra-capacitors to discharge. Toyota calls for a wait of at least five minutes. On the vehicles with very high voltages, longer waiting times are safer. The basic procedure for de-powering the high-voltage system includes:

1. Remove the key from the ignition switch. If the vehicle has a "smart" key, turn the smart key system OFF. This may be done by applying pressure to the brake pedal while depressing the start button for at least two seconds. If the READY lamp goes off, continue. If it does not, diagnose the problem before continuing.
2. Disconnect the negative (-) terminal cable from the auxiliary 12-volt battery. This should turn the high-voltage system off, but does not complete the de-powering process.
3. Make sure you are wearing insulated gloves and remove the service plug from the battery module (Figure 9-5). To gain access to the plug, you may need to move the carpeting in the rear of the vehicle.
4. Put the service plug in your pocket to prevent others from reinstalling it before the system is ready or while you are working on the vehicle.
5. Put electrical insulating tape over the service plug connector.
6. Wait at least five minutes before proceeding or doing any work on or around the high-voltage system.

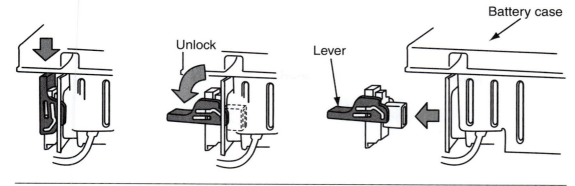

Figure 9-5 Removal of the service plug on a Toyota hybrid to disconnect the high-voltage circuits.

7. Prior to handling any high-voltage cable or part, check the voltage at the terminals. There should be less than 12 volts.
8. If a high-voltage cable must be disconnected for service, wrap its terminal with insulating tape to prevent a possible short.

Ford Motor Company Hybrids

The hybrids from Ford have a similar procedure as Toyota. However, it is not the same. The service disconnect plug can be installed in different positions, rather than simply be in or out. Ford also recommends that whenever service is being performed on a hybrid that a buffer zone be set around the vehicle. Their recommendation is to place four orange traffic cones about 3 feet (1 meter) from the corners of the vehicle to establish the buffer zone. This zone marks off the area where customers or others should not enter unless they are trained to work on the hybrid vehicle. Once this zone has been established, follow this procedure:

1. Place the gear selector into the PARK position and remove the ignition key. This should isolate the high-voltage system, but for safety reasons continue through the procedure.
2. Disconnect the negative cable at the 12-volt battery; this should also isolate the high voltage, but continue through this procedure.
3. Lift up the carpeting behind the rear seat.
4. Locate and turn the service disconnect plug from the LOCK position to the UNLOCK position. Then lift it out. This disconnects the high-voltage battery pack from the vehicle.
5. Reinstall the plug with its arrow at the SERVICING/SHIPPING mark (Figure 9-6).

■ CAUTION:
Although the removal of the plug disconnects the battery pack, it should be returned to its bore but in the SERVICING/SHIPPING position to prevent debris from entering into the battery pack while services are being performed.

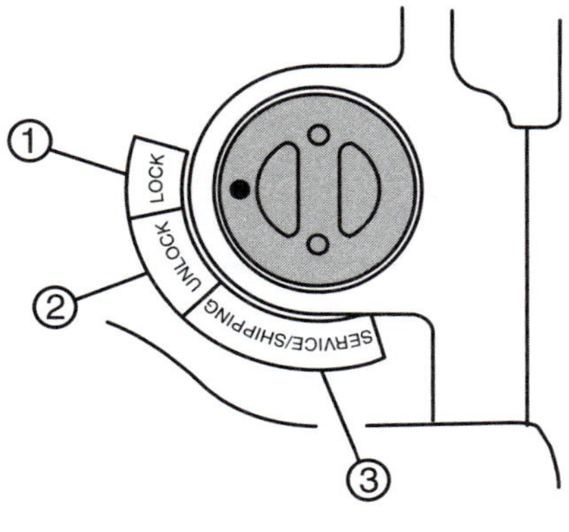

Figure 9-6 To isolate the high-voltage system, turn the service disconnect plug from the LOCK position (1) to the UNLOCK position (2). Then lift it out and reinstall it in the SERVICING/SHIPPING position (3).

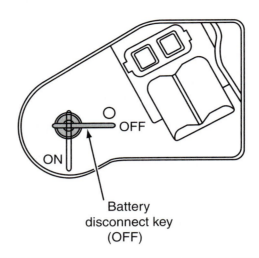

Battery
disconnect key
(OFF)

Figure 9-7 When the service disconnect switch for a GM hybrid switch is in the horizontal position, the switch is OFF

GM's Silverado/Sierra Hybrids

General Motors is currently using different hybrid systems. The most common of these are found in the Silverado and Sierra pickups. These systems have a 42-volt starter/generator located between the engine and transmission. These vehicles also have a portable generator feature that can provide 110 volts of AC at power outlets. To safely work on these vehicles, both the 42-volt battery pack and the auxiliary power outlet system need to be de-powered. All 42-volt and 110-volt circuits are labeled and colored orange.

These vehicles have a stop-start feature that is easily disabled by opening the hood. There is a hood ajar switch that prevents the engine from starting when the hood is open. However, if the engine is running and then the hood is opened, the engine will continue to run. To safely work under the hood of these vehicles, simply turn off the vehicle and open the hood.

To control the **auxiliary power outlet (APO)** circuit, there is an APO button on the dash and an APO indicator lamp. When the lamp is lit, the circuit is active. To turn off the system, depress the button, turn off the engine, and remove the key.

To disconnect the 42-volt system from the vehicle, there is a switch on the side of the battery pack, called the **energy storage box (ESB)**. The battery pack is located under the rear passenger seat. The service disconnect switch is located behind a removable cover on the battery pack. Once the cover is removed, the switch is turned to its horizontal or OFF position (Figure 9-7). This disconnects the 42-volt system from the rest of the vehicle.

General Motors also advises technicians to disconnect the 12-volt battery whenever working under the vehicle's hood and when working on or around any part related to the vehicle's hybrid system.

Safety Features

All hybrid vehicles have certain safety features that are designed to isolate the high-voltage system in the case of an accident or electrical fault. You should be aware of these features, as they may cause the hybrid system to be inoperable. Some of these safety features are unique to specific hybrid models, whereas others are universally applied.

On all hybrids, the high-voltage cables and connectors are colored orange, and warning labels are affixed to all high-voltage components. In addition, the battery packs are placed in

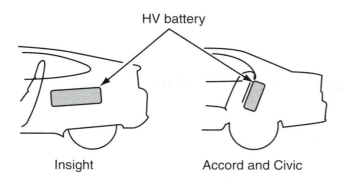

HV battery

Insight   Accord and Civic

Figure 9-8 Battery packs are placed in areas where they are protected.

protective zones (Figure 9-8) and are enclosed in containers. These containers are designed to withstand a degree of impact and to contain any battery chemicals that may leak from the batteries. The cables are also routed in "safe" areas and if they are not routed through the vehicle's frame, they are protected with extra metal.

The damage that results from a collision can result in damage to the high-voltage cables and/or components, which could result in high-voltage shorts. These could be disastrous, for the vehicle, the passengers inside, and emergency responders. To immediately isolate or shut down the high-voltage system, hybrids are fitted with many sensors (Figure 9-9). These sensors respond to impact or the activation of the airbags. In most accidents, the high-voltage system is automatically and immediately disabled.

Every hybrid is designed to open the circuit from and to the battery pack whenever the airbag(s) are deployed. In many models, high-voltage is shut down before the airbags are deployed. Impact sensors are placed strategically in the areas where the high-voltage system is located. On GM hybrid pickups, the battery pack is located on the passenger side. A side impact sensor will trigger the battery cut-off switch to isolate the 42-volt battery pack and the auxiliary power outlet circuit if an impact to the passenger door is detected.

Ford hybrids have impact switches at the front and rear of the passenger side of the vehicle. If a force or jolt causes either of these switches to open, the high-voltage system and the circuit to the fuel pump are disconnected. On Toyota hybrids, the impact sensor is located inside the inverter. Again, when an impact is detected, the high-voltage system is shut down.

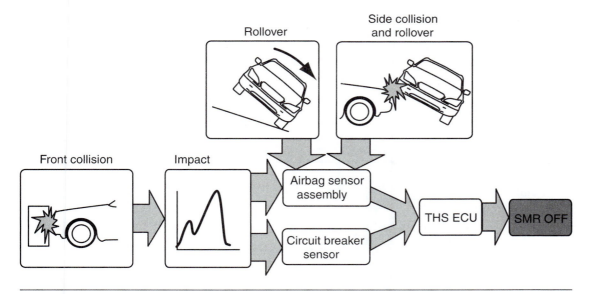

Figure 9-9 Through the use of sensors. High-voltage systems are shut down when a collision is detected.

Other sensors are also used to prevent high voltage from leaking into the passenger compartment. These sensors monitor the condition of the high-voltage wiring and cables. If there is any evidence of a short or high current through the circuit, the high-voltage system is immediately shut down. The high-voltage system will also be shut down if the circuit opens, which can be caused by such things as a broken wire or disconnected connector. Also, the system will shut down if battery temperatures rise beyond a specified temperature (normally 140°F).

On GM pickups with the auxiliary power outlets, the same techniques are used to protect the passengers from faults within that system. The APO circuit is fitted with a ground fault detection circuit. This is similar to those found in the kitchens and bathrooms of most houses. These circuits monitor the current flow through the wires of the circuit. If there is more than a few milliamps difference between the two wires (neutral and hot), the APO circuit will be shut down. The system will also shut down if there is an opening in the circuit.

Batteries

Most hybrids have two separate battery packs. One is the high-voltage pack and the other is a 12-volt battery. The high-voltage battery pack typically supplies the power to start the engine, assist the engine during times of heavy load, and in full hybrid it supplies the energy to move the vehicle without the engine's power. The battery pack is the one that is most associated with the hybrid system. The 12-volt battery is associated with the rest of the vehicle, such as the lights, accessories, and power equipment. The 12-volt battery also supplies the power for the electronic controls that monitor and regulate the operation of the hybrid system. If this power source is not working correctly, the hybrid system will not. Therefore, this low-voltage power source should never be ignored when working on a hybrid or a conventional vehicle.

12-Volt Batteries

⚠ WARNING:
Batteries contain sulfuric acid. Wear eye protection when working near or with the battery. If the acid gets on your skin or in your eyes, flush immediately with water for a minimum of 15 minutes and get immediate medical attention.

The auxiliary battery used in many hybrid vehicles is the same type found in conventional vehicles; however, some are AGM batteries. It is important that you properly identify the type of battery in the vehicle so that you correctly service the battery. For the most part, inspection of and service to the auxiliary battery in a hybrid is no different than if it were not in a hybrid. It is important to note that if the battery's voltage drops below a specific level, the emissions MIL and/or hybrid warning lights may illuminate. It is also important to know that if the vehicle has not been driven for more than a month; both the low-voltage and high-voltage batteries will be low on charge. Most manufacturers recommend that hybrids should be started and run for at least ten minutes every month. This will keep the high-voltage battery charged enough to operate the vehicle but may not be enough to keep the low-voltage battery charged.

The 12-volt batteries are located in the trunk or under the hood (Figure 9-10). They should be inspected (Figure 9-11) on a regular basis. Make sure the cable connectors are tight. Also,

Figure 9-10 Note that the auxiliary battery is under the hood and the battery pack is in the rear of a Honda Civic.

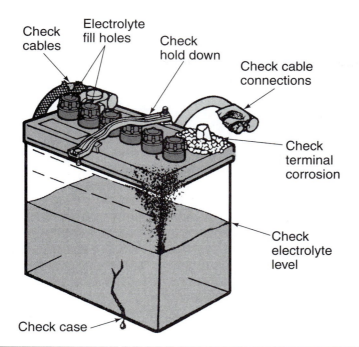

Check cables

Electrolyte fill holes

Check hold down

Check cable connections

Check terminal corrosion

Check electrolyte level

Check case

Figure 9-11 Carefully inspect the auxiliary battery, especially if its voltage is low.

▲ **WARNING:**
Do not conduct this test on a lead-acid battery that has recently been recharged. The gases released by the battery during charging may ignite during this test. Also, if the vehicle has been recently driven, wait at least 40 minutes to allow the gases to dissipate and for all control modules to power down.

make sure the electrolyte level is satisfactory. The battery, its terminals, and cables should be clean. If there is dirt or corrosion on the battery or cable ends, clean them. If the battery cables are disconnected for cleaning, all control systems will lose their memory. This is not desirable in hybrid vehicles; therefore, steps must be taken to preserve the memory. Memory savers are available and should be used. Some manufacturers also give specific instructions for disconnecting the battery while maintaining memory. When disconnecting battery cables, always remove the ground cable from the negative terminal first and reinstall it last.

Battery voltage is quickly checked with a voltmeter. Typically, a good 12-volt battery will have 12.6 to 14.2 volts, although anything above 12 volts would indicate the battery is near fully charged. An auxiliary battery can be low on charge for a number of reasons. It could be bad battery, poor connections, or the charging system is not functioning properly. However, a common problem is parasitic drain.

Vehicles should have no more than a 50 mA current draw with the ignition off. To measure the drain on the battery and to identify the source of the excessive current draw, use an in-line digital ammeter rated at 10 amps with the capability of measuring in milliamps. Make sure the battery is clean and check for voltage leakage through the battery case. Then connect a jumper wire fitted with a 10-amp circuit breaker or fuse between the negative battery cable and the negative battery post. This will maintain the memories in the vehicle. Then disconnect the negative cable at the battery. Be careful not to disconnect the jumper wire while doing this. Connect the in-line ammeter between the end of the negative cable and the battery's negative post. Once the meter is in place, remove the jumper wire. Observe the reading on the meter.

If the reading is excessive, pull one fuse at a time from the fuse box. Check the current after each one is removed. When the current drops significantly after a fuse is removed, the cause of the excessive drain is most likely in the circuit being protected by that fuse. If all of the fuses have been removed and there is still excessive draw, check the wiring diagram to identify the circuits that are not protected by the fuses that were removed. Then disconnect these circuits, one at a time, until the problem circuit is identified. Once the problem is identified and repaired, replace all fuses and reconnect everything. Do not remove the ammeter until the fused jumper wire is placed between the battery and the cable end. After removing the meter, reconnect the battery's negative cable, making sure that the jumper wire stays in place. Once the cable terminal is tightened, the jumper wire can be removed.

Battery Capacitance Test The auxiliary battery can be load tested using a conventional battery tester. However, many manufacturers recommend that a battery capacitance or **conductance test** be performed. Conductance describes a battery's ability to conduct current. It is a measurement of the plate surface available in a battery for chemical reaction. Measuring conductance provides a reliable indication of a battery's condition and is correlated to battery capacity. Conductance can be used to detect cell defects, shorts, normal aging, and open circuits that can cause the battery to fail.

A fully charged new battery will have a high conductance reading, anywhere from 110% to 140% of its CCA rating. As a battery ages, the plate surface can sulfate or shed active material which will lower its capacity and conductance. When a battery has lost a significant percentage of its cranking ability, the conductance reading will fall well below its rating and the test decision will be to replace the battery. Because conductance measurements can track the life of the battery, they are also effective for predicting end of life before the battery fails.

To measure conductance, the tester (Figure 9-12) creates a small signal that is sent through the battery, and then measures a portion of the AC current response. The tester displays the service condition of the battery. The tester will indicate that the battery is good, needs to be recharged and tested again, has failed, or will fail shortly.

Removal and Installation The procedure for removing the auxiliary battery from a hybrid vehicle is the same as for a conventional vehicle. However, in some hybrids with the auxiliary battery in the trunk, a Prius for example, it may be impossible to get to the battery if it is dead.

Figure 9-12 A battery conductance tester.

The lack of battery power will prevent the rear hatch from opening. To gain access to the battery, the vehicle will need to be jump-started. Toyota has a special jump-starting terminal under the front hood. Once the car is started, the trunk can be opened normally. Always adhere to all safety precautions when removing and installing a battery.

Because of the advanced electronics in hybrids, steps must be taken after installation to allow the computers to re-learn. The regenerative braking system needs to relearn the initial position of the brake pedal. After the battery is reconnected, slowly depress and release the brake pedal one time. The engine also needs to relearn its idle and fuel trim strategy. If this is not done immediately after reconnecting the battery, the engine will idle and run poorly until it sets up its strategy. A typical procedure begins with turning off all accessories and starting the engine. The engine is idled until it reaches normal operating temperature, then it should be allowed to run at idle for one minute. After that time, the air conditioning is turned on and the engine again is allowed to idle for at least one minute. Now the vehicle should be driven for about ten miles. All manufacturers have their own sequence for doing this, so be sure to follow their procedure.

On some vehicles, such as the Honda Civic Hybrid, the battery level gauge will not display the battery's state of charge when the engine is first started after the auxiliary battery has been replaced or disconnected. To reactivate the indicator, start the engine and run it at 3000–4000 rpm with all accessories off. Keep the engine at that speed until the battery level gauge displays at least three bars or segments.

Recharging The correct method for recharging the auxiliary battery varies with the type of battery. Most manufacturers recommend a slow charge of less than 3.5 amperes. However, others recommend the use of an intelligent charger. In most cases, the vehicle's engine is the best and safest charger. If the engine runs and the charging system works normally, drive the vehicle to charge all of the batteries. When recharging, always consider the following precautions:

❏ Never recharge the battery when the hybrid system is on.
❏ Turn off all accessories before charging the battery and correct any parasitic drain problems.
❏ Make sure the charger's power switch is off when you are connecting or disconnecting the charger cables to the battery.
❏ Always charge the battery in an unconfined and well-ventilated area.
❏ Keep all flames, sparks, and excessive heat away from the battery at all times, especially when it is being charged.

High-Voltage Batteries

Any discussion about high-voltage battery and energy systems must include warnings regarding the need to isolate the high-voltage from the rest of the vehicle and reminders to adhere to all precautions stated by the manufacturer. This is important for all services, not just electrical. Air-conditioning, engine, transmission, and bodywork can require services completed around and/or with high-voltage systems. If there is any doubt as to whether something has high voltage or not, or if the circuit is sufficiently isolated, test it before touching anything.

To test high-voltage systems you need a **Category 3 (CATIII)** digital volt ohmmeter (Figure 9-13) and, of course, a good pair of insulating gloves. Although the high-voltage system can be isolated from the rest of the vehicle, high-voltage is still at and around the battery pack. When checking for the presence of high-voltage, make sure to check the inverter assembly. This is typically where the large or ultra-capacitors are, and until they are discharged, they are lethal.

The high-voltage battery packs in nearly all HEVs are made up of several NiMH cells. There are certain characteristics about these cells that must be understood and remembered, especially for diagnostic purposes and safety.

▲ **WARNING:**
When lifting a battery, excessive pressure on the end walls can cause acid to leak through the vent caps. The acid can cause burns and/or skin irritation. Always lift a battery with a battery carrier or with your hands on opposite corners of the battery.

Figure 9-13 Only meters with this symbol should be used on the high-voltage systems in a hybrid vehicle.

❏ NiMH batteries do not store well for long lengths of time, so if the vehicle has not been driven for a while, the battery may lose quite a bit of its normal capacity.
❏ These batteries have a limited number of charging cycles (the number of times a battery can be charged and discharged).
❏ All hybrids charge the battery as you drive.
❏ The capacity of NiMH batteries decreases with an increase in heat.
❏ If the battery is exposed to intense heat, it is possible that hydrogen will be released from the battery, and hydrogen can be explosive when introduced to flame or sparks.
❏ The battery cells contain a base electrolyte comprised mostly of potassium hydroxide.
❏ Exposure to the electrolyte can cause skin/eye irritation and/or burns.
❏ The battery packs are very heavy.
❏ Used batteries should be disposed of according to local and federal laws.

Whenever a battery pack, cooling system components (Figure 9-14), or other hybrid parts must be removed, serviced, and/or replaced, also follow the specific procedures given by the manufacturer. If these are not followed exactly, injury to you, the vehicle, and others can result.

Recharging Recharging the high-voltage battery pack is best done by the vehicle itself; however, there are times when it may be necessary to recharge the battery in the shop. Doing so is not a typical procedure. Chances are your shop will not have the correct charger. For example,

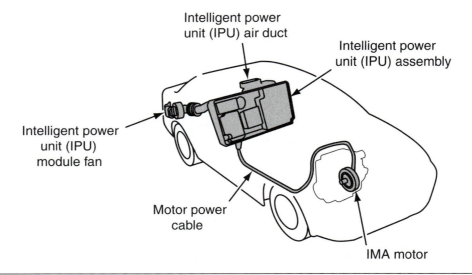

Figure 9-14 The battery pack and associated components in a Honda Civic.

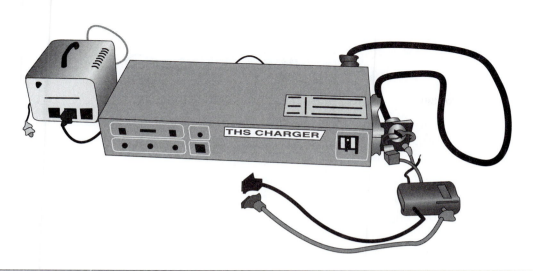

Figure 9-15 The high-voltage battery charger for Toyota hybrids. There are two output cables, one for the battery pack and the other operates the 12V system for the battery-cooling fans and computer.

Toyota hybrid batteries require a special charger that is not sold to its dealerships (Figure 9-15). If there is a need for one, the dealership must contact the regional office and have one delivered and only someone from that office is allowed to operate it. This charger has the normal connections plus a cable to power the battery's cooling system. The charger is designed to bring the battery pack to a 40-50% state-of-charge within three hours. This is enough to start the vehicle and allow the engine to bring the battery back to full charge.

Jump-Starting

If the vehicle will not start, several things can be the cause. Like conventional vehicles, the vehicle must have fuel, there must be ignition, intake air, compression, and exhaust. Before proceeding with a no-start diagnosis, make sure the immobilizing system is working properly. Most hybrid vehicles cannot be push- or pull-started. If the auxiliary or high-voltage battery is discharged, the engine will not start nor will the vehicle be able to operate on electric power only. Manufacturers have built in ways to jump-start these vehicles, if and when the batteries go dead. The basic connection from a booster battery to the dead battery is the same but the connecting points may be different and there are certain precautions to consider when jump-starting. There are also separate procedures for jump-starting with the low- and high-voltage systems. Some of the precautions are:

❏ Be sure the connections are done correctly and tightly.
❏ Make sure all electrical accessories of both vehicles are off.
❏ When making the connections, do not lean over the battery or accidentally let the jumper cables or clamps touch anything except the correct battery terminals.
❏ Use only a 12-volt supply as the booster battery.
❏ The gases around the battery can explode if exposed to flames, sparks, or lit cigarettes.

To jump-start a typical hybrid vehicle with the battery of another vehicle, follow this procedure:

1. Park the booster vehicle close to the hood of the disabled vehicle making sure the two vehicles do not touch.
2. Set the parking brake on both vehicles.
3. Turn off the hybrid system and remove the ignition key.
4. Open the hood or trunk to gain access to the auxiliary battery. On some hybrid vehicles with the battery in the trunk, there is a special jump-starting terminal under the hood.

CAUTION:
Never connect the end of the second cable to the negative terminal of the battery that is being jumped.

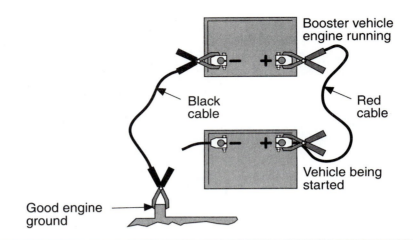

Booster vehicle
engine running

Black
cable

Red
cable

Vehicle being
started

Good engine
ground

Figure 9-16 The correct cable hook-up for jump-starting.

5. Connect the clamp of the positive (red) jumper cable to the positive terminal on the battery or the jump-starting terminal.
6. Connect the clamp at the other end of the positive (red) jumper cable to the positive (+) terminal on the booster battery.
7. Connect the clamp of the negative (black) jumper cable to the negative (-) terminal on the booster battery.
8. Connect the clamp at the other end of the negative (black) jumper cable to an exposed metal part of the dead vehicle's engine (Figure 9-16), away from the battery and the fuel injection system. Never use a fuel line, engine camshaft or rocker arm covers, or the intake manifold as the grounding point for making this negative connection.
9. Make sure the jumper cables are clear of fan blades, belts, moving parts of both engines, or any fuel delivery system parts.
10. Run the engine of the booster vehicle at a medium speed for about five minutes.
11. Now, attempt to start the hybrid vehicle.
12. If the hybrid vehicle did not start, check the connections of the jumper cables and if a problem was found, correct it and try again. If the engine still does not start, the low- or high-voltage battery should be replaced.
13. Once the disabled vehicle has started, allow both engines to run for about five minutes.
14. Then disconnect the negative cable from the hybrid vehicle, and then from the booster battery.
15. Next, disconnect the positive cable from the hybrid vehicle, and from the booster battery.

High-Voltage Batteries Some hybrid vehicles have separate procedures for jump-starting, one for the low-voltage battery and the other for the high-voltage battery. If the vehicle cranks but the engine does not start, the high-voltage battery may need to be jump-started. Also, the "Service Soon" lamp may be illuminated to indicate a problem.

To jump-start, or get the engine running, on a Ford Escape or Mariner Hybrid, make sure the ignition is OFF before proceeding. Then, open the access panel on the driver's side foot well and press the jump-start button (Figure 9-17). Wait at least eight minutes before continuing. This is important! If you continue sooner, the energy from the auxiliary battery will not be able to supply enough power to start the engine by the battery pack. When you depress the button, the

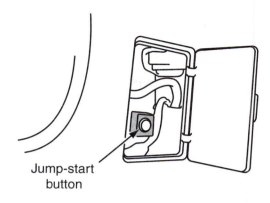

Jump-start
button

Figure 9-17 To jump-start a Ford Escape or Mariner Hybrid high-voltage battery, turn the ignition OFF, open the access panel on the driver's side foot well, and press the jump-start button.

system sends energy from the auxiliary battery to the battery pack. If the auxiliary battery has ample energy, it will be enough to start the engine. However, if the auxiliary battery is weak, it should be jump-started rather than the high-voltage battery pack. After pressing the button, you should wait eight minutes before attempting to start the engine; otherwise, the high-voltage battery may not receive sufficient charge to start the engine.

After the wait time, the warning lamp may blink for up to two minutes. After it stops, attempt to start the engine. If the engine still does not start, try again in couple of minutes. If the engine still does not start, the low-voltage battery must be recharged or the vehicle jump-started through the 12-volt battery.

An Atypical Service

Some early Prius models had a problem with electrolyte seepage from the battery that caused resistance at the positive terminals for the battery pack. This problem caused the illumination of the master and hybrid warning lamps. The technical service bulletin and recall notice issued for this problem included detailed steps for correcting this problem. Although there is no attempt to duplicate the procedure here, it is important to summarize the corrective steps. This is a good example of things that can go wrong with NiMH batteries. These batteries are relatively new for automotive technicians and require different thinking, as well as different service considerations.

The procedure involves resealing the battery pack, which requires the removal of the pack. Before beginning, check the system for any related electrical problems and correct these first. This check includes checking for DTCs, which will be discussed in greater detail later in this chapter. It also includes the de-powering of the high-voltage system and a thorough inspection of the hybrid system. Also, remember to adhere to all safety precautions. You should be wearing electrically insulated gloves when you de-power the system and should continue wearing them during this entire procedure. Once the high-voltage system has been isolated, make sure it is by checking the voltage at the cable terminals. There must be less than 12 volts present before it is safe to continue. If there is more, diagnose the problem!

To gain access to the battery pack, the carpeting, some panels, and the rear seatback need to be removed (Figure 9-18). Then remove the service plug if it has not been removed already. Make sure you are wearing insulated gloves that are in good condition. Double check to make sure there are no more than 12 volts at the power cable terminals, then disconnect them. Wrap the ends of the cable with electrical tape.

⚠ WARNING: This is just a summary of the procedure and should not be followed to correct the problem. ALWAYS follow the instructions given by the manufacturer.

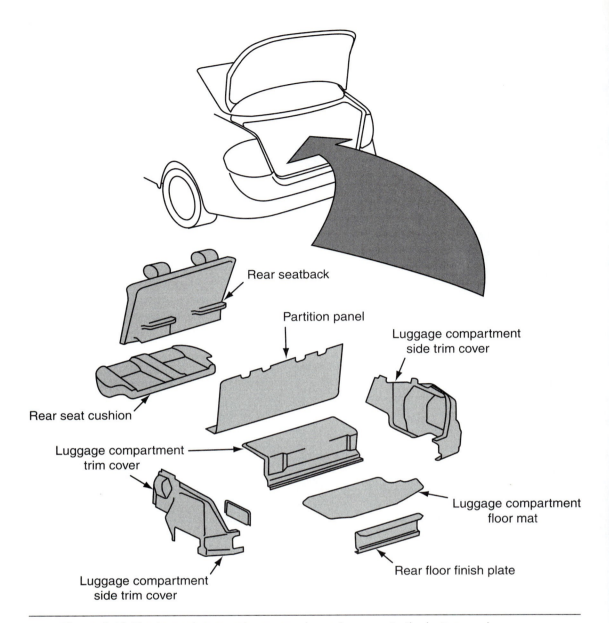

Figure 9-18 The items that must be removed to gain access to the battery pack.

Remove the battery brackets and disconnect the connectors from the battery's control unit, the SMRs, and blower motor controller. The male connector of the SMR should also be loosened from the cooling air duct, as should the wiring harness for the rest of the battery pack.

The entire cooling air ductwork system should be removed and placed aside. At this point, the battery pack can be unbolted and removed. This pack is heavy and you should seek help in lifting it out. Make sure your assistant is wearing the correct protective gear. As the battery pack is lifted from its cavity, support it on wood across the spare tire. This protects you, the battery pack, and the tire. It also provides a resting spot before the battery is moved to a working surface covered with a non-conductive material.

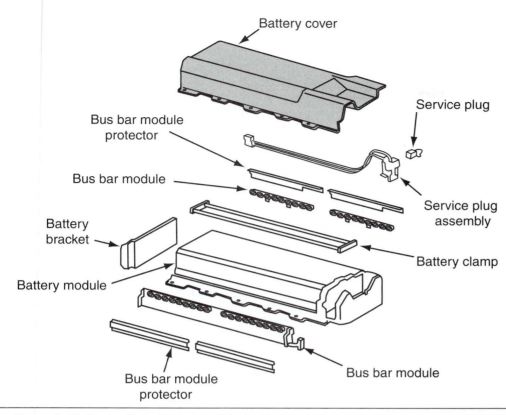

Battery cover

Bus bar module protector

Bus bar module

Battery bracket

Battery module

Service plug

Service plug assembly

Battery clamp

Bus bar module protector

Bus bar module

Figure 9-19 The basic components of the battery pack.

Once on the work surface, remove the remaining parts of the air ductwork and the vent hoses. Then remove the cover over the batteries (Figure 9-19). Then remove the insulators or protectors that are placed over the bus bars for the batteries. These protectors can be removed by carefully separating their retainers and lifting them off. Again (and this will be repeated several times), keep your gloves on throughout this procedure. Remove the plastic cover from the service plug assembly and the HV fuse.

Now, loosen and remove the retaining screws or bolts for the bus bar modules. Wrap the ends of your tools with insulating tape or use good insulated tools when doing this—never use power tools to loosen or tighten these screws. Once removed, cover the battery terminals with electrical tape. Then, clean the bus bars in a boric acid solution. Do not use a wire brush; in most cases a paper towel or a nylon wire will be able to completely clean the surface. If the modules cannot be cleaned, they should be replaced.

The seepage problem is corrected by placing resin plates between the cells of the battery and the case (Figure 9-20). To insert them, the thermistors in the case must be removed. It is important that these be reinstalled in their exact location, so mark their mounting point before removing them. The resin plates can now be installed. Also, Toyota has technicians place absorbent sheets below the positive terminals of the battery. The technical service bulleting

■ CAUTION:
Any residue on the side surfaces of the battery module is probably from the electrolyte. If this gets in your eyes, it may cause vision loss. If this happens, immediately flush your eyes and seek medical attention. If the residue comes in contact with your skin, it may burn or cause a serious skin irritation. Wash the exposed skin with large amounts of water.

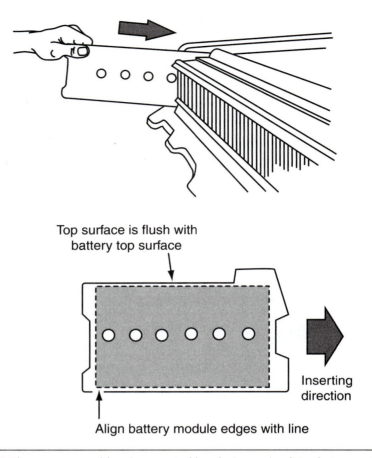

Top surface is flush with
battery top surface

Inserting
direction

Align battery module edges with line

Figure 9-20 The seepage problem is corrected by placing resin plates between the cells of the battery and the case.

details how the plates and sheets should be trimmed and placed; these details must be followed exactly to prevent further damage and to make sure the placement of these does not affect the operation of the hybrid system.

After installation, the battery pack is reassembled. New power cables are installed and the bus bar modules are installed with masking tape along the edges of the two bus bars. The masking tape serves as a limiter for the sealing compound that is applied next. If the masking tape is not correctly applied, the sealant will not fill the cavity behind the bus bar module, which would negate the efforts of doing this service.

The sealant is injected in the same way as other caulking, although specific points needs to be filled with the sealant. Toyota instructs that a hole be made in the masking tape so that the sealant can be injected into each port above the positive terminal of the battery pack. After this is done, all excess sealant needs to be immediately cleaned from the edges of the bus bar and other areas of the battery case. It is recommended that the sealant be dried with a hair dryer. If this is not possible, time should be allowed for the sealant to set. After the sealant has been applied, an ohmmeter should be used to verify that there is no continuity between the battery module's positive terminal and the battery case (Figure 9-21).

After the resin plates, absorbing sheets, and sealant have been installed or applied, the battery pack can be totally reassembled and installed back in the vehicle. After the high-voltage system is reactivated, and the vehicle turned on, the system should be checked for DTCs.

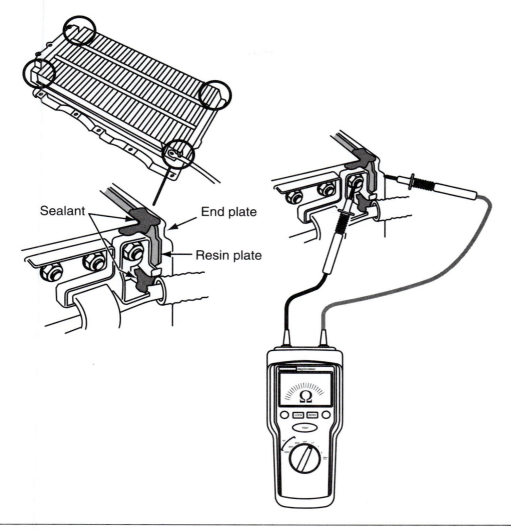

Sealant

End plate

Resin plate

Figure 9-21 After the sealant has been applied and dried, make sure the positive terminal is not shorted to the battery case.

Maintenance

Maintenance of a hybrid vehicle is much the same as a conventional vehicle. The manufacturers list the recommended service intervals in their service and owner's manuals. Nearly all of the items are typical of a conventional vehicle. Care needs to taken to avoid anything orange while carrying out the maintenance procedures.

The computer control systems are extremely complex, especially in assist and full hybrids, and are very sensitive to voltage changes. This is why the manufacturers recommend a thorough inspection of the auxiliary battery and connections every six months.

The engines used in hybrids are modified versions of engines found in other models offered by the manufacturer. Other than fluid checks and changes, there is little maintenance required on these engines. However, there is less freedom in deciding the types of fluids that can be used and the parts that can replace the original equipment. Hybrids are not very forgiving. Always use the exact replacement parts and the fluids specified by the manufacturer.

Typically, the weight of the engine oil used in a hybrid is very light. If the weight is increased, it is possible that the computer system will see this as a problem. This is simply caused by the extra current needed to turn over the engine. If the computer senses very high current draw while attempting to crank the engine, it will open the circuit in response.

Special coolants are required in most hybrids because the coolant not only cools the engine, but also cools the inverter assembly. Cooling the inverter is important, and checking its coolant condition and level is an additional check during preventative maintenance. The cooling systems used in some hybrids feature electric pumps and storage tanks. The tanks store heated coolant and can cause injury if you are not aware of how to carefully check them. The battery cooling system may need to be serviced at regular intervals. There is a filter in the ductwork from the outside of the vehicle to the battery box. This filter needs to be periodically changed. If the filter becomes plugged, the temperature of the battery will rise to dangerous levels. In fact, if the computer senses high temperatures it may shut down the system.

A normal part of preventative maintenance is the checking of power steering and brake fluids. The power steering systems used by the manufacturers varies, some have a belt-driven pump, some have an electrically driven pump, and other have a pure electric and mechanical steering gear. Each variety requires different care; therefore always check the service manual for the specific model before doing anything to these systems. Also, keep in mind that some hybrids use the power steering pump as the power booster for the brake system.

Hybrids are all about fuel economy and reduced emissions. Everything that would affect these should be checked on a regular basis. Items such as tires, brakes, and wheel alignment can have a negative effect, and owners of hybrids will notice the difference. These owners are constantly aware of their fuel mileage, due to the displays on the instrument panel.

Diagnostics

A hybrid vehicle is an automobile and as such is subject to many of the same problems as a conventional vehicle. Most systems in a hybrid vehicle are diagnosed in the same manner as well. However, a hybrid vehicle has unique systems that require special procedures and test equipment. It is imperative that you have good information before attempting to diagnose these vehicles. Also, make sure you follow all test procedures precisely as they are given.

Hybrids present unique considerations when they have a driveability problem. The problem can be caused by the hybrid system, engine, or transmission. Determining which system is at fault can be difficult. On some hybrids, it is possible to shut down the hybrid system and drive the vehicle solely by engine power. On others, such as Toyota and Ford hybrids, this is not possible. If electric power can be shut off and the vehicle still drives poorly, the problem is the engine or transmission. If it is not possible to shut down either power source, your diagnosis must be based on the symptom and information retrieved with a scan tool.

Gathering Information

Diagnosis of a hybrid vehicle should follow a logical approach. The first step is gathering as much information as possible from the customer about the concern. This is followed by a thorough inspection of the vehicle. A road test is then taken to verify the problem, as well as to define the problem. After having a good understanding of what is and is not working properly, all service bulletins that relate to the problem should be read and the appropriate procedures followed. If the cause is still unknown, specific tests should be conducted to pinpoint it. Once the cause is identified, repairs should be made and then the repair should be verified.

In order to properly diagnose a problem, you must totally understand the customer's concern or complaint. It is essential that you gather as much information from the customer as possible (Figure 9-22). Get a good description of the concern from the customer. Find out when the problem was first noticed and if the problem is evident now. Find out as much as you can about the conditions that exist when the problem occurs. The conditions to consider are weather conditions, vehicle load, city or highway driving, acceleration, coasting, and braking.

It is very important that you experience how a normal hybrid reacts to certain situations. There are many characteristics of a normal operating system that may seem strange and can be labeled as abnormal when it is actually a characteristic of the system. Start and drive a similar

Figure 9-22 Diagnostics should begin with getting as much information as you can from the customer.

model hybrid before making any conclusions about the vehicle you are diagnosing. This is also important to the customer, as most have never had a hybrid vehicle and their concerns may not be the result of a problem, but may be just part of the system!

Conducting a thorough inspection of the entire vehicle is extremely important. Inspect the battery and all related wiring and connectors. Make sure you are wearing lineman's gloves while inspecting the system, and make sure the high-voltage system is isolated. Check the wheels and tires of the vehicle. Make sure they are the correct size for the vehicle; a change in tire diameter can affect hybrid operation. Also, identify any and all non-factory installed electrical accessories. Make sure these have been properly installed. If these were incorrectly connected to the system or connected into a control module's harness, the system will not operate properly.

All warning lamps in the instrument panel should be checked (Figure 9-23). If any of these remain on after the engine is started, the cause should be identified and corrected before continuing with diagnosis. Lastly, a scan tool should be used to retrieve any fault codes held in the computers' memory. In many cases, a manufacturer-specific scan tool is required to test hybrids. Aftermarket scan tools may be able to retrieve codes and display some data, but may have limited capabilities. Also, follow the prescribed sequence for retrieving and responding to all diagnostic trouble codes (DTCs). Details about using a scan tool on the common hybrids are covered later in this chapter.

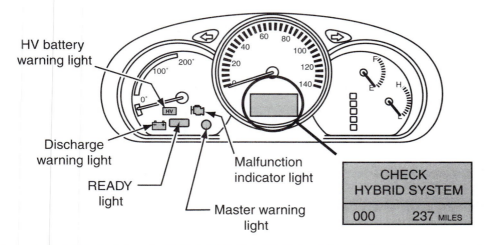

Figure 9-23 An example of some of the warning lights on a hybrid vehicle.

Test Equipment

An important diagnostic tool is a DVOM. However, this is not the same DVOM you use on a conventional vehicle. The meter used on hybrids (and EVs and FCEVs) should be classified as a category III meter. There are four categories for low-voltage electrical meters, each built for specific purposes and to meet certain standards. Low voltage, in this case, means voltages less than 1000 volts. The categories define how safe a meter is when measuring certain circuits. The standards for the various categories are defined by the American National Standards Institute (ANSI), the International Electrotechnical Commission (IEC), and the Canadian Standards Association (CSA). A CAT III meter is required for testing hybrid vehicles because of the high voltages, three-phase current, and the potential for high-transient voltages. Transient voltages are voltage surges or spikes that occur in AC circuits. To be safe, you should have a CAT III-1000 V meter. Within a particular category, meters have different voltage ratings. These reflect a meter's ability to withstand higher transient voltages. Therefore, a CAT III 1000 V meter offers much more protection than a CAT III meter rated at 600 volts.

Another tool that will save much time and effort during diagnosis is an **insulation resistance tester**. These meters can check for voltage leakage from the insulation of the high-voltage cables. Obviously no leakage is desired and any leakage can cause a safety hazard as well as damage to the vehicle. Minor leakage can also cause hybrid system-related driveability problems. This meter is not one commonly used by automotive technicians, but should be for anyone who might service a damaged hybrid vehicle, such as doing body repair. This should also be a CAT III meter, for the same reasons as the DVOM. In fact, these meters often have the capability of checking resistance and voltage of circuits, like a DVOM.

To measure insulation resistance, system voltage is selected at the meter, and the probes placed at their test position (Figure 9-24). The meter will display the voltage it detects. Normally, resistance readings are taken with the circuit deenergized and this is true for resistance checks

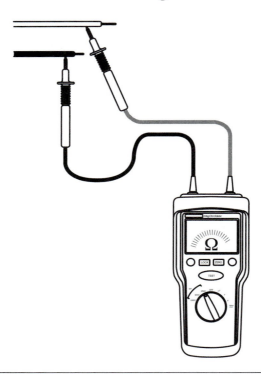

Figure 9-24 Connecting a MegOhmMeter to check the resistance of the wire insulation in a circuit.

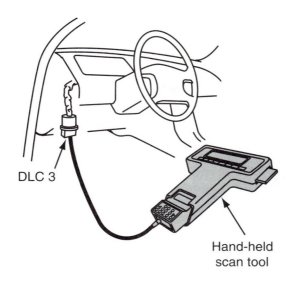

DLC 3

Hand-held
scan tool

Figure 9-25 The scan tool should be connected to the DLC.

with this meter unless you are checking the effectiveness of the cable or wire insulation. In this case, the meter is measuring the insulation's effectiveness and not its resistance.

The probes for the meters should have safety ridges or finger positioners. These help prevent physical contact between your fingertips and the meter's test leads.

Toyota Hybrids

Diagnosis of Toyota hybrids depends on common sense, and of course, what the control modules tell you. After the customer interview, visual inspection, and review of information, Toyota instructs you to connect the scan tool to DLC3, which is the diagnostic connector that ties into the multiplex system (Figure 9-25). You should record all codes as they appear. If the codes are hybrid system related, they should be divided by their power source. All engine- and emission-related DTCs should be handled first, and then move on to the other systems.

Toyota's instructions continue with a test drive. Starting or driving a hybrid is much different than driving a conventional vehicle. Remember, when the car is ready to be driven, the engine may not be running, all that happens is the "READY" lamp will be lit. If the "READY" lamp does not come on, the vehicle cannot be driven. Typical causes for this are:

❏ The service plug is removed.
❏ The inverter unit cover is not closed securely.
❏ There is excessive current in the inverter circuit.
❏ There is a problem with the hybrid system
❏ One or more electric motors/generators are drawing excessive current.
❏ One or more of the impact sensors have been tripped

During the test drive, note the operation of the vehicle at different speeds and conditions, and pay attention to the speed of the vehicle, electric motors, and engine. The recommended test speeds are acceleration to 20 mph, and then steady speeds of 35 and 45 mph. During each of the transitions between speeds, the system should be monitored to identify what is providing the propulsion power and what the SOC of the battery is. Fortunately, Toyota has a test drive sheet available that will help you identify what needs to be observed and recorded. The basic premise of the test drive is to determine what system may be causing the problem.

Some of the DTCs retrieved with the scan tool will also display three-digit information codes that will help your diagnosis (Figure 9-26). They provide additional information and freeze frame data. As an example, the DTC-P0A4B indicates there is a problem with the generator position

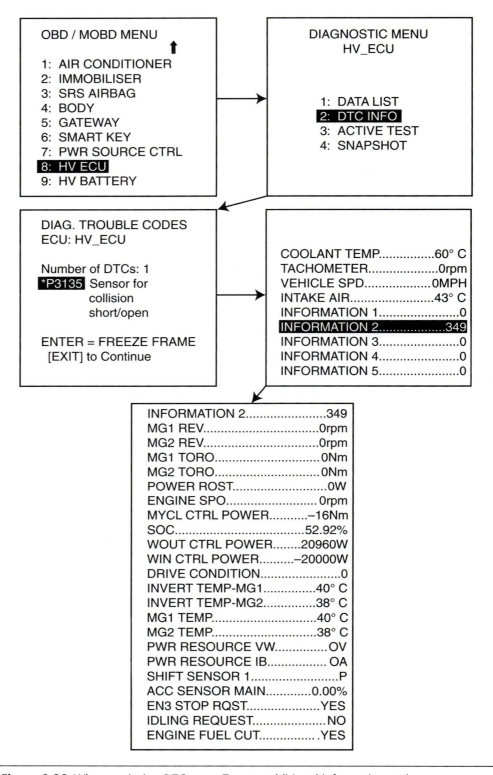

Figure 9-26 When retrieving DTCs on a Toyota, additional information codes may appear that will aid in diagnostics. *Courtesy of Toyota Motor Company*

(resolver) sensor circuit. To help identify the problem, three information codes are also displayed on the scan tool: code 253 (Interphase short in resolver circuit), 513 (Resolver output is out of range), and 255 (Open or short in resolver circuit). These codes further define the problem, as well as give guidance as to what should be checked. All DTCs and information codes should be interpreted according to the charts given in the service manual. Each code has specific procedures for further testing; these should be followed.

The scan tool also allows for some active tests that enable you to excite or disable certain outputs so their operation can be monitored. These "inspection modes" can crank the engine to conduct a compression test, turn the traction control on and off, and turn the inverter on and off. The value of these modes is the ability to isolate systems, which will definitely help in diagnosis.

Honda Hybrids

Honda hybrids are more like a conventional vehicle in that they have a normal transmission. Driveability problems are caused by either the hybrid system or the engine. If there is a problem with the hybrid system, the IMA warning lamp will be lit (Figure 9-27). When the lamp is lit, that system should be checked first. Remember to take all necessary precautions when working around the high-voltage system.

If the IMA warning lamp stays on when the engine is running, turn the ignition off. Then connect the scan tool to the DLC. Once it is securely connected, turn the ignition on. Select the IMA system on the scan tool and record all DTCs. Also, check the freeze frame data to get a complete picture of what was happening when the computer noted the fault.

Using the DTC chart in the service manual, match the code with the appropriate troubleshooting procedure. Make sure to follow all sequences exactly as they are presented. If the code was set by an intermittent fault, you may not be able to identify it. Therefore, the codes should be cleared and the vehicle driven to see if the codes are reset. Codes can be cleared through the commands of the scan tool or by removing the No. 9 Backup fuse from the underhood fuse/relay box for 60 seconds. Using the latter technique will cause the IMA battery level indicator to show zero bars. To bring a voltage reading back to the meter, start the engine and run it at about 4000 rpm until at least three or four bars are displayed on the gauge.

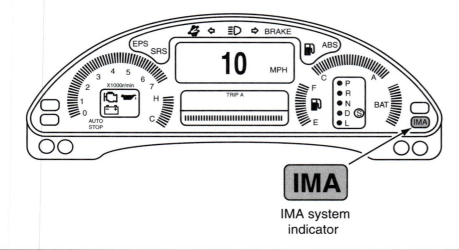

Figure 9-27 When the IMA warning lamp remains lit after the engine is started, the control system has detected a problem.

Prior to the road test, perform a wiggle test on all connectors to see if this resets the codes. Make sure you are wearing lineman's gloves when moving wires and connectors around.

Once diagnostics are complete, clear any existing codes from the memory. Then turn off the ignition and disconnect the scan tool. Again, the IMA battery gauge will need to be reset, so start the engine and run it until at least three bars appear on the gauge. Once this is complete, the PCM must re-learn the idle strategy if the battery was disconnected or the codes were cleared by any means other than the scan tool.

The procedure for conducting "PCM idle learn" begins with turning off all accessories. Then start the engine and run it at about 3000 rpm until the electric cooling fan comes on. Then the engine should idle for at least five minutes before it is turned off.

The scan tool is also used to reset or calibrate the electric motor's rotor. This must be done whenever the following components have been removed, and/or replaced:

❏ Motor Control Module (MCM)
❏ Motor rotor position sensor
❏ Stator or rotor of the IMA motor
❏ Engine
❏ Transmission

To calibrate the rotor position sensor, connect the scan tool with the ignition off. Then turn the ignition on. Select the IMA System on the scan tool. Then scroll the adjustment menu and select Motor Rotor Position Calibration. The sensor should now be reset. Turn the ignition off and disconnect the scan tool.

Ford hybrids require specific troubleshooting procedures that must be followed. Like other hybrid models, these hybrids are diagnosed primarily with a scan tool. The DTCs will lead to a general area where additional tests should be made. The basic diagnostic procedure begins with gathering information from the customer to define the problem. The vehicle should then be taken on a test drive to verify the customer's concerns. The concerns should then be matched to any related technical service bulletins, and the action prescribed by these should be followed.

Diagnostics continue with a thorough visual inspection of the mechanical and electrical systems. This inspection should include the transmission, battery pack and its components, the jump-start switch, and junction and fuse boxes. Make sure you are wearing lineman's gloves, and follow all manufacturer specific precautions. If you discover a likely cause for the customer's concerns, correct the problem before continuing.

Check the auxiliary battery; it should have at least 12 volts. If it does not, identify and correct that problem before continuing. If the battery is okay, make sure the ignition is off and connect the scan tool to the DLC. Record all DTCs (Figure 9-28) and match them to the pinpoint tests given in the service manual. All engine-related DTCs should be corrected first. Follow the steps given for each test exactly as they are listed. Failure to do so may lead to a misdiagnosis or damage to the electrical circuit. If no DTCs related to the customer's concerns are retrieved, diagnose the problem with the symptom charts given in the service manual (Figure 9-29).

B1016	Jump Start Control Module Fault
B1143	Excessive Battery Contractor Close Requests
B1239	Air Flow Blend Door Driver Circuit Failure
B1342	ECU is faulted
B2950	Air Conditioning System Fault
C1862	Contractor Circuit Failure
P0535	A/C Evaporator Temperature Sensor Circuit
P0A0A	High Voltage System Inter-Lock Circuit
P0A1F	Battery Energy Control Module
P0A27	Hybrid Battery Power Off Circuit
P0A7D	Hybrid Battery Pack State Of Change Low
P0A7E	Hybrid Battery Pack Over-Temperature
P0A80	Replace Hybrid Battery Pack (End of Useful Life)
P0A81	Hybrid Battery Pack Cooling Fan 1 Control Circuit
P0A8B	14 Volt Power Module System Voltage
P0A8D	14 Volt Power Module System Voltage Low
P0A8E	14 Volt Power Module System Voltage High
P0A95	High Voltage Fuse
P0A96	Hybrid Battery Pack Cooling Fan 2 Control Circuit
P0A9B	Hybrid Battery Temperature Sensor Circuit
P0AA6	Hybrid Battery Voltage System Isolation Fault
P0AA7	Hybrid Battery Voltage Isolation Sensor Circuit
P0AAC	Hybrid Battery Pack Air Temperature Sensor A Circuit
P0AB1	Hybrid Battery Pack Air Temperature Sensor B Circuit
P0ABF	Hybrid Battery Pack Current Sensor Circuit Open
P0AC0	Hybrid Battery Pack Current Sensor Circuit Range/Performance
P0AE1	Hybrid Battery Precharge Contractor Circuit
P2533	Ignition Switch Run/Start Position Circuit
P2612	A/C Refrigerant Distribution Valve Control Circuit Low
P2613	A/C Refrigerant Distribution Valve Control Circuit High
U0001	High Speed CAN Communication Bus
U0073	Control Module Communication Bus Off
U0100	Lost Communication With PCM
U0101	Lost Communication with eCVT Transmission Control Module (TCM)
U0300	Internal Control Module Software Incompatibility
U0401	Invalid Data Received From PCM
U0402	Invalid Data Received From TCM

Figure 9-28 A list of hybrid-related DTCs for a Ford Escape and Mercury Mariner Hybrid.

Condition	Possible Sources
No communication with the traction battery control module (TBCM)	Fuse(s) Circuitry
The charging system warning indicator is on with the engine running and the charging system voltage does not increase — battery icon ON, low-voltage charging system at the DC/DC converter	Electronically controlled continuously variable transmission (eCVT) DC/DC converter interlock circuit(s) High-voltage cable(s) Traction battery control module (TBCM) Powertrain control module (PCM)
The charging system warning indicator is on with the engine running and the charging system voltage does not increase — battery icon ON	DC/DC converter interlock circuit(s) High-voltage cable(s) Low-voltage cable(s) 12-volt battery DC/DC converter
The charging system warning indicator is on with the engine running and the charging system voltage does not increase — battery icon ON	Instrument cluster Climate control system Traction battery control module (TBCM) Powertrain control module (PCM)
The charging system warning indicator is off with the ignition switch in the RUN position and the engine is off	Instrument cluster
The high-voltage traction battery (HVTB) is noisy	HVTB internal cooling fans (part of the HVTB) HVTB Air leakage from the climate control system
Radio interference	Audio unit Generator (part of eCVT) DC/DC converter
The low-voltage battery is discharged or battery voltage is low	12-volt battery 12-volt cables DC/DC converter

Figure 9-29 A typical symptom chart that should be used when diagnosing a Ford Escape and Mercury Mariner Hybrid.

■ CAUTION:
Always follow the precautions given by the manufacturer when conducting a compression test or other engine-related tests. Also, wear the proper protective gear when working near the high-voltage system.

Engines

Most engines used in hybrid vehicles are based on an engine used in other model vehicles, however they are usually modified to increase fuel mileage and reduce emissions. Testing these engines is much the same as doing so in conventional vehicles, however, because any time the engine rotates it may be rotating the generator, there is a potential safety hazard. Keep this mind when conducting a compression test on the engine. In most hybrids, the engine is cranked by a high-voltage motor. Because this motor is required to run the test, the high-voltage system cannot be isolated. Therefore, extreme care must be taken.

The procedure for conducting a compression test on Honda hybrids is much the same as done on a conventional vehicle. Before beginning the test, use the scan tool to shut off the fuel injection and ignition systems. To run a compression test on a Honda Hybrid:

1. Warm up the engine to normal operating temperature.
2. Turn the ignition switch OFF.
3. Connect the scan tool to the DLC.
4. Following the menu of the scan tool, turn off all fuel injectors.
5. Disconnect and remove all of the ignition coils.
6. Remove all of the spark plugs.
7. Tighten the hose for the compression tester into a spark plug bore (Figure 9-30).
8. Open the throttle to its wide-open position and hold it there.
9. Crank the engine with the starter motor.
10. Record the compression readings on the gauge.
11. Repeat the process at each cylinder.

The results of the compression test can be interpreted in the same manner as a conventional engine. The pressure should be greater than 135 psi (930 kPa) and there should less than 28 psi (200 kPa) difference between cylinders. Once this test is completed and all parts reinstalled, the PCM must be reset. If this step is not done, the injectors will not work.

Ford and Toyota hybrids use Atkinson cycle engines. If you recall, these engines delay the closing of the intake valve. As a result, the overall compression ratio and displacement of the engine is reduced. Therefore, when conducting a compression test on these engines, expect a slightly lower reading than what you would expect from a conventional engine. Intake manifold vacuum will also be lower. Normally a healthy engine produces at least 17 in/Hg (57 kPa) during cranking. The expected reading from an Atkinson cycle engine is at least 15 in/Hg (51 kPa).

To conduct a compression test on a Ford Escape, you must use a scan tool, and the one from Ford is preferred. The scan tool allows you to enter into the engine cranking diagnostic mode. This mode allows the engine to crank with the fuel injection system disabled. It also

Figure 9-30 A compression tester connected to a cylinder.

makes sure the starter motor/generator is not activated (except for activating the starter motor to crank the engine), which is not only good for safety purposes, it is also good because the load of the generator cannot affect the test results because it is not energized. Always follow the sequence as it is stated in the service manual. Failure to do so will result in bad readings. NOTE: If the battery pack's has a SOC of less than 35%, the engine will not crank for the test.

1. Apply the parking brake and start the engine. Allow it to run until it reaches normal operating temperature.
2. Turn the ignition off.
3. Connect the scan tool.
4. Remove all spark plugs.
5. Install the hose for the compression tester in cylinder #1.
6. Set the scan tool to the cranking diagnostic check.
7. Place the gear selector in the PARK position.
8. Depress and hold the brake pedal.
9. Move the ignition to the on position but do not start the engine.
10. Within 5 seconds, press the throttle pedal to the floor and hold it there for ten seconds.
11. Release the pedal, shift the transmission into NEUTRAL, and fully depress the throttle pedal again.
12. Hold the throttle pedal down for ten seconds.
13. Release the pedal and move the gear selector to the PARK position. If this sequence is done properly, the hazard indicator on the dash will flash once per second.
14. Turn the ignition to the START position, crank the engine through a minimum of five compression strokes, and record the highest reading.
15. Put the ignition switch back into the ON position.
16. Release the accelerator and brake pedals.
17. Repeat the sequence on each cylinder and record the results. Make sure the engine is cranked through five compression strokes.
18. Position the key to the OFF position to deactivate the cranking diagnostic mode.
19. Clear all DTCs.
20. Compare your results with the specifications.

Cooling Systems

Normal maintenance and service to engine cooling systems in hybrid vehicles can become a hazardous, frustrating event. For example, late-model Toyotas have a system that heats a cold engine with retained hot coolant to provide reduced emissions levels. This is a great idea but can cause problems for technicians servicing the cooling system.

Hot coolant is stored in a container (Figure 9-31) and it will circulate through the engine immediately after start-up. The fluid also may circulate through the engine many hours after it

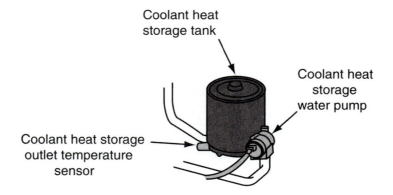

Figure 9-31 The hot coolant storage tank for Toyota hybrids.

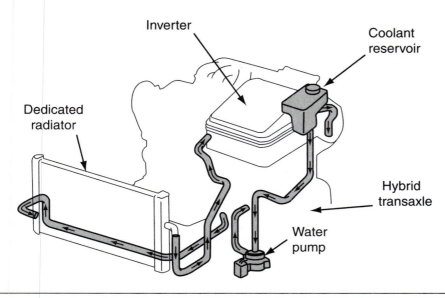

Figure 9-32 Air is easily trapped after servicing the cooling system because of the route the coolant takes, therefore, bleeding the system may be necessary.

was shut off. This fluid is under pressure and can cause serious burns to anyone who opens the system for inspection and/or repairs. To safely service the cooling system, the pump for the storage tank must be disconnected. The cooling system also is tied into the inverter assembly. This also presents a potential problem, as it is easy to trap air in the cooling system due to the path of coolant flow (Figure 9-32). To purge the system of air, there is a bleeder screw and a scan tool is used to run the electrical pump. The recommended procedure for draining and refilling the cooling system on a Toyota hybrid includes the following steps.

1. Remove the radiator's top cover and cap.
2. Disconnect the connector for the coolant heat storage tank's water pump to prevent circulation of the coolant. (To gain access to the connector, either remove the left headlamp assembly or pull down on the front portion of the left front fender liner).
3. Connect a drain hose to the drain port on the bottom of the coolant heat storage tank, and then loosen the yellow drain plug on the tank.
4. Connect a drain hose to the drain port on the lower left corner of the radiator, and then loosen the yellow drain plug on the radiator.
5. Connect a drain hose to the drain port on the rear of the engine and loosen the drain plug.
6. After the coolant is drained, tighten the three drain plugs.
7. Reconnect the connector to the coolant heat storage tank's water pump.
8. Connect a hose to the radiator's bleeder valve port and place the other end of the hose into the coolant reservoir tank.
9. Loosen the radiator's bleeder plug.
10. Fill the radiator with the specified coolant.
11. Tighten the radiator's bleeder plug and install the radiator cap.
12. Connect the scan tool to DLC3.
13. Using the scan tool, run the water pump for the storage tank for 30 seconds.
14. Then, loosen the radiator's bleeder plug.
15. Remove the radiator cap and top off the coolant in the radiator.
16. Repeat the refilling and bleeding sequence as often as necessary. Normally, when no additional coolant is needed after the sequence, the system is bled.
17. Start the engine and allow it to run for one to two minutes.
18. Turn off the engine and top off the fluid, if necessary.

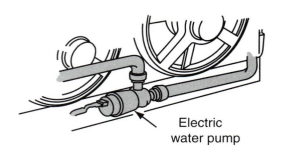

Electric
water pump

Figure 9-33 The electric water pump for the M/E cooling system on a Ford Escape Hybrid.

Ford hybrids have two separate cooling systems, one is for engine cooling and the other is for hybrid system components, called the **Motor Electronics (M/E) cooling system**. The engine cooling system is conventional. The M/E cooling system uses an electric water pump (Figure 9-33) to move coolant through the inverter, transmission, and a separate radiator mounted next to the conventional radiator (Figure 9-34). The M/E coolant reservoir is located behind the engine coolant reservoir. Although the two systems operate similarly, the M/E cooling system typically operates at lower pressures and temperatures. The fluid levels in both cooling systems must be maintained.

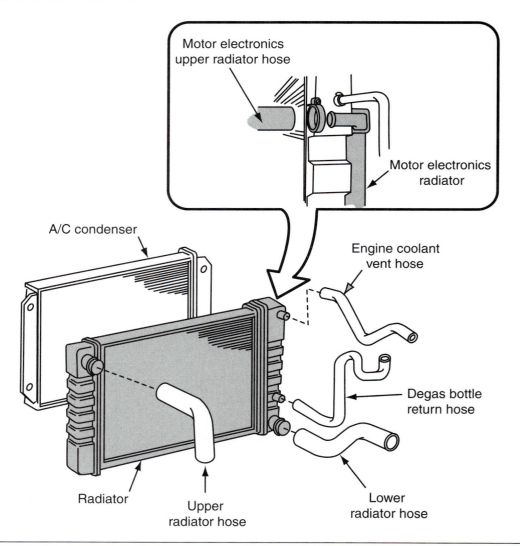

Figure 9-34 The radiator assembly for a Ford Escape Hybrid.

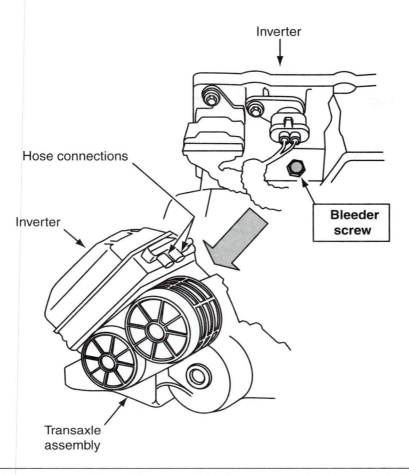

Figure 9-35 Location of the cooling system bleeder screw on a Ford Escape Hybrid.

It is easy to trap air in the M/E cooling system when filling and/or flushing the system. The system is fitted with a bleeder screw at the top of the inverter (Figure 9-35). When servicing the system, make sure the high-voltage system is isolated by having the service connector in the SERVICING/SHIPPING position. Also, wear lineman's gloves because the bleeder screw is very close to the high-voltage cables.

To drain, refill, and bleed the system, put the vehicle on a hoist. Remove the splash shield from the left front of the vehicle. Then loosen the hose clamps and remove the coolant hoses at the transaxle and allow the coolant to drain into a catch can. Once the fluid is drained, reinstall the hoses and tighten the clamps. Reinstall the splash shield and lower the vehicle. Loosen the bleeder screw and pour the specified type of coolant into the coolant (degas) reservoir. Fill the reservoir until coolant starts to leak from the bleeder screw, then tighten the bleeder screw. Turn the ignition on to allow the M/E water pump to run. Add coolant to the reservoir as the level drops. Once the level stays at the FULL mark, loosen the bleeder screw slightly to allow any air to escape. Tighten the bleeder screw and bring the coolant level back to the correct level.

Battery Cooling System Filter The battery has its own cooling or air conditioning system. The control module monitors the temperature of the cells and activates fans and/or the rear air conditioning system when the temperature rises. The battery pack's cooling system draws in outside air from a vent built into the rear side window. Within the ductwork there is an air filter that requires periodic replacement (every six months in normal conditions). If the filter is dirty or restricts airflow, the battery can overheat, which may cause the hybrid system to shut down. To inspect and/or change the filter, remove the access panel in rear trim panel on the driver's side

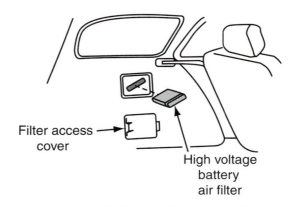

Filter access → cover

High voltage battery air filter

Figure 9-36 Changing the high-voltage battery air filter element on a Ford Escape Hybrid.

of the vehicle. The cover for the filter is retained by small tabs. To remove the cover, push on the tabs while pulling on the cover (Figure 9-36). Remove the filter. When installing the filter, make sure it is positioned correctly. Then reinstall the filter's cover and the access panel.

Transmission

Transmission service in a hybrid vehicle is no different from that in a conventional vehicle for about half of the hybrids on the road. Honda's use conventional-type transmissions and many GM hybrids do also. The only real difference is that the flywheels on these hybrids bolt to the electric motor's rotor. This means the high-voltage systems must be isolated prior to performing any transmission service. In addition, because the rotor is a strong permanent magnet, all precautions regarding the rotor must be adhered to (Figure 9-37).

The hybrids made by Toyota and Ford, as well as the two-mode transmissions used in some GM and Chrysler hybrids, are unique units. The only services that can be done to these transmissions are fluid checks, fluid changes, and replacing the entire unit. The required precise position of the motors in a two-mode transmission make them very unlikely candidates to be rebuilt in a regular service department, ever.

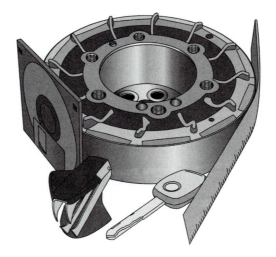

Figure 9-37 The permanent magnet rotor in the IMA is very strong. All iron, magnetic, and electronic materials must be kept away from the rotor. People with pacemakers should not handle an IMA rotor.

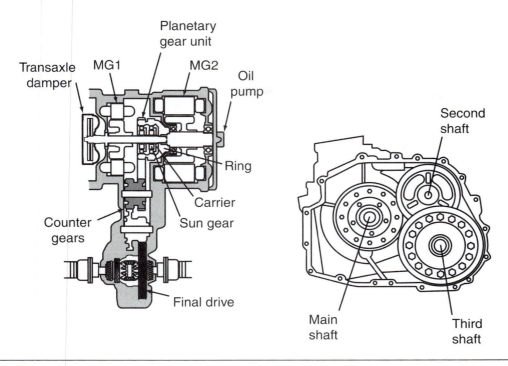

Figure 9-38 A compound gear set-equipped Toyota transaxle.

The transaxles used in Toyota and Ford hybrids are also not rebuildable and are replaced as units if they fail. These transaxles contain a single or compound planetary gear set and two motors/generators (Figure 9-38). Some designs have an additional set of reduction gears in the housing. The transaxles not only provide different forward drive ratios and a reverse, they also blend the torque output from the engine and the traction motor. The transaxles are CVTs, but are unlike CVTs used in conventional vehicles.

These CVTs can be diagnosed with a scan tool, which can retrieve codes pertaining to the transaxle. Keep in mind, problems with the motors and engine can also cause the transaxle not to operate properly. Normal diagnostic procedures should begin with a customer interview, verifying the customer's concerns, researching all service information, and conducting a thorough inspection of vehicle, especially the components of the hybrid system.

Part of the inspection should include checking the condition of the auxiliary battery and observing all warning light activity. If the cause of the concern is not found during these checks, the scan tool should be connected to retrieve all DTCs. All codes that are directly related to emissions should be diagnosed first. Match all codes with the appropriate chart, and follow the specific procedures for identifying the exact cause of the problem. If no DTCs are retrieved, diagnostics should continue according to the symptoms of the problem.

The rear transaxle assemblies in 4WD Toyota hybrids are also diagnosed with the scan tool.

The fluid level in all transmissions and transaxles must be periodically checked. Many designs also require periodic fluid and filter changes. To check the fluid on an automatic transmission equipped with a dipstick, place the vehicle on a level surface. Wipe all dirt off the filler or dipstick tube and the dipstick handle, before removing the dipstick. Most often, the fluid level can only be accurately checked when the transmission is at operating temperature. Most manufacturers recommend running the engine while checking the fluid level. Always refer to the service manual to identify the correct procedure. Also, make sure that the parking brake is engaged and take all necessary safety precautions while working under the hood.

273

Remove the dipstick and wipe it clean with a lint-free cloth or paper towel. Reinsert the dipstick, remove it again, and note the reading. Markings on a dipstick indicate ADD levels, and on some models, FULL levels for cool, warm, or hot fluid. If the fluid level is low and/or off the crosshatch section of the dipstick, the problem could be external fluid leaks. Check the transmission case, oil pan, and cooler lines for evidence of leaks.

Low fluid levels can cause a variety of problems. Air can be drawn into the oil pump's inlet circuit and mix with the fluid. This will result in aerated fluid, which causes slow pressure build-up, and low pressures, which will cause slippage between shifts or delayed shifts.

Excessively high fluid levels can also cause aeration. As the planetary gears rotate in high fluid levels, air can be forced into the fluid. Aerated fluid can foam, overheat, and oxidize. All of these problems can interfere with normal valve, clutch, and servo operation. Foaming may be evident by fluid leakage from the transmission's vent.

The condition of the fluid should be carefully examined while checking the fluid level. The normal color of automatic transmission fluid (ATF) is pink or red. If the fluid has a dark brownish or blackish color and/or a burned odor, the fluid has been overheated. A milky color indicates that engine coolant has been leaking into the transmission's cooler in the radiator. If there is any question about the condition of the fluid, drain out a sample for closer inspection.

After checking the ATF level and color, wipe the dipstick on absorbent white paper and look at the stain left by the fluid. Dark particles are normally band and/or clutch material, whereas silvery metal particles are caused by the wearing of the transmission's metal parts. If the dipstick cannot be wiped clean, it is probably covered with varnish, caused by fluid oxidation. Varnish will cause the spool valves to stick, causing improper shifting speeds. Varnish or other heavy deposits indicate the need to change the transmission's fluid and filter. Contaminated fluid can sometimes be felt better than be seen. Place a few drops of fluid between two fingers and rub them together. If the fluid feels dirty or gritty, it is contaminated with burned frictional material.

The fluid level of most CVTs (including those used in Toyota and Ford hybrids), manual transmissions, and the rear transaxle in 4WD Toyota hybrids is checked by removing the filler plug on the side of the transaxle. Each manufacturer specifies a distance below the filler plug bore that the fluid level should be at (Figure 9-39). If the level is low, fluid should be added. However, before adding any fluid, make sure it is the type specified for that transmission. If internal damage is suspected, the fluid from these transmissions should be drained through an absorbent white paper into a catch can. The paper should then be carefully inspected for residue.

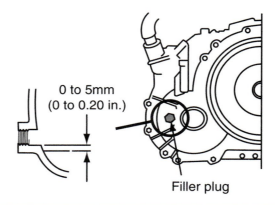

0 to 5mm
(0 to 0.20 in.)

Filler plug

Figure 9-39 Each manufacturer specifies a distance below the filler plug bore where the fluid level should be.

Brakes

The brake systems used in hybrids are quite conventional. The most noticeable difference is the regulation of the hydraulic system to allow for regenerative braking. All this means is the action of the normal brake system is delayed to allow the generators to work. Basically, the control module determines how much regenerative braking there should be and how much hydraulic pressure is needed to stop the vehicle. For the most part, there are no physical changes to the hydraulic brake system, except for an electronic control that interrupts the flow of pressurized fluid from the brake master cylinder to the wheel units (Figure 9-40).

The brake system in Ford hybrids has a feature that can become a safety hazard to the unknowing. The system basically checks for leaks by applying the brakes without being commanded to so by the driver. The hydraulic control unit (HCU) controls the brake force distribution and the anti-lock brake system. The HCU has an accumulator and an electric pump (Figure 9-41). The pump charges the accumulator when the brakes are applied, and whenever the vehicle's doors are opened, the ignition switch is turned on and the brake pedal is depressed. The latter conditions put the system into a self-test. During this time, brake fluid is drawn from the master cylinder and sent to the wheel units. The pressure in the system is monitored. If there is a leak, the HCU will sense the drop in pressure and turn on the appropriate warning light. If there is no leak after four minutes, the fluid is returned to the master cylinder. To prevent an unwanted application of the brakes while working on the brakes, the auxiliary battery should be disconnected and two fuses (numbers 24 and 31) removed from the fuse box under the hood.

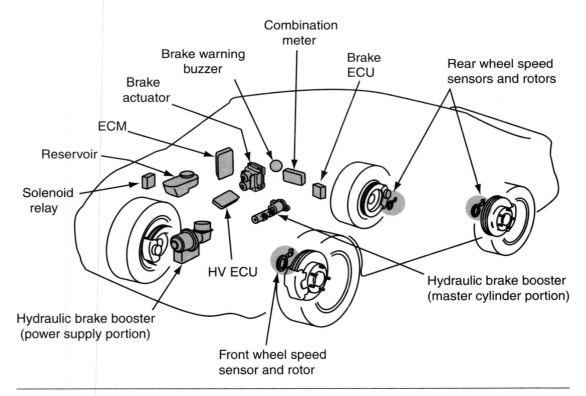

Figure 9-40 The major components in the brake system for an early Prius.

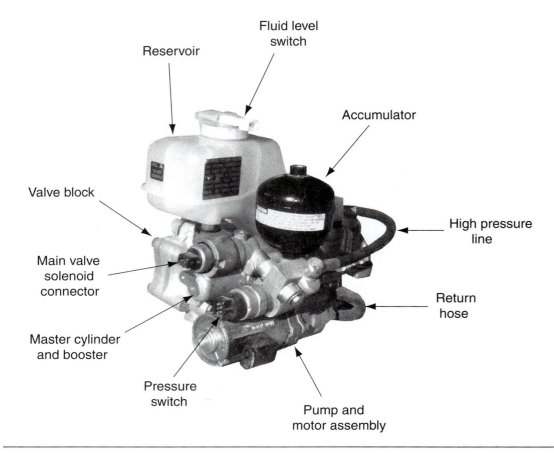

Reservoir

Fluid level switch

Accumulator

Valve block

High pressure line

Main valve solenoid connector

Return hose

Master cylinder and booster

Pressure switch

Pump and motor assembly

Figure 9-41 A typical electronically controlled hydraulic control unit.

Electronically Controlled Brake System

Toyota's electronically controlled brake (ECB) system was introduced on late-model hybrids along with the introduction of more powerful motors/generators. Because these can provide more propulsion power, they are also very efficient at capturing energy during regenerative braking. The ECB system (Figure 9-42) controls power brake assist, ABS, traction control, and on some models, vehicle stability control. The ECB calculates the required braking force based on the amount of effort and force applied to the brake pedal. The ECB then appropriately activates the regenerative brakes and the hydraulic brakes.

The system does not use a conventional brake power booster. Also, during normal braking, the fluid pressure developed in the brake master cylinder does not go directly to the wheel units. The pressure is applied to the brake actuator, which controls and sends the pressure to the wheels. The brake actuator houses an accumulator, electric pump, solenoids, and several valves. The pump supplies pressurized fluid to the accumulator, which supplies pressure to the hydraulic brake system. Brake assist is provided by the pressure in the accumulator, which varies according to current conditions, the required braking force, and driver demands.

On late-model hybrids, Toyota has added a power source backup unit (Figure 9-43) to allow the ECB system to operate when the auxiliary battery fails. This backup power source is only designed to allow the vehicle to stop, and cannot provide continuous power. The backup unit is made up of individual capacitors. The capacitors discharge when the ignition switch is turned OFF. If you need to replace this unit, use a voltmeter to see if it is discharged before handling it.

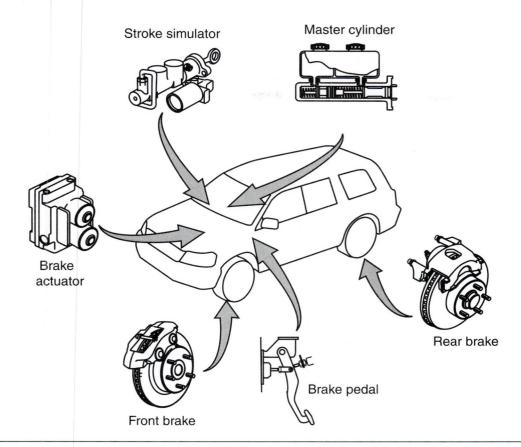

Figure 9-42 The main hydraulic components in Toyota's ECB system.

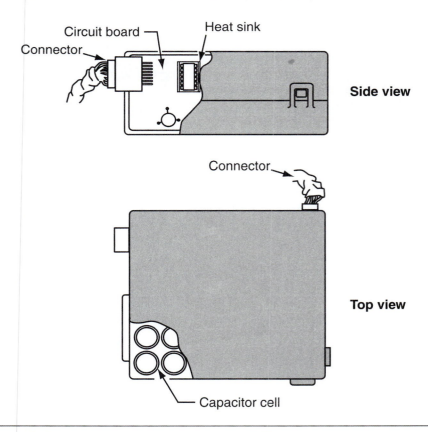

Figure 9-43 The power source backup unit for the ECB system.

The system is diagnosed with a scan tool. It is important to understand that this electronic system controls the action of the hydraulic brake system. Therefore, when the vehicle has a brake problem, diagnosis should lead to defining the problem as being electrical or hydraulic.

Bleeding The system must be disabled whenever the hydraulic brake system needs to be bled and when other brake work is being done. The system is disabled by removing the system's two relays from the junction/fuse box or by using the scan tool to shut the system down. Check the service manual for the correct procedure.

Steering

Most hybrid vehicles have electric power steering (EPS). These systems do not have a power steering pump and require no fluid services. A 12-volt motor (Figure 9-44) is fit into the steering column and provides steering assist according to commands given by the EPS control unit. Because these systems are electronically controlled, diagnosis is performed in the typical way. There is a warning lamp on the dash that comes on when a problem is detected. A DTC will also be set. In most cases, when the warning lamp is lit, there will be no power assist available.

The systems are constantly (from key on to key off) monitored by the control unit. If a problem is detected, the EPS lamp may stay lit after the ignition is turned on, or it may come on while it is being driven. To identify the reason for the illumination of the lamp, interview the customer to find out when and where it first came on. Try to duplicate the situation during a road test and then retrieve any and all DTCs (Figure 9-45). If the problem cannot be duplicated, do a careful inspection of all associated wiring and connectors.

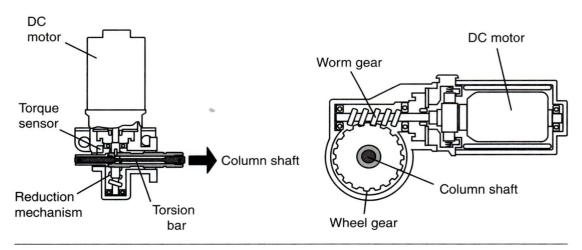

Figure 9-44 The DC motor used in an EPS system.

```
TRQ1...................................2.53V
TRQ2...................................2.51V
TRQ3...................................2.34V
SPD.....................................0MPH
MOTOR ACTUAL......................0A
COMMAND VALUE...................0A
THERMISTOR TEMP.............21°C
PIG SUPPLY.........................12.1V
IG SUPPLY...........................11.9V
TRQ1 ZERO VAL.................2.51V
TRQ2 ZERO VAL.................2.49V
TRQ3 ZERO VAL.................2.37V
MTR TERMINAL (+)...............5.8V
MTR TERMINAL (−)...............5.8V
MTR OVERHEAT...................Unrec
MTR LOW POWER..............Unrec
CONTROL MODE...............$010E
IG ON/OFF TIMES..........255times
#CODES.....................................0
ASSIT MAP...............................02
ECU I.D. .....................................01
TEST MODE STAT.........NORMAL
READY STATUS....................OFF
```

Figure 9-45 This screen print from a scan tool represents a normal condition for the EPS system. *Courtesy of Toyota Motor Company*

Air Conditioning

Hybrids are equipped with air conditioning systems with either a belt-driven or an electrically powered A/C compressor (Figure 9-46). The electrical units are powered by high-voltage and all precautions should be taken to work safely with these units. Always wear lineman's gloves when inspecting or servicing high-voltage air conditioning systems.

All air conditioning diagnosis and service on a hybrid vehicle are the same as that for a conventional vehicle. The exceptions are the special needs of the electric compressor, if the vehicle has one. Like all high-voltage systems, the cables to the compressor will be orange and there will be caution labels affixed to the compressor. (Respect the voltage!) These compressors require a different type of refrigerant oil. The oil not only serves as a lubricant but it also must be able to insulate the motor's electrical components from the compressor's housing. Always refer to the service manual to identify the proper oil for the compressor.

The operation of the air conditioning system may affect the stop-start feature. If the A/C is turned on, the engine may not shut down when the vehicle comes to a stop. This occurs mostly on hybrids with a belt-driven compressor, but can occur on others if battery voltage is low and the engine is needed to drive the generator and recharge the battery pack. When the compressor is driven by the engine, there will be no cooling if the engine is shut down.

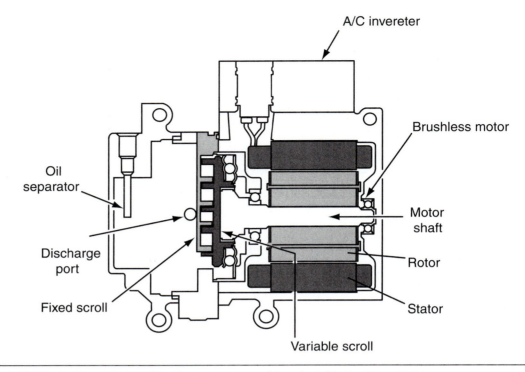

Figure 9-46 *An air conditioning compressor driven by a high-voltage motor.*

Review Questions

1. What is the correct procedure for testing the integrity of lineman's gloves, and why is this an important thing to do?
2. List five commonsense rules that should be followed when working on a hybrid vehicle.
3. Describe the procedure for checking for parasitic battery drains.
4. What is the best way to recharge a high-voltage battery pack?
5. After isolating the high-voltage system, what is the minimum time you should wait before beginning to work on or around the hybrid system?
   A. one hour
   B. thirty minutes
   C. fifteen minutes
   D. five minutes
6. True or False: All hybrids use the engine's cooling system to cool the inverter assembly.

7. List and simply describe three important tools that should be used when troubleshooting a hybrid system.
8. When using a scan tool on Toyota hybrids, which of the following will be displayed after the DTCs to help guide the diagnostic procedure?
   A. Pinpoint diagnostic charts
   B. Related technical bulletins
   C. Information codes
   D. Intermittent codes
9. Which of the following is not powered by the auxiliary battery in Ford and Toyota hybrids?
   A. Hybrid control module
   B. Engine starter
   C. Exterior lights
   D. Power steering motor
10. Why should only CAT III meters be used to test the circuits and components in a hybrid system?

FUEL CELL AND OTHER POSSIBLE VEHICLES

After reading and studying this chapter, you should be able to:

❏ Describe the basic configurations for the power train in a fuel cell vehicle.

❏ Describe the major components of a fuel cell vehicle.

❏ Explain how a fuel cell works.

❏ Describe the different types of fuel cells currently being considered for use in vehicles.

❏ Explain why fuel cell vehicles are not yet practical for the average consumer.

❏ Describe what hydrogen is and where it can be found.

❏ Explain the various processes used to produce hydrogen.

❏ Describe the different ways hydrogen can be stored.

❏ Explain the various technologies that can be incorporated into an internal combustion engined vehicle to make it more efficient.

❏ List and describe the power plants for a zero-emission vehicle.

❏ Describe the technologies that may make diesel automobiles more common on U.S. roads.

❏ Describe how steam can be used as an auxiliary power source in an automobile.

Introduction

The main topics of this chapter are fuel cells, fuel cell vehicles, and hydrogen. Also discussed are some of the alternate propulsion systems that may be used in the future. This chapter looks at the future; how much in the future is a good question. Most of what is covered exists in concept vehicles and systems that have been produced. Some have already been leased to businesses and individuals to serve only as "proof of concept" vehicles.

How soon you will see a fuel cell vehicle is anyone's guess and there are many guesses. These vehicles are the topic of much news and, for many, a topic of interest.

Fuel cell electric vehicles are the result of the many years of research and development on electric and hybrid vehicles. EVs, HEVs, and FCEVs are electric drive vehicles. They share many of the same technologies but differ greatly in the source of energy used to power the electric motors that are used to move the vehicle.

An EV relies on the energy stored in batteries as the sole energy source. The energy that is stored comes from readily available electrical power from external power lines. To refill or replenish the energy, or fuel, the batteries are charged by the normal source of electricity: an electrical outlet. It takes many hours to refill the batteries. An EV also stores electricity that is captured during braking. EVs offer the most economical and cleanest alternative to an ICEV. However, they have very limited driving range and require very long refilling (recharging) times. Both of these are contrary to the driving habits and desires of most consumers and are the primary reasons EVs are not commonly seen on the roads today.

Solar-powered EVs are also being tested. Currently, these have not proven to be very practical due to the required size and location of the solar panels. However, further development can lead to more efficient traction motors and additional applications of solar energy.

An HEV has two different energy sources for energy: the battery and the engine. The batteries are charged by using some of the energy from the engine to turn a generator. The energy in the batteries can be used to propel the vehicle. An HEV battery also stores electricity

Key Terms

Air engine
Alkaline fuel cell (AFC)
Direct-methanol fuel cell (DMFC)
Electrolysis
Fuel cell stack
Hydrogen
Molten Carbonate fuel cell (MCFC)
Phosphoric acid fuel cell
Photo biological
Photo electrolysis
Proton Exchange Membrane (PEM) fuel cell
Selective Catalytic Reduction (SCR)
Steam reforming
Solid oxide fuel cell (SOFC)

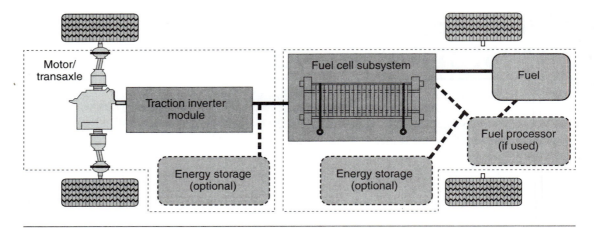

Figure 10-1 The basic layout for a fuel cell vehicle.

that is captured during braking. Although no external means is necessary to refill or charge the batteries, the vehicle must be refilled with fuel for the engine. HEVs achieve excellent fuel mileage and lower emissions when compared to an ICEV. When compared to an EV, they offer a much longer driving range but emit some pollutants and still rely on a fossil fuel for energy.

FCEVs have electric motors, but the energy source for those motors is not necessarily batteries. Some FCEVs use an ultra-capacitor in place of the battery pack. Regardless of where the energy is stored, all FCEVs rely on the electricity generated by an on-board fuel cell assembly. The energy from the fuel cell can directly power the motors or be sent to the storage device (Figure 10-1). Some FCEVs also have regenerative braking. An external energy source is not required to refill the electrical storage unit; however, the fuel used in the fuel cell must be refilled. Pure water and heat are the only emissions from a fuel cell. This technology is not new, nor is it unproven, NASA (National Aeronautics and Space Administration) has been using this technology in its spacecraft for years. Fuel cells provide the energy for the various electronic devices on-board the spacecraft.

Fuel Cell Vehicles

Fuel cell vehicles use hydrogen as their fuel or energy source (Figure 10-2). The supply of hydrogen can be stored in tanks in the vehicle or can be provided by a reformer that extracts hydrogen from another fuel, such as gasoline, methanol, or natural gas. A major obstacle in the practicality of a fuel cell vehicle is the absence of an infrastructure for supplying pure hydrogen. A reformer answers that concern, as the required fuels are readily available. However, the cost of the reformer adds to the already high cost of a fuel cell.

There are many obstacles that must be overcome before there is general use of a fuel cell vehicle. A Honda spokesperson predicted that fuel cell vehicles could have a market share of 5% by 2020. A Toyota spokesperson said that the introduction of a fuel cell vehicle for consumers will not happen before 2010. This is the same year GM has said it plans to have a production-ready fuel cell vehicle. They also plan to sell one million FCEVs by the year 2020. We do not know when FCEVs will be a common sight on the road, but we do know that many companies are working to get them there.

Many manufacturers have joined together in this effort, whereas others are working alone. Ford and DaimlerChrysler are majority owners of Ballard Power Systems Inc., which is a leading developer and manufacturer of fuel cell stacks. Ballard also has supplied fuel cells to Mitsubishi, Nissan, Volkswagen, and Honda. A fuel cell vehicle is much like a battery-operated electric vehicle. It operates like one and has many of the same characteristics as one: electricity powers an

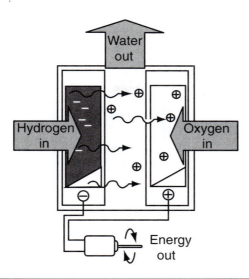

Figure 10-2 The basics of fuel cell operation.

electric motor to drive the vehicle, the vehicle operates very quietly, and the output of CO_2 and other harmful emissions is zero. The main powertrain components in a typical fuel cell vehicle are:

- ❏ Fuel cell stack—An electrical generation device made up of several individual fuel cells.
- ❏ High-pressure hydrogen supply system or reformer with a fuel tank.
- ❏ Air supply system—An air pump to supply the fuel cells with air.
- ❏ Humidification system—Recycles water vapor generated in the FC stack to humidify the hydrogen and air, so the fuel cell's membrane does not dry out.
- ❏ Fuel cell cooling system.
- ❏ Storage battery or ultra-capacitor.
- ❏ Traction motor and transmission.
- ❏ Control module and related inputs and outputs—includes a DC/DC converter.

The energy generated by a fuel cell can directly power the traction motor of the vehicle and/or can be stored in a battery or ultra-capacitor. If there is no storage (battery or capacitor) in the system, regenerative braking does not exist. However, when equipped with a battery or capacitor, regenerative braking is used and the energy stored in either one can provide power boosts for the vehicle.

Fuel Cells

A fuel cell produces electricity through an electrochemical reaction that combines hydrogen and oxygen to form water. The basic principle is the opposite of electrolysis. **Electrolysis** is the process of separating a water molecule into oxygen and hydrogen atoms by passing a current through an electrolyte placed between two electrodes (Figure 10-3). In an internal combustion engine, fuel is combined with oxygen and this causes combustion. In combustion, the chemical energy in the fuel is changed to heat energy. In a fuel cell there is no combustion; the reaction is purely chemical. Catalysts are used to combine the fuel (hydrogen) with oxygen. The reaction releases electrons or electrical energy. Fuel cells have no moving parts and can continue to work until the fuel supply is depleted. In other words, the driving range of a fuel cell vehicle is largely dependent on the amount of fuel it can carry.

Fuel cell technology is not new. In 1839, a British scientist, William Robert Grove, proved it was possible to reverse electrolysis and produce electricity. However, the first successful fuel cell was not developed until 1959. Since then, using fuel cells as power generators has increased rather

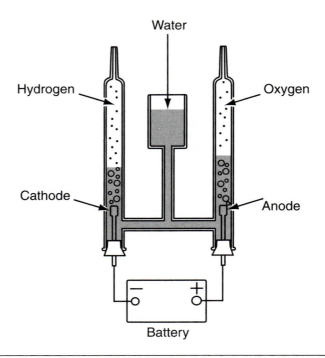

Figure 10-3 The process of electrolysis converts water into hydrogen and oxygen using electricity as the source of energy to cause the reaction.

rapidly. However, fuel cells are expensive to make. Therefore, they have been mostly used where practicality and necessity outweigh cost, such as space programs. The idea of using fuel cells in vehicles did not make sense until new, low-cost materials could be used to make a fuel cell.

A single fuel cell produces very low voltage, normally less than 1 volt. To provide the amount of power needed to propel a vehicle, several hundred fuel cells are connected in series. This assembly is the **fuel cell stack** (Figure 10-4), called this because the cells are layered or stacked next to each other. Each fuel cell produces electricity and the combined output of the cells is used to power the vehicle.

A fuel cell has two electrodes coated with a catalyst. The electrodes are separated from each other by an electrolyte and separators (Figure 10-5). One of the electrodes has a positive polarity, the anode, and the other is negative and is the cathode. The electrolyte is most often a polymer membrane, called the proton or ion exchange membrane. Polymers can be very resistant to chemicals and can serve as an electrical insulator or separator. The polymer membrane in a fuel cell does both. The catalyst, normally platinum, on the electrodes causes the chemical reaction in the fuel cell, but it does not materially take part in the reaction. Therefore, the catalysts are not consumed during the operation of a fuel cell. A fuel cell consumes only hydrogen and oxygen. The oxygen is delivered to the cell by an air compressor that draws air in from outside the fuel cell. Hydrogen is fed into the fuel cell from a pressurized tank or from a reformer. In a direct hydrogen fuel cell vehicle, there are zero emissions. However, if the hydrogen is extracted from a fuel by a reformer, there will some vehicle emissions.

When hydrogen is delivered to the anode, the catalyst causes the hydrogen atoms to separate into electrons and protons. Electrons always move to something more positive but cannot pass through the membrane. Therefore, their only path to the positive side of the fuel cell is through an external circuit. The movement of the electrons through that circuit results in direct current flow. It is this current flow that powers the vehicle's electric propulsion motors.

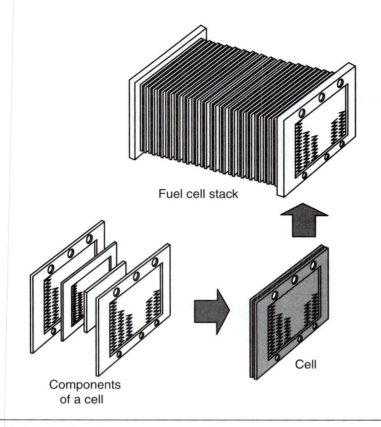

Fuel cell stack

Components
of a cell

Cell

Figure 10-4 A fuel cell stack.

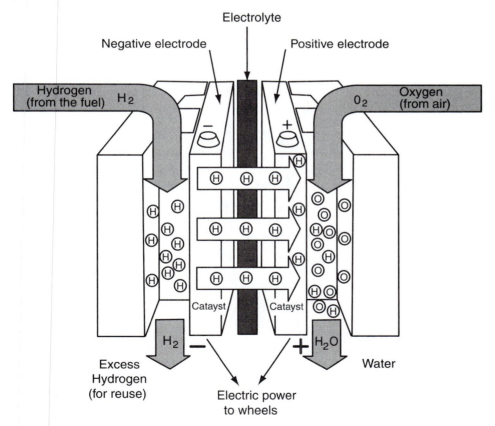

Figure 10-5 All fuel cells contain two electrodes—one positively and one negatively charged— and an electrolyte sandwiched between them.

Oxygen enters the other side of the fuel cell and reacts to the catalyst on the cathode. This reaction splits the oxygen molecules into oxygen ions. The protons (hydrogen ions) that were released from the hydrogen at the anode move toward the oxygen ions. The membrane that separates the two electrodes will only allow protons to pass through, and they do. At the cathode, two hydrogen ions bond with each oxygen ion to form water.

To function, the ion exchange membrane must be kept moist. Therefore, some of the water produced by the fuel cell is used to humidify the incoming hydrogen and oxygen. The remaining water is emitted as exhaust from the fuel cell. Some heat is also emitted by the fuel cell. The heat is either released to the outside air or captured and used to heat the fuel cell or can be used to heat the passenger compartment.

Types of Fuel Cells

The fuel cell described above is currently the most commonly used fuel cell in concept vehicles, the PEM. There are many other designs; some are impractical for use in an automobile whereas others show promise. The following descriptions of the different types only includes those that exist or are being developed at the time of this writing. At this point, it is hard to tell what design will actually be used on the roads of tomorrow, but it seems certain that a fuel cell will power some of the vehicles of tomorrow, eventually. Most fuel cell designs vary by size, weight, fuel, cost, and operating temperature. However, ALL have two electrodes and an electrolyte between them. In addition, all types of fuel cells are more efficient than an internal combustion engine.

Proton Exchange Membrane The **Proton Exchange Membrane (PEM) fuel cell** (Figure 10-6), or derivatives of it, is a favored design because it allows for adjustable outputs, which are necessary for driving. The speed of the vehicle can be controlled by controlling the output of the fuel cell. Although it is quite compact, it is capable of providing high outputs. When compared to other fuel

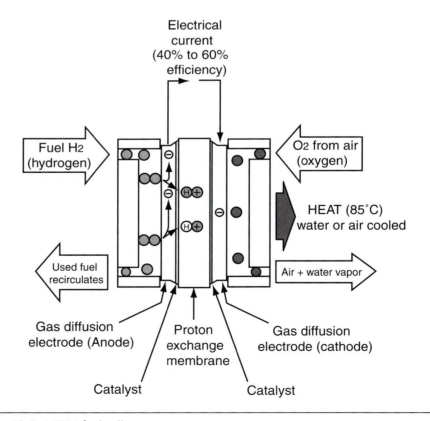

Figure 10-6 A PEM fuel cell.

cell designs, it operates at a relatively low temperature of 176°F (80°C). However, it is expensive to manufacture. Much of the cost is because the catalysts are platinum based. One of the biggest disadvantages of the PEM cell is the need to keep the membrane moist. In cold temperatures, the water can freeze, making the fuel cell very difficult to get started. Also, carbon monoxide (CO) can weaken the platinum catalysts. Because outside air is delivered to one side of the cell, the presence of CO in that air will reduce the output of the cell. Much research and development is taking place to alleviate these obstacles.

Solid Oxide The **solid oxide fuel cell (SOFC)** may be the first design to be used in a mass-produced automobile. However, it will not be used to power a traction motor. Rather it may be used to replace the belt-driven generator (alternator) on internal combustion engines. Current alternators are not very efficient and their output is dependent on rotational speed. They also rely on engine power to operate, which means they contribute to an engine's fuel consumption. Removing the alternator and using an SOFC will increase the efficiency of the engine. The SOFC can also provide much higher power levels, which means more accessories can be electrically driven. This again will increase the efficiency of the engine. Using an SOFC will also allow accessories to operate when the engine is not running and without draining the battery. In addition, the heat generated by the cell can be used to heat the passenger compartment.

These cells have a ceramic anode, ceramic cathode, and a solid electrolyte (Figure 10-7). The cells operate at very high temperatures from 1290 to 1830°F (700 to 1000°C). Although these operating temperatures restrict the type of materials used in the cells to ceramics, they also eliminate the need for expensive catalysts. Lower production cost is one of the reasons SOFCs are considered as a likely choice for replacing the alternator. These fuel cells can operate with a simple, single-stage, built-in reformer because of the high operating temperatures. Also, the high temperatures eliminate the chances of CO poisoning the electrodes. Efficiency estimates for this type fuel cell vary from 40 to 45%, as compared to 20 to 30% for an internal combustion engine.

The high operating temperature is also a reason these fuel cells may not be used to power a vehicle. When this high heat is generated, it must be released. Releasing a large quantity of high heat can cause many problems in other automotive systems. When used to replace the alternator, the quantity of heat is far less and it can be easily moved away from the vehicle.

Molten Carbonate This design of fuel cell is unlikely to ever be used in an automobile; it is best suited as a power generator to supply factories and perhaps cities. It has the ability to generate electricity from coal-based fuels or natural gas. The **Molten Carbonate fuel cell (MCFC)** uses a

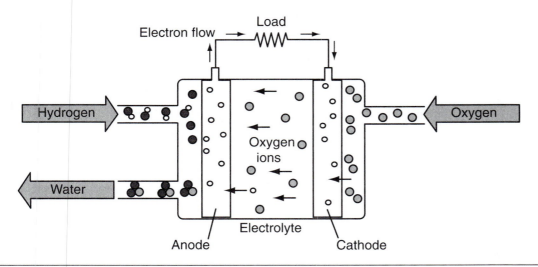

Figure 10-7 A solid oxide fuel cell (SOFC).

liquefied carbonate salt as its electrolyte. This fuel cell operates at between 1110°F (600°C) and 1200°F (650°C). However, its output is strongly dependent on operating temperatures, as a drop of just 100°F (50°C) will drop its output by as much as 15%. It also needs to recirculate carbon dioxide (CO_2), which is one of its by-products. The other end product is water. Recirculating the CO_2 and making sure all of the hydrogen fed into the cell is used are the two major issues with this design. Research is ongoing to develop a membrane that would recirculate the unused hydrogen back into the fuel intake.

The anode is typically made from a highly sintered nickel powder alloyed with chromium, and the cathode is a porous nickel oxide alloyed with lithium.

Direct Methanol The **Direct-methanol fuel cell (DMFC)** is a type of PEM fuel cell. Liquid methanol, rather than hydrogen, is the fuel oxidized at the anode and the oxygen from the outside is reduced at the cathode. Methanol is considered an ideal hydrogen carrier because it takes little energy to cause it to release its hydrogen. A mixture of methanol/water is delivered directly into this modified PEM cell and releases the hydrogen needed by the fuel cell. This cell uses a thin membrane lightly covered on both sides with a layer of a platinum-based catalyst. A reformer is not needed and, therefore, the cost of these cells is lower than the PEM. This is because less platinum is needed. Liquid methanol is also easier to store than hydrogen and has a much higher energy density than compressed hydrogen.

These cells are also simple and compact units that can provide a good amount of energy for a long period of time. However, these cells are not as efficient as PEMs and the response of these fuel cells is slower than a PEM. Also, they have emissions that are not present with other fuel cells designs. As the hydrogen is removed from the methanol, carbon is released. The carbon and hydrogen combines with the oxygen at the outlet of the cell to form carbon dioxide and water.

These cells use a mixture of methanol and water that is introduced to a negatively charged electrode. This electrode immediately reacts and breaks the methanol molecules apart. The carbon and oxygen atoms from the methanol combine to form CO_2. The hydrogen atoms are separated into protons and electrons. The electrons move through an external circuit and return to the cell at a positive electrode. Once in the external circuit, they combine with the protons, which arrived there by passing through the membrane. At this point water is formed and exhausted from the cell.

Phosphoric Acid The **phosphoric acid fuel cell** is the most commercially used fuel cell and can operate at 37 to 42% efficiency. These fuel cells use liquid phosphoric acid (Figure 10-8) as the electrolyte with electrodes made of carbon paper coated with a platinum catalyst. The use of platinum means they are costly to manufacture. They also operate at a relatively high temperature (anywhere between 300 to 400°F [150 to 205°C]). The operation of the fuel cell is much the same as others; the catalyst separates the fuel into electrons and protons. The electrons move through an external circuit and the protons move to the cathode. At the cathode, the (electrons) oxygen and (protons) hydrogen ions join to form water. The water is emitted as steam, which can be used to power another electrical generating device. When this occurs, the efficiency of the fuel cell doubles. It is unlikely that this design of fuel cell will be used in an automobile because of the high temperatures, reluctance to have varying outputs, and cost. It works fine when power output demands are constant.

Alkaline The **alkaline fuel cell (AFC)** is the one used primarily by NASA. It is expensive, but highly efficient. In a spacecraft, the water (its exhaust) is used as drinking water for the space travelers. This fuel cell will undoubtedly never be used in automobiles because of its cost. It is also very sensitive to carbon dioxide, which means it does best where all CO_2 can be removed from the incoming supply of air. This fuel cell operates in the same manner as a PEM.

FUEL CELL

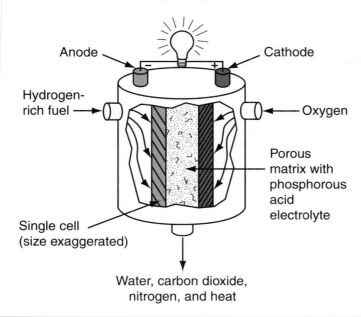

Figure 10-8 A phosphoric acid fuel cell.

Alkaline fuel cells use a water-based solution of potassium hydroxide (KOH) as the electrolyte. The electrodes are coated with a catalyst, although due to the operating temperature and the purity of the incoming gases, platinum is not required.

Obstacles for Fuel Cell Vehicles

Fuel cells provide clean energy and are quite efficient, which is the main reason for considering their use in future automobiles. However, before fuel cells will be used on any large scale, certain obstacles must be overcome. The lack of a hydrogen infrastructure is one of biggest obstacles. This is an obstacle of practicality and consumer acceptance, not of engineering. In order for a fuel cell vehicle to be practical, its fuel must be readily available.

Another problem related to fuel supply is storage. To be practical, any vehicle must have a decent driving range, one of at least 300 miles (483 km). To accomplish this, fuel cell vehicles must be able to store a lot of hydrogen. Many different methods of storage are being researched. Storing hydrogen in pressure tanks may be the answer; however, high pressures are required and high-pressure tanks are very expensive. The typical fuel cell vehicle stores hydrogen at 5000 psi (352 kg/cm) and has a driving range of about 150 miles (241 km). Doubling the pressure would nearly double the driving range. To double the pressure, stronger tanks are required, which adds to the cost of the tanks. Hydrogen storage is a major area for research and will be discussed in more detail later in this chapter.

Some FCEVs have on-board reformers (Figure 10-9) that extract hydrogen from gasoline, ethanol, or methanol. Storing these fuels requires less space and is much simpler than storing pure hydrogen. The objections to using a reformer are plentiful. A reformer has undesirable emissions, such as carbon dioxide. Using reformers does not reduce our dependence on fossil fuels. Reformers are slow and require long run times before they can provide enough hydrogen to move a vehicle a few feet.

The cost of a fuel cell and its supporting systems is extremely high. It is estimated that the cost of one fuel cell vehicle is one million dollars. Obviously, as more FCEVs are built, the cost will come down. Also, the advances made with hybrid vehicles that can be shared with FCEVs will also lower future costs. As it stands right now, very few consumers could afford to purchase a fuel cell vehicle. To gain public acceptance, the cost must be drastically reduced.

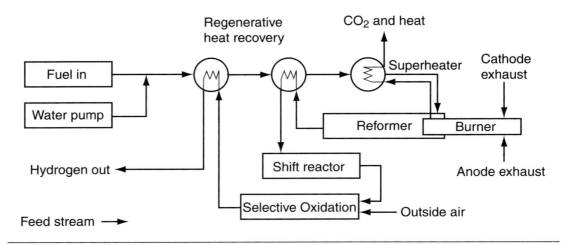

Figure 10-9 The flow diagram for converting methanol to hydrogen with a reformer.

An issue that may seem to be an odd concern is noise. FCEVs are too quiet—so quiet that there are safety concerns. If the vehicles can be seen but not heard, their approach to the rear of pedestrians and bicyclists can present dangerous situations. This is an area that must be studied.

To control the output of a fuel cell and therefore the speed of the vehicle, advanced electronics are necessary. Much of this technology is already used in hybrid vehicles but the uniqueness of the fuel cell demands additional new controls. FCEVs have high- and low-voltage systems and electronic controls are necessary to allow the fuel cell to power both. These controls are in addition to the typical computer systems of other vehicle types. The traction motors used in a FCEV must be very efficient and able to respond to changing driving conditions. Hybrid motor technology is an area of constant study and that technology can be shared with FCEVs.

Most fuel cells take some time to start, especially when they are cold. As mentioned before, freezing temperatures can kill a fuel cell. Ice on the membrane can destroy it or at least stop the fuel cell from working. An exhaust system plugged with ice will shut down a fuel cell. This is an area of much research and some manufacturers have had some success dealing with the problem. The basic thrust has been making sure all water is removed from the fuel cell after it has been shut down. This requires energy from a storage device. There is also research being done on mixing special coolants in the water. The fact that a fuel cell does not generate electricity until it has a temperature of 32°F (0°C) is an obstacle that needs to be overcome.

On the other side of the temperature scale, heat must be carefully controlled. Fuel cells become heated while they operate and operate best within a particular temperature range. That range depends on the type of fuel cell. PEM cells operate best at a lower temperature than conventional ICEs. This presents a major challenge as it is more difficult to get rid of low heat than it is high heat. This means the cooling system must be more efficient than those used in conventional vehicles. Typically a PEM cell requires larger and/or more radiators. This means more space is needed in the vehicle just for the cooling system. This results in less useable space for passengers and luggage. When the space for the cooling system is added to required space for the fuel cell stack and other components, very careful planning of space is necessary (Figure 10-10). This becomes more of a challenge when one considers that the electronics and traction motors must also be kept cool. The cooling of these requires an additional cooling system because they operate at a different temperature range than the fuel stack. An additional cooling problem enters when the vehicle is equipped with a high-voltage battery pack and/or ultra-capacitors.

An obstacle that pertains to other fuel cell designs is the isolation of the extremely high temperatures of the fuel cell. For example, the solid oxide fuel cell operates at temperatures from 1290 to 1830°F (700 to 1000°C). If this heat is not totally insulated from the passenger compartment, it could bake everyone and everything inside the vehicle.

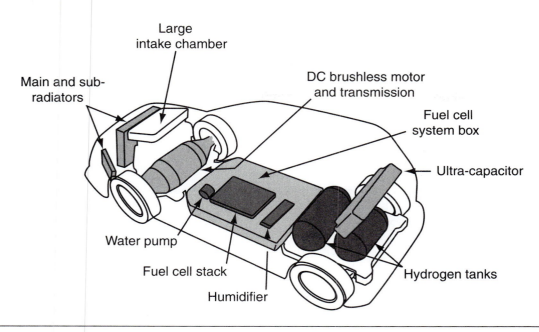

Large
intake chamber

Main and sub-
radiators

DC brushless motor
and transmission

Fuel cell
system box

Ultra-capacitor

Water pump

Fuel cell stack

Humidifier

Hydrogen tanks

Figure 10-10 The layout of the components for the fuel cell system in a Honda FCX.

Another heat-related problem is generated by the air compressor that feeds outside air into the fuel cell. As air is compressed, its temperature increases. Because the fuel cell works best with a specific temperature range, the compressed air can heat up the cell beyond that range. To eliminate this, intercoolers must be added to the air compressor system. These, again, occupy space. There is also the problem of filtering the incoming air. Ideally, the incoming air would be free of all dirt and other contaminants. A filtering system occupies space and has an impact on the overall layout and design of the vehicle.

As you can see, fuel cell vehicles are very complicated machines.

Hydrogen

The fuel for a fuel cell is hydrogen. **Hydrogen** is full of energy because of its atomic structure. Hydrogen is also the most abundant natural resource in our world. Hydrogen has been used as a fuel for spacecraft and electrical generators. It is also used to manufacture reformulated gasoline, ammonia for fertilizer, and many different food products. And in the next step, hydrogen may be the fuel for the future for personal transportation. The automotive industry has long used the energy released by separating hydrogen from a substance and recombining it to oxygen. In a gasoline ICE, gasoline is forced, by heat, to combine with oxygen. The result is combustion, which releases energy. That energy is used as mechanical energy. In a fuel cell, the same basic thing happens, but the chemical energy is released as electrical energy.

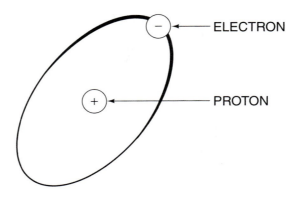

Figure 10-11 A hydrogen atom.

In a fuel cell, hydrogen can be delivered or can be extracted from commonly available sources, such as water, gasoline, diesel fuel, and methanol. Because hydrogen gas is rarely present in its natural state (hydrogen by itself), it must be manufactured or processed by a reformer.

What is Hydrogen?

Hydrogen is the simplest and lightest of all elements. It is made up of one proton and one electron (Figure 10-11). Hydrogen is a colorless and odorless gas. It is one of the most abundant elements on earth. However, it is only found in compound form. The combination of hydrogen and oxygen forms water. Fossil fuels are combinations of carbon and hydrogen, which is why they called hydrocarbons.

Sources of Hydrogen

Hydrogen is extracted from various substances through a process that pulls hydrogen out of its bond with another element or elements. Hydrogen is commonly extracted from water, fossil fuels, coal, and biomass. The two most common ways hydrogen is produced are steam reforming and electrolysis. Although FCEVs are considered zero-emission vehicles, there are some related emissions due to the production process. Hydrogen production is commonly done, but it is very costly. Currently it costs much more to produce hydrogen than it does to produce other fuels, such as gasoline. This, again, is an obstacle and the focus of much research.

Steam reforming is the most common method used to produce hydrogen. This process uses high-temperature steam to extract hydrogen from natural gas or methane (Figure 10-12). Methane is the simplest of all hydrocarbons and is readily available. Methane is the primary component of natural gas, which is found in oil fields, natural gas fields, and coal beds. This method is used to produce about 95% of the hydrogen that is available today. Steam reforming is currently the most cost effective way to produce hydrogen. However, it relies on fossil fuels to create the steam and uses a fossil fuel as the source for hydrogen. Therefore, it does not reduce our dependence on fossil fuels and releases emissions during the process.

A cleaner, but more costly, method for producing hydrogen is **electrolysis**. In this process, electrical current is passed through water. The water then separates into hydrogen and oxygen. The hydrogen atoms collect at a negatively charged cathode and the oxygen atoms collect at the positively charged anode. Producing hydrogen by electrolysis costs approximately ten times more than using steam reforming. However, the process does result in pure hydrogen and oxygen.

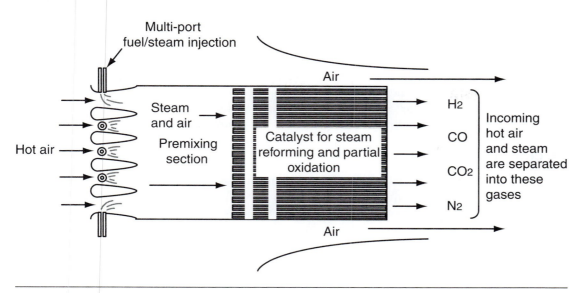

Figure 10-12 Basic view of how a steam reformer produces hydrogen.

Using solar energy to separate the atoms in water is being heavily researched. Water is a renewable resource and light from the sun is readily available. Basically, the sun's light is collected at photovoltaic cells and converted to electricity. That electrical energy is then passed through the water to begin electrolysis. This process is called **photo electrolysis**. Another photolytic process is also being researched. **Photo biological** methods rely on the activities of some algae and bacteria that produce hydrogen when exposed to light (Figure 10-13).

Biomass (plants or agricultural waste) may provide an economical alternative to fossil fuel-based hydrogen production. By gasifying or burning biomass with high heat, the biomass separates into hydrogen and other gases. Biomass can also be used to provide the heat to cause the separation. This means the process uses no fossil fuels.

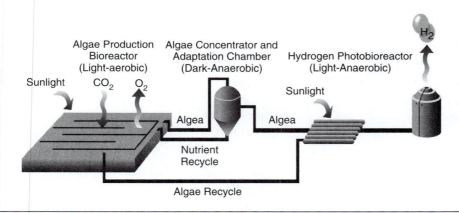

Figure 10-13 The principle of photo-biological hydrogen production.

Some concept FCEVs have reformers that extract hydrogen from a fossil fuel directly on the vehicle. Three fuels are commonly used: gasoline, methanol, and natural gas. Reformers make the vehicle more practical because the fuel supply is easily replenished. However, reformers have some emissions issues, are costly, and consume valuable vehicle space. There is also an issue of the purity of the fuels that will be reformed. Many of these fuels have a substantial amount of sulfur. The sulfur can contaminate the catalysts used in the fuel cell and may not be totally filtered out of the hydrogen during the reforming process.

Hydrogen Fuel for ICEs

There is a lot of energy available in hydrogen. Some automobile manufacturers are experimenting with fueling internal combustion engines with pure hydrogen. Research is also being done with adding hydrogen to other fuels. In both of these cases, exhaust emissions are reduced without a great decrease in power output; in some cases the power actually increased. In addition to these benefits, hydrogen-fueled ICEs may be the impetus for building a solid infrastructure for dispersing hydrogen as a fuel.

Three major auto manufacturers have developed and tested hydrogen-fueled internal combustion engines. These vehicles actually have bi-fuel capabilities. Of interest is BMW's bi-fueled V-12 engine, which uses liquefied hydrogen or gasoline as its fuel (Figure 10-14). When running on hydrogen, the engine emits zero carbon dioxide emissions. To prove the feasibility of the engine, BMW took a specially equipped car (Figure 10-15) out to its high-speed test track in Miramas, France. There the car set nine international speed records for hydrogen-driven vehicles. To store the liquefied hydrogen, the storage tank is kept at a constant temperature of −423°F (−253°C). At this temperature, the liquid hydrogen has the highest possible energy density.

Figure 10-14 BMW's hydrogen powered V-12 ICE. *Photo courtesy of BMW of North America, LLC.*

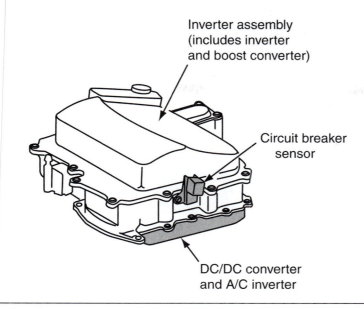

Inverter assembly
(includes inverter
and boost converter)

Circuit breaker
sensor

DC/DC converter
and A/C inverter

Figure 10-15 BMW's hydrogen powered racecar.

Ford and Mazda have also developed vehicles with hydrogen power. Mazda is using its rotary engine, which it claims is ideal for running on hydrogen. The concept vehicles from both manufacturers are also bi-fuel vehicles. One of the engines converted by Ford is its 2.3-liter, I-4 engine used in the Ford Ranger. Engine modifications include a higher compression ratio, special fuel injectors, and a modified electronic control system. When running on hydrogen, the engine is more than 10% more efficient than when it runs on gasoline. Another benefit is emissions levels that are near zero. Because the fuel contains no carbon, there are no carbon-related emissions (CO, HC, or CO_2). Typically, an engine running on hydrogen produces less power than a same-sized gasoline powered engine. So, Ford added a supercharger with an intercooler to the engine to compensate for the loss of power.

Infrastructure and Storage

Other than manufacturing costs, the biggest challenge for hydrogen-powered vehicles, whether with a fuel cell or an engine, is the lack of an infrastructure. These vehicles need to be able to be refueled quickly and conveniently. The use of reformers may be the short-term answer to this problem. However, reformers add more weight and technical complexity to a vehicle. Also, the reformed hydrogen is not free of contaminants. Some of what could be in the hydrogen may poison the fuel cell and cause it to underperform.

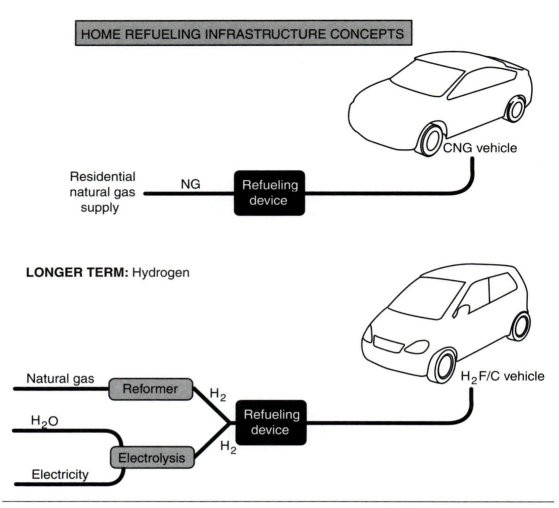

Figure 10-16 Creating a home based hydrogen-refueling center.

Honda is developing a home energy station that extracts hydrogen from natural gas. These stations will allow owners of hydrogen-powered vehicles to refill their tanks at home (Figure 10-16). The stations are also capable of supplying electricity, hot water, and heat for the home. The stations use electrolysis to extract the hydrogen and solar energy as the source for electrical current.

In-Vehicle Storage

Hydrogen contains more energy per weight than any other fuel, but it contains much less energy by volume. This makes storing enough hydrogen for an acceptable driving range very difficult. Naturally, you can store more in a larger container but that container would consume more space and add considerable weight to the vehicle. In-vehicle hydrogen storage is another area of much attention for researchers and engineers.

In most concept FCEVs, hydrogen is stored either as a liquid or as a compressed gas. When stored as a liquid, hydrogen must be kept very cold. Keeping it that cold adds weight and complexity to the storage system. At cryogenic (icy cold) temperatures, more hydrogen can be stored in a given space. Cryogenic fuels are used in the rockets of NASA's space shuttle. However, liquid storage has some safety issues that are not present with compressed hydrogen storage. The

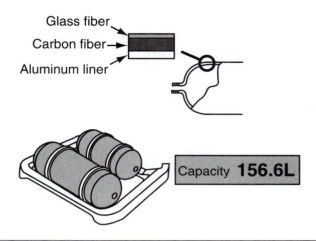

Figure 10-17 In-vehicle storage tanks for compressed hydrogen.

tanks required for compressed hydrogen need to be very strong, which translates to very heavy (Figure 10-17). Also, higher pressures mean more hydrogen can be packed into the tank but the tank must be made stronger before the pressure can be increased.

Other storage technologies are being developed. Two of these technologies getting the most attention are systems based on metal hydrides and carbon nanotubes. The use of metal hydrides offers the possibility of storing three times more hydrogen in a given volume than when it is compressed. The metal hydrides, normally powdered magnesium-based alloys, collect hydrogen atoms and hold them at low temperatures. When the metal hydrides are heated, the hydrogen is released until heat is removed from the metal. Carbon nanotubes are microscopic tubes of carbon that can store hydrogen in their pores. Because the surface of these tubes is quite irregular, the actual surface area is larger than the size of the tubes. The use of metal hydrids and carbon nanotubes may help solve the hydrogen storage problem for the future.

Prototype FCEVS

Fuel cell vehicles for everyday use are still years away; however, there are many fuel cell prototypes and concept vehicles on the road all over the world. All of these are part of the ongoing research that taking is place. Every major manufacturer has built at least one type of FCEV and many have developed a new model nearly every year. These vehicles are testing different technologies in a real-world setting. It is difficult to predict exactly how a mass-produced FCEV will be equipped, but it is certain that some of the technology used in today's prototypes will be part of that final design. In addition, much of what was learned from the experiences of EVs and HEVs will also be part of the future FCEV.

There are three basic configurations that describe the design of a FCEV powertrain (Figure 10-18). The powertrain of a basic fuel cell vehicle is referred to as the direct-supply system. With this design, the energy from the fuel cell is delivered directly to the electric traction motor(s). Vehicles with this configuration do not have regenerative braking, and propulsion power depends entirely the output of the fuel cell.

In a battery hybrid powertrain system, the energy from the fuel cell is sent to the motor(s), the battery pack, or both. This configuration can use regenerative braking. The battery can also

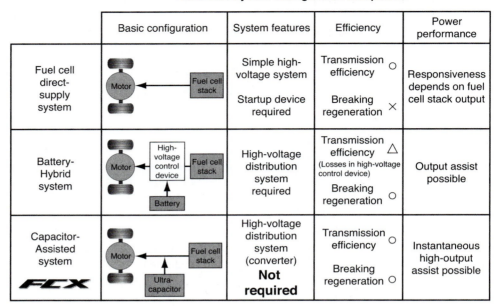

	Basic configuration	System features	Efficiency	Power performance
Fuel cell direct-supply system	Motor ← Fuel cell stack	Simple high-voltage system Startup device required	Transmission efficiency ○ Breaking regeneration ×	Responsiveness depends on fuel cell stack output
Battery-Hybrid system	Motor — High-voltage control device — Fuel cell stack; Battery	High-voltage distribution system required	Transmission efficiency △ (Losses in high-voltage control device) Breaking regeneration ○	Output assist possible
Capacitor-Assisted system *FCX*	Motor ← Fuel cell stack; Ultra-capacitor	High-voltage distribution system (converter) **Not required**	Transmission efficiency ○ Breaking regeneration ○	Instantaneous high-output assist possible

Figure 10-18 Honda's view of the different power train configurations for FCEVs. *Courtesy of American Honda Motor Co., Inc.*

supplement the fuel cell's energy to improve performance. This system requires more electronic controls than the direct-supply system. The third configuration uses ultra-capacitors rather than a battery. The ultra-capacitors are charged by the fuel cell and regenerative braking. Ultra-capacitors charge and discharge quickly, which allows the powertrain to respond quicker to changing conditions. Complex electronic systems are also required for this type of system.

Some of the manufacturers have developed their own fuel cell stacks, whereas others use a fuel cell manufactured by either Ballard Power Systems or United Technologies Company. Both companies are continuously working on new fuel cell designs. Much of this development is centered on the use of new materials to reduce cost and increase the efficiency of their fuel cells.

Fuel supply is also a factor in the design of an FCEV. Storage tanks and reformers take up space and are costly. It is the capability to feed a fuel cell for a long time that increases the vehicle's driving range. There are also variations in the electric traction motors used in the prototypes. Some use a motor from an EV or HEV, whereas others have motors located at the wheels. Some have a combination of both.

All FCEVs have a low-voltage system to energize and operate lights, accessories, and the various control modules. This means that all FCEVs need a DC/DC converter to reduce the high-voltage from the fuel cell. In addition, because a fuel cell generates DC voltage, an inverter is not needed unless the traction motors and the accessories require AC voltage.

To illustrate the constant development of fuel cell vehicles, examples of some of the vehicles from some manufacturers will be looked at. The examples are just a sampling of what has been developed. Many of the manufacturers that are working on fuel cell vehicles are also developing hybrid vehicles and other alternatives that will reduce our dependency on fossil fuels.

DaimlerChrysler

DaimlerChrysler started developing fuel cell vehicles in 1994 and has produced well over 100 vehicles for testing purposes. These vehicles include cars, buses, and vans. Their vehicles are part of the NECAR (New Electric Car) and NEBUS (New Electric Bus) series. Its FCEVs are on the roads in the United States, Europe, China, Australia, Japan, and Singapore. In each location, feasibility studies are being made while the vehicles are being used in

YEAR	VEHICLE	CONFIGURATION	FUEL CELL MANUFACTURER	RANGE	FUEL STORAGE
1994	NECAR 1 – Mercedes-Benz van	Direct-supply	Ballard – Twelve stacks (PEM)	81 mi (130km)	Compressed at 4300 psi
1996	NECAR 2 – Mercedes-Benz van	Direct-supply	Ballard (PEM)	155 mi (150km)	Compressed at 3600 psi
1997	NECAR 3 – Mercedes-Benz A-class car	Direct-supply	Ballard – Two stacks Mark 700 series (PEM)	250 mi (400km)	Methanol reformer
1999	NECAR 4 – Mercedes-Benz A-class car	Direct-supply	Ballard – Mark 900 series (PEM)	280 mi (450km)	Liquid hydrogen
2000	Jeep Commander 2	Battery hybrid	Ballard – Two stacks Mark 700 series (PEM)	118 mi (190km)	Methanol Reformer
2000	NECAR 5 – Mercedes-Benz A-class car	Direct-supply	Ballard – Mark 900 series (PEM)	280 mi (450km)	Methanol Reformer
2001	NECAR 5.2 – Mercedes-Benz A-class car	Battery hybrid	Ballard – Mark 900 series (PEM)	300 mi (482km)	Methanol Reformer
2001	Natrium – Town & Country Mini-van	Battery hybrid	Ballard – Mark 900 series (PEM)	300 mi (482km)	Extracted from sodium borohydride
2002	F-Cell - Mercedes-Benz A-class car	Battery hybrid	Ballard – Mark 900 series (PEM)	90 mi (145km)	Compressed at 5000 psi
2005	F-Cell - Mercedes-Benz B-class car	Battery hybrid	Ballard – Mark 900 series (PEM)	250 mi (400km)	Compressed at 5000 psi

Figure 10-19 The chronology of FCEVs from DaimlerChrysler.

varying driving and climate conditions. Figure 10-19 shows the chronological development of many of DaimlerChrysler's fuel cell vehicles.

Through continuous research, DaimlerChrysler has been able to extend driving range, minimize the space required for the fuel cell components, and improve cold weather starting and operation. To minimize the space requirement, engineers have fit the entire fuel cell drive system in the floor. Doing this allowed them to convert a light and small car, the Mercedes-Benz A-Class, into a FCEV and still have room for passengers and luggage (Figure 10-20). They have also experimented with the location and number of electric motors. For example, the Jeep Commander 2 is driven by front and rear electric motors to provide full-time, four-wheel drive.

General Motors Corporation

General Motors's has made a huge commitment to the development of FCEVs. In fact, it has a stated goal to design and validate a fuel cell propulsion system by 2010, one that is competitive with current engines with regard to cost, reliability, and performance. Its venture began in 1997 with the introduction of a fuel cell powered Opel Sintra. This is a European-only mini-van. The fuel cell Sintra was followed up, in 1998, by a fuel cell powered Zafira, which was the replacement vehicle for the Sintra. This model vehicle was the initial test bed for early GM fuel cell vehicles. The first Zafira FCEVs used a methanol reformer and a Ballard fuel cell. It had a decent driving range of 300 miles (483km). This original FCEV, and the Zafira, served as the basis for three generations of what GM called its HydroGen series. Each generation used a GM developed PEM fuel cell. The HydroGen1 and 3 (there was no true HydroGen 2) relied on liquid cryogenic hydrogen and had a range of 250 miles (400km). The HydroGen1 was based on the battery hybrid configuration, whereas the HydroGen3 was direct supply. The advanced HydroGen3 was also based on the direct supply configuration (sent to testing one year after the original

Figure 10-20 Mercedes-Benz A-Class F-Cell passenger cars, like the one seen here, are being operated by customers in Singapore, Japan, Germany, and the United States. *Courtesy of Ballard Power Systems*

HydroGen3), which relied on high-pressure (10,000 psi [703 kg/cm]) hydrogen storage. Its range was lower than the previous generations, but used less fuel overall.

Using what it learned from the HydroGen series, GM tested a fuel-cell-powered Chevrolet S-10 pickup in 2001, which combined several of the technologies used previously. This S-10 pickup had a low power fuel cell from GM and used the battery hybrid configuration. It also had a gasoline reformer to supply the required hydrogen. The driving range for this truck was 240 miles (386km). Due to the low output from the fuel cell, the vehicle had a lower top speed and was less inclined to accelerate than previous GM FCEVs.

After these experiments, GM worked to minimize the required space and increase fuel economy. In 2002, it introduced its "skateboard" concept (Figure 10-21). In this design, all of the fuel cell-related components are packed into a carbon fiber structure that also served as the chassis for the vehicle. The first of such vehicles were the AUTOnomy and the Hy-wire. These vehicles featured many futuristic concepts, including total drive-by-wire systems. The concept was based on the idea of developing a propelled chassis that any body style or configuration could be placed upon.

The latest concept GM fuel cell vehicle is the Sequel. The Sequel uses the technologies that worked well in the AUTOnomy, Hy-wire, and the advanced HydroGen3. This battery hybrid fuel cell vehicle uses a lithium ion battery to provide extra electrical energy to the three electric motors during acceleration and to capture energy during braking. A transverse-mounted, three-phase AC motor drives the front wheels, and 2 three-phase AC wheel hub motors drive the rear wheels. There is a separate inverter for each motor. The electrical system includes three separate systems with three different voltages. A high-voltage system provides energy for the traction motors; the 42-volt system supplies energy for the brakes, steering, air conditioning, and other by-wire systems; and the 12-volt system is used for the conventional accessories and lights.

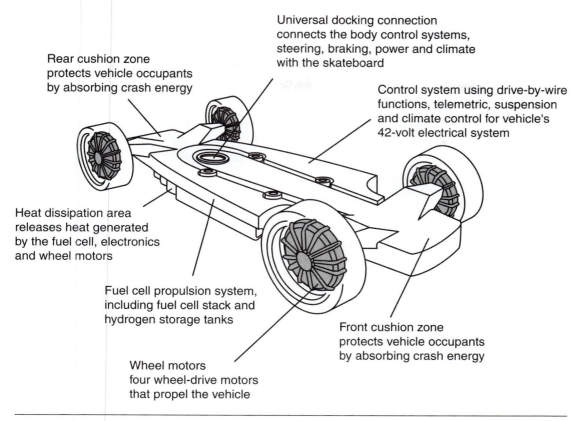

Rear cushion zone
protects vehicle occupants
by absorbing crash energy

Universal docking connection
connects the body control systems,
steering, braking, power and climate
with the skateboard

Control system using drive-by-wire
functions, telemetric, suspension
and climate control for vehicle's
42-volt electrical system

Heat dissipation area
releases heat generated
by the fuel cell, electronics
and wheel motors

Fuel cell propulsion system,
including fuel cell stack and
hydrogen storage tanks

Front cushion zone
protects vehicle occupants
by absorbing crash energy

Wheel motors
four wheel-drive motors
that propel the vehicle

Figure 10-21 The basic layout of General Motors's skateboard chassis.

The fuel cell stack, hydrogen and air processing subsystems, high-voltage distribution system, and hydrogen storage tanks are housed in the skateboard. The storage tanks are designed to hold compressed hydrogen at 10,000 psi (703 kg/cm).

Toyota

Toyota has developed some FCEVs, along with its continuous work with hybrid technology. Much of what it learned with its hybrids is transferable to its fuel cell vehicles. In fact, many of the same components can be transferred as well. Toyota also has experience with battery electric vehicles and its first FCEV, in 1996, was based on the RAV 4 EV. The fuel cell prototype was simply the EV with a fuel cell. The RAV 4 FCEV used a Toyota developed PEM fuel cell and was configured as a battery hybrid FCEV. There were two generations of the RAV FCEV; the first stored the hydrogen in metal hydrides and the other had a methanol reformer. The second generation of this FCEV had a range of 310 mi (500km).

As Toyota continued to make advancements in hybrid technology, it applied the technology to new fuel cell prototypes and concept vehicles. To take a quick look at how hybrid technology and components are used in a fuel cell vehicle, consider the FCHV (Toyota Fuel Cell Hybrid Vehicle), which is based on the Highlander (Figure 10-22). The power train is entirely based on the hybrid technology used in the Prius.

The vehicle is a battery hybrid FCEV and is propelled by a single electric motor. The compressed fuel is stored in four hydrogen fuel tanks at 5000 psi (352 kg/cm). The vehicle uses the same nickel-metal hydride battery pack as the Prius, including all of the associated electronics. A power control unit is a slightly modified version of that used in the Prius. It monitors the current operating conditions and determines when to use the battery, fuel cell, or both to propel the vehicle and to charge the battery. This is the same strategy used in hybrid vehicles. However, in the FCEV the fuel cell and its output replace the engine.

Figure 10-22 A Toyota FCHV-4. *Courtesy of Dewhurst Photography and Toyota Motor Sales, U.S.A., Inc.*

Honda

Honda, the other major manufacturer of hybrid vehicles, also has been busy with fuel cell vehicle research. Unlike Toyota, the hybrid system used in Honda's does not easily adapt to a FCEV. However, many of the controls and features do transfer rather nicely. Plus, Honda also has much experience with battery electric vehicles and DC brushless motors. Honda started its venture into fuel cell vehicles in 1989 and has been road testing vehicles since 1999, when it introduced the FCX. Through the years, Honda has had four versions of this original model, the FCX-V1, V2, V3, and V4 (Figure 10-23). After the FCX-V4, it continued with new models but those are referred to as the FCX.

The original two versions were released the same year and were quite different from each other. V1 was a battery hybrid fuel cell vehicle that used a Ballard Mark 700 series PEM fuel cell. Hydrogen was stored in metal hydride. V2 was a direct supply FCEV and used a Honda developed PEM fuel cell with a methanol reformer. These two versions were developed to test the two different configurations and storage systems. What Honda learned from these two was then applied to V3, which is the basis for all subsequent fuel cell prototypes.

	V1	V2	V3	V4
Fuel	Pure Hydrogen	Methanol	Pure Hydrogen	Pure Hydrogen
Energy storage	Metal Hydride	Reformer	Compressed Hydrogen gas (25MPa)	Compressed Hydrogen gas (35MPa)
Motor power	49 kW	49 kW	60 kW	60 kW
Fuel cell stack	Ballard	Honda original	Honda original and Ballard	Ballard
Passengers	2	2	4	4 plus trunk

Figure 10-23 A comparison of the original FCX series from Honda.

The FCX-V3 was equipped with a conglomeration of components from other Honda vehicles. It used the basic body and chassis from the EV-Plus, including its electric motor, transmission, and braking system. It was fitted with a modified version of Honda's hybrids' electronic control system. To store the hydrogen, the V3 was fitted with high-pressure tanks used in Honda's natural gas-powered Civic. Some components are unique to the FCX and have since been used in other models. One of the components unique to the FCX is an ultra-capacitor, which replaces the battery that was used in the V1. The benefits for using an ultra-capacitor are many, but the major benefits are the instantaneous response times for discharge and charge and a capacitor's ability to directly feed a DC motor without regulation or inverter. Both an inverter and voltage regulator consume electrical energy and by eliminating them, driving range is increased.

A Ballard Mark 700 series fuel cell was used with the ultra-capacitor. The capacitor was used to boost performance and to capture energy from regenerative braking. Hydrogen was stored at 3600 psi (253 kg/cm).

A year later in 2001, the FCX-V4 was introduced. This was an ultra-capacitor hybrid. It used a Ballard Mark 900 series fuel cell and stored hydrogen at 5000 psi (352 kg/cm). This resulted in an improvement in driving range: it could travel 185 miles (300 km) on a tank of fuel.

In 2002, the FCX was displayed. It was based entirely on the FCX-V4. However, it had several improvements, enough to increase the driving range by nearly 40 miles (64 km). This extended range was largely due to increased storage space for the compressed hydrogen. There were also other refinements that brought about the increase in range, as well as an increase in top speed and acceleration. The FCX became the first fuel cell vehicle certified by the Environmental Protection Agency (EPA) and during this certification, the FCX was rated at the equivalent of 52 mpg on the highway and 49 mpg in the city.

As the FCX evolved, the biggest changes, until this year, have been in the overall packaging of the fuel cell system. In 2003, Honda consolidated all components of the fuel cell system into the chassis (Figure 10-24) and developed several editions based on that. All designs used an ultra-capacitor, Ballard Mark 900 Series fuel cell, and compressed hydrogen storage. The fuel cell system box, as it sits in the chassis, contains the fuel cell stack, humidifier units, a water pump, and many electronic components. Because the PEM fuel cell needs to run at relatively low temperatures, the vehicles are fitted with three separate radiators. The traction motor and PCU with related electronic controls are under the hood. The assembly of ultra-capacitors is behind the rear seat. All of these packaging changes make the FCX a more practical automobile.

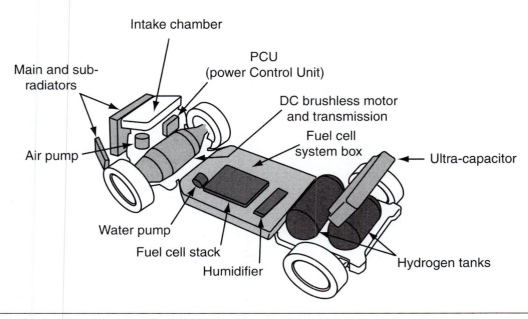

Figure 10-24 The layout of the fuel cell powertrain in a late-model Honda FCX.

In 2006, Honda announced that it will begin production, in Japan, of its next-generation FCX in three to four years. This vehicle uses a Honda fuel cell that is quite compact but delivers more power than the previously used fuel cells. The fuel cell has been developed to control water flow, because this is critical to fuel cell efficiency and start-up times. This new fuel cell is a PEFC (polymer electrolyte fuel cell). Oxygen and hydrogen flow from the top to the bottom of the fuel cell stack and the fuel cells are arranged vertically to achieve efficient packaging. However, the biggest improvement to the fuel cell is that it is designed to allow gravity to get rid of the unwanted water in the system. By disposing of the water built up during operation, Honda's fuel is capable of starting in temperatures as low as –4°F (–20°C).

The fuel cell stack has metal separator structures that are easier to make, thereby reducing overall costs. Costs are further reduced by the use of an aromatic electrolyte membrane that also increases the fuel cell's efficiency through a broad range of temperatures. This new FCX uses Honda's ultra-capacitor and three separate electric motors. One motor drives the front wheels and there is one motor at each rear wheel. This arrangement minimizes the space required for the drive system. The new FCX also has a special material in its hydrogen storage tanks. This material, not described at this time, doubles the storage capability of a tank and allows the new FCX to have a range of nearly 350 miles (563 km).

These vehicles are also equipped with a satellite-linked navigation system that displays the locations for all existing hydrogen-fueling stations within its database.

Others

All of the major auto manufacturers have been developing fuel cell and/or hybrid vehicles. It is impractical to describe or list all of the prototypes they have developed, or are in process of developing. The following is simply a description of where some of them have focused their attention and what they have introduced.

Ford Motor Company Ford's venture into FCEVs is best defined by its Ford Focus FCV (Figure 10-25). The latter edition is a battery hybrid that uses a Ballard Mark 900 series fuel

Figure 10-25 A Ford Focus FCV. *Courtesy of Ford Motor Company*

Figure 10-26 A 2003 Nissan X-TRAIL FCV. *Courtesy of Nissan North America, Inc.*

cell. It relies on a nickel metal-hydride battery pack to capture energy during regenerative braking. The vehicle runs on compressed hydrogen gas and its power train uses an electric motor that was first used in the Ranger pickup truck. It has a range of approximately 100 mi (160 km).

Nissan The first fuel cell prototype from Nissan was the 2002 Nissan X-TRAIL (Figure 10-26). This vehicle used a PEM fuel cell manufactured by UTC. It had a 100 mi (161 km) driving range and was a battery hybrid. Hydrogen storage was held at 5000 psi (352 kg/cm). This model was followed by the Nissan Xterra-FCV, which is a direct supply system that uses a Ballard Mark 900 stack. Nissan announced in 2005 that it would be producing its own fell cells and its own high-pressure storage tanks. The fuel cell features new electrodes with geometric designs to increase efficiency and reduce the size of the unit.

Hyundai Motor Company In June 2003, UTC Fuel Cells and Hyundai agreed to jointly develop a new PEM fuel cell capable of starting and operating during very cold temperatures. This fuel cell will be used in battery hybrid FCEV models of the Hyundai Santa Fe and Tucson sports utility vehicles. The average range for the concept vehicles is about 186 miles (300 km).

Volkswagen Volkswagen has a fuel-cell vehicle called the Bora HyMotion. Hydrogen is stored at cryogenic temperatures, which allows the vehicle to have a driving range of about 220 mi (355 km). The top speed of this vehicle is 90 mph (145 km/h) and it is capable of accelerating from 0 to 60 mph (0 to 97 km/h) in less than 12.5 seconds, both of which are good for a FCEV.

Other Alternatives

In an attempt to reduce our dependency on fossil fuels and to decrease the harmful emissions from our transportation systems, manufacturers are exploring many different avenues other than alternative fuels, hybrid vehicles, and fuel cells. Many of these are very futuristic but others are here now or will be in the near future. This look at the future is certainly not inclusive, as many manufacturers chose not to share their current projects and there are many different technologies being studied.

The future of the automobile (how it will look, what fuel it will use, what will supply the power, etc.) is totally undecided. The only thing that is certain is that it will emit much less emissions, rely less on fossil fuels, go farther on a gallon (or equivalent) of fuel, and cost more.

ICE Modifications

As far as alternative fuels are concerned, Ford has had many available model vehicles that can run on E-85 fuel (85 percent ethanol and 15 percent gasoline), which is primarily a renewable fuel. Other manufacturers, such as GM, are also introducing E-85 engines. Other manufacturers, particularly Honda, are working on compressed natural gas vehicles. CNG vehicles still depend on fossil fuels but their emission levels are lower than gasoline-powered vehicles. And of course, there are a few manufacturers fueling their engines with hydrogen.

In the very near future, more vehicles will be equipped with systems already found in some vehicles, such as direct injection, variable valve timing, cylinder cutoff systems, stop-start capability, and 42-volt systems. The higher voltage allows more components to be electrically powered rather than belt driven. This reduces the load on the engine and makes it more efficient.

An old technology may be used in the future. Although it has been around for many years, electronics may allow it to be practical today. Homogeneous Charge Compression Ignition (HCCI) for gasoline engines relies on cylinder heat and pressure to ignite the air/fuel mixture. In this system, there are no spark plugs and therefore no ignition system. With HCCI, the air fuel mixture is the same throughout the entire combustion chamber. In typical ICEs, there are pockets within the cylinder where the mixture is leaner or richer than the rest of the mixture. Because the mixture is uniform, a cleaner combustion takes place, which results in lower emissions and more efficient fuel usage. The mixture is self-ignited, similar to the ignition in a diesel engine. Once ignition begins, nearly all of the mixture ignites immediately. Without an ignition system, the timing of ignition is difficult to control. Advances made in electronics may provide the precise control required to use HCCI in the future.

Zero-Emissions Vehicles

Many believe the world would be a much better place if the roads were filled with zero-emission vehicles. To the unknowing this is possible and the only thing that stops us is "Big Oil." As we have seen throughout this book, achieving zero emissions from an automobile is no easy task and the obstacles do not seem to come from the oil companies. The obstacles are primarily cost and practicality. Currently there are four different powerplant systems that can provide a zero-emission vehicle:

- ❏ Battery electric vehicles
- ❏ Flywheel energy storage vehicles
- ❏ Fuel cell electric vehicles
- ❏ Air engined vehicles

Keep in mind that some of these do have associated emissions in the production of the fuel or the electrical energy required to recharge batteries.

The basics of three of these powerplant systems has already been covered in this book. Air engines have not. However, before describing an air engine, it should be stated that due to advances in technology, two manufacturers are again testing battery electric vehicles. The manufacturer of Subaru, Fuji Heavy Industries Ltd., recently announced it would produce ten electric cars for feasibility testing. This announcement comes at a time when nearly all other manufacturers are focusing on hybrid and fuel cell vehicles. The new EVs will use a lithium ion battery that will, it is hoped, have a minimum driving range of 50 miles (80 km) and a recharging time of 15 minutes for 80% of the battery's capacity. The battery should also last twice as long as those currently in hybrid vehicles. Mitsubishi Motors Corp. is also making plans for a battery electric vehicle, which will also use a lithium ion battery. Mitsubishi claims to have a prototype that has a range of 93 miles (150 km) and its goal is to extend that range to 155 miles (250 km) by the year 2010.

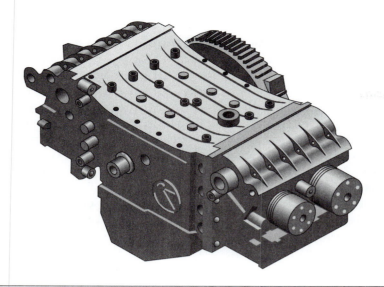

Figure 10-27 An air engine.

Air Engine An **air engine** is an emission-free piston engine that uses compressed air as its fuel (Figure 10-27). The engine relies on the expansion of compressed air to drive the pistons in a modified engine (Figure 10-28). One leading company in the development of air engines is Moteur Development International (MDI). They are based in Spain and have offices in the United Kingdom, Spain, Portugal, Latin America, and Canada. The MDI engine was developed in 2001.

The actual operation of the engine can be somewhat confusing, but the basics are simply that compressed air is used to force a piston down in the same way as the pressure increase in an ICE cylinder. The expansion of the compressed air relies on outside heat and the increase and decrease of the temperature of that air. The engine uses a unique connecting rod and crankshaft assembly that allows the pistons to be held at Top Dead Center for 70° of the cycle. Doing this, pressure is allowed to increase in the cylinder and the pressure on the piston is increased. The exhaust from this system is very cold air, which can be used to cool the passenger compartment. This is a zero emissions engine and has a driving range that is about twice that of a battery electric vehicle. These engines can be used with a conventional ICE (two separate engines). The ICE can drive an air compressor to replenish the pressure in the air tanks and extend the driving range of the vehicle.

Figure 10-28 Compressed air from a tank is used to move the pistons in an air engine.

Diesel Engines

Diesel engines in cars and light trucks will become more common soon. There are many reasons for this, one of which is that low sulfur diesel fuel will be available in the United States. Diesel vehicles are very common is Europe and other places where cleaner fuels are available. Diesel engines achieve better fuel economy than gasoline engines of the same size. With new technologies and the cleaner fuels, their emissions levels can be comparable to the best of gasoline engines. Plus, diesel hybrids could achieve fuel mileage ratings well beyond today's best.

Emissions have always been an obstacle with having diesel cars and new stricter emissions standards will go into effect shortly. Many of the new diesel vehicles will have an assortment of traps and filters to clean the exhaust before it leaves the tailpipe; others will use **Selective Catalytic Reduction** (SCR) systems. In fact, most diesel cars will be using SCR within the next few years. SCR is one of keys for providing cleaner running diesel engines.

SCR is a process in which a reductant is injected into the exhaust stream and then absorbed onto a catalyst (Figure 10-29). This action breaks down the exhaust's NOx to form H_2O and N_2. A reductant removes oxygen from a substance and combines with the oxygen to form another compound. In this case, oxygen is separated from the NOx and is combined with hydrogen to form water. The common reductants used in SCR systems are ammonia and urea water solutions. The problem with these reductants is again the lack of an infrastructure. The tanks that hold the reductant must be able to be refilled. When the tanks are empty, the emission levels will not be satisfactory. Alternatives to ammonia and urea are being studied in hopes that a suitable reductant that is readily available can be found.

The reductant is injected in the exhaust stream over a catalyst. These special catalytic converters work well only when they are within a specific temperature range. The engine's control unit is programmed to keep the temperature of the exhaust within that range. Also, the amount of reductant sprayed into the exhaust must be proportioned to amount of exhaust flow. The amount of reductant sprayed is controlled by the engine's control module.

Steam Hybrid

Another older technology that is being looked at for the future is steam power. BMW has built and tested an auxiliary drive system that uses steam to assist the engine. The heat to create the steam comes from the heat of the exhaust. This idea was first patented in 1914 and steam engines were around years before that. The original ideas had many disadvantages, one being the space required to hold the water and to house the heat-recovery system. BMW seems to have solved these problems and has been able to achieve more than a 10% increase in power and a decrease of 15% in fuel consumption. The real advantage of this system is that the heat has no cost associated with it. The required heat is normally wasted and transferred to the atmosphere.

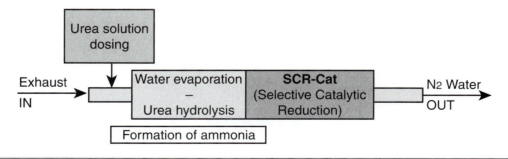

Figure 10-29 Selective Catalytic Reduction is a process where a reductant is injected into the exhaust stream and then absorbed onto a catalyst.

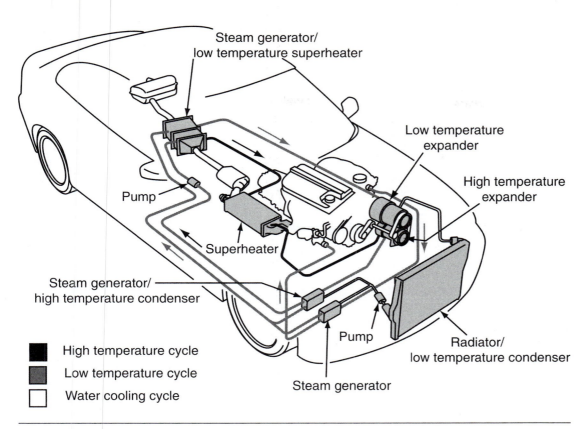

Figure 10-30 BMW's Turbosteamer.

The steam auxiliary drive system, called the Turbosteamer, can be installed on any engine as long as there is room in the engine compartment. There are two fluid circuits in the system (Figure 10-30). The circuit used to assist the engine is the high-temperature circuit. The fluid moves around the exhaust system and heat exchangers.

The primary high-temperature circuit moves the fluid through heat exchangers positioned behind the catalytic converter. By capturing more than 80% of the exhaust's heat, the water is heated to 1,022°F (550°C). The resultant steam is transferred to an expansion unit, which changes the steam into mechanical energy. The expansion unit then drives pulleys that are connected to the engine's crankshaft. There the mechanical energy assists the rotation of the crankshaft. The remaining steam moves to the second circuit, which also serves as the engine's cooling system. This circuit is filled with engine coolant and absorbs the steam's heat and transfers it and the engine's heat outside the vehicle, just as a traditional cooling system would.

Review Questions

1. *True or False* All fuel cells vehicles have regenerative braking and 12-volt auxiliary systems.
2. Which of the statements about hydrogen is true?
   A. A hydrogen atom is one proton and two electrons.
   B. Hydrogen is one of the heaviest elements known and is full of energy.
   C. Fossil fuels are combinations of carbon and hydrogen.
   D. Hydrogen is produced in a fuel cell when water is broken down into its basic elements.
3. *True or False* A PEM fuel cell needs to operate at very high temperatures.
4. What characteristic of an ultra-capacitor allow it to work well in a fuel cell drive system?
5. Why is water control so important to the effectiveness of a PEM fuel cell?
6. Which of the following CANNOT be used as a source for the production of hydrogen?
   A. Gasoline
   B. Methanol
   C. Carbon dioxide
   D. Natural gas
7. The type of fuel cell that will undoubtedly be used first in an automobile is the:
   A. Proton exchange membrane fuel cell.
   B. Solid oxide fuel cell.
   C. Molten carbonate fuel cell.
   D. Alkaline fuel cell.
8. Currently there are four different powerplant systems that can provide a zero-emissions vehicle. Name them.
9. What two substances are commonly used as a reductant in SCR systems for diesel engines?
10. Which of the following statements about fuel cells is NOT true?
    A. A single fuel cell produces very low voltage, normally less than one volt.
    B. A fuel cell produces electricity through an electro-chemical reaction that combines hydrogen and oxygen to form water.
    C. A fuel cell is composed of two electrodes coated with a catalyst and separated from each other by an electrolyte and from the case by separators.
    D. In a fuel cell, catalysts are used to ignite the hydrogen; this causes a release of electrons or electrical energy.

RESOURCES

The following websites have information about electric drive vehicles. Some of these are very technical, while others are simply fun and informational.

http://evalbum.com
http://geocities.com/ev_list
http://mit42v.mit.edu
www.acpropulsion.com
www.acterra.org
www.afdc.nrel.gov
www.avere.org
www.cafeelectric.com
www.calcars.org
www.dcpowersystems.com
www.driveclean.ca.gov
www.eaaev.org
www.electricdrive.org
www.etikkit.com
www.evadc.org
www.evco.ca
www.evparts.com
www.evproject.com
www.evsupersite.net
www.evworld.com
www.hybridcars.com
www.kta-ev.com
www.manzanitamicro.com
www.megawattmotorworks.com
www.metricmind.com
www.MixedPower.com
www.nedra.com
www. priuschat. com
www. priusplus. org
www.shepinc.com

GLOSSARY

A

A/F (air/fuel ratio) A ratio expressing the amount of fuel mixed with air, by weight, that enters the combustion chamber in an internal combustion engine.

Absorbed (or absorptive) glass mat A technique for sealed lead-acid batteries. The electrolyte is absorbed in a matrix of glass fibers, which holds the electrolyte next to the plate and immobilizes it, preventing spills.

AC See Alternating current.

Accelerator pedal A device used to control the throttle opening, and thereby intake air, on an internal combustion engine.

Accessory power module (APM) Part of GM's EV1 that contains a power supply, which works with the auxiliary battery to energize various accessories.

Active control engine mount (ACM) A powerplant mounting system that is designed to suppress the natural vibrations of the driveline by instantly changing the damping of the mounts.

Active mass The substance used in a positive or negative electrode of a battery or fuel cell, which creates current flow by an electrochemical reaction.

Active material The specific material at the positive or negative electrode of a battery or fuel cell that takes part in the charge and discharge reactions of the device.

Active noise control (ANC) A system that monitors low-frequency engine noise in the passenger compartment and sends out an equal but opposite noise to effectively cancel the engine noise.

Additives Chemicals added to fuel in very small quantities to improve and maintain fuel quality. Detergents and corrosion inhibitors are examples of gasoline additives.

Advanced technology vehicle (ATV) A vehicle that combines engine/power/ drivetrain systems to improve fuel economy. This includes hybrid power systems and fuel cells, as well as some specialized electric vehicles.

AFC (alkaline fuel cell) A fuel cell with an alkaline electrolyte that can only be operated with pure oxygen.

Aftermarket Broad term that applies to any change after the original purchase, such as adding equipment not a part of the original purchase.

Air conditioning (A/C) The process of adjusting and regulating, by heating or refrigerating, the quality, quantity, temperature, humidity, and circulation of air in a space or enclosure; to condition the air.

Air engine An emission-free piston engine that uses compressed air as its fuel.

Air pollution Unwanted particles, mist, or gases put into the atmosphere as a result of motor vehicle exhaust, the operation of industrial facilities, or other human activity.

Air toxics Toxic air pollutants defined under Title II of the Clean Air Act (CAA), including benzene, formaldehyde, acetaldehyde, 1 -3 butadiene, and polycyclic organic matter (POM).

Alcohol fuels A class of liquid chemicals that have certain combinations of hydrogen, carbon, and oxygen, and that are capable of being used as fuel.

Alcohols Organic compounds that are distinguished from hydrocarbons by the inclusion of a hydroxyl group. The two simplest alcohols are methanol and ethanol.

Aldehydes A class of organic compounds that can be produced from the oxidation of an alcohol (removing the hydrogen atoms from the alcohol).

Alkaline battery A primary battery that uses an alkaline aqueous solution as its electrolyte.

Alkaline fuel cell (AFC) The type of fuel cell used by NASA. These cells use a water-based solution of potassium hydroxide (KOH) as the electrolyte and electrodes coated with a catalyst.

Alternating current (AC) An electric current that reverses its direction of flow from positive to negative at regular intervals, typically 60 times per second.

Alternative fuel Fuel for internal combustion engines that can be used in place of gasoline or diesel fuel. The term includes blends of other fuels with gasoline or diesel fuel. The most common alternative fuels are methanol, ethanol, compressed natural gas (CNG), liquefied natural gas (LNG), liquefied petroleum gas (LPG), and hydrogen.

Alternative fuel vehicle (AFV) A vehicle that does not use a typical form of energy for operation: electricity, E85, propane, natural gas, and others.

Ambient The surrounding atmosphere; encompassing on all sides; the environment surrounding a body but undisturbed or unaffected by it.

Ambient air temperature Surrounding temperature, such as the outdoor air temperature around a building.

American wire gauge (AWG) A standard method of denoting the diameter of electrically conducting wire.

Ampere/Amperage (Amp) Standard unit used to measure electric current; proportional to the quantity of electrons flowing through a conductor past a given point in one second. Amperage is calculated by dividing watts by volts.

Anode The positive electrode at which oxidation (loss of electrons) takes place in a liquid solution. Depending on the current direction, each of the two electrodes can become an anode in secondary cells. The negative electrode is then the anode when discharging.

ANSI American National Standards Institute is the national organization that coordinates development and maintenance of consensus standards and sets rules for fairness in their development. ANSI also represents the United States in developing international standards.

Aromatics Hydrocarbons based on the ringed six-carbon benzene series of related organic groups. Benzene, Toluene, and Xylene are the principal aromatics, commonly referred to as the BTX group. They represent one of the heaviest parts of gasoline.

Ash A non-organic, non-flammable substance left over after combustible material has been completely burned.

Assist hybrid vehicle A vehicle that cannot be powered only by the electric motor. The electric motor helps or assists the engine to overcome increased load. At all other times, the vehicle is powered by the ICE.

AT–PZEV Advanced technology–partial zero emissions vehicle.

Atkinson cycle During the Atkinson cycle, the intake valves are kept open for a while during the compression stroke; this reduces the actual displacement and the power output of the engine.

Atom The smallest unit of an element consisting of a dense positively charged nucleus (of protons and neutrons) orbited by negatively charged electrons.

Auxiliary power outlet (APO) An AC power outlet available in some hybrid pickup trucks and SUVs. They are designed to power tools and other electrical appliances.

B

Battery A container, or group of containers, holding electrodes and an electrolyte for producing electric current by chemical reaction and storing energy. The individual containers are called "cells." Batteries store direct current (DC) voltage.

Battery ECU Monitors the charging condition of the HV battery.

Battery life Number of cycles or miles an EV will travel on one battery pack before the pack must be replaced.

Battery pack monitor (BPM) Monitors state of charge, temperature, and other vital battery readings.

Battery smart unit A device used to monitor the voltage, current flow, and temperature of the battery pack. It may also have a leak-detection circuit that watches for excessive current drain. Serial communication is used to transfer the digital signals from the battery smart unit to the electronic control unit.

BEAN Body Electronic Area Network. A form of communication used in a multiplexed system.

Belt alternator starter (BAS) A combination motor/generator that is driven by the engine's crankshaft via a drive belt. It replaces both the engine's alternator and starter motor.

Benzene A type of colorless liquid hydrocarbon that can be used as a fuel.

Bi-fuel vehicle A vehicle with two separate fuel systems designed to run on either an alternative fuel and gasoline or diesel, using one fuel at a time; may be referred to as a dual-fuel vehicle.

Biodiesel A biodegradable fuel for use in diesel engines. It is produced by organically derived oils or fats. It may be used as a replacement for or as a component of diesel fuel.

Biomass Renewable organic matter that can be used as a source of energy, such as plants, wood chips, bales of straw, liquid manure, organic wastes, agricultural crops, crop-waste residues, wood, animal and municipal waste, aquatic plants, fungal growth, etc.

Brake system control module (BSCM) A control unit that calculates the required braking force for slowing down and stopping a vehicle. Based on these calculations, it controls regenerative braking and the hydraulic brake system.

British thermal unit (Btu) The standard measure of heat energy. It takes one Btu to raise the temperature of one pound of water by one degree Fahrenheit at sea level. One Btu is equivalent to 252 calories, 778 foot-pounds, 1055 joules, and 0.293 watt-hours.

Busbar A conductor that serves as a common connection for two or more circuits. It may be in the form of metal bars or cables.

Butane A hydrocarbon gas found in the earth along with natural gas and oil. Butane turns into a liquid when it is put under pressure.

Button cell A miniature cell. A button- or coin-shaped battery with a diameter greater than its height.

C

California Air Resources Board (CARB) A state agency that regulates the air quality in California. Air quality regulations established by CARB are often stricter than those set by the federal government.

CAN (controller area network) bus A commonly used multiplexing protocol for serial communication The communication wire is a twisted-pair wire.

Capacitor A device for holding and storing a surge of current.

Capacity Amount of electrical energy a cell or battery contains expressed in ampere/hours.

Capacitance The unit of measure for a capacitor's ability to store an electric charge.

Carbon dioxide (CO_2) A colorless, odorless, nonpoisonous gas that is a normal part of the air. Carbon dioxide is exhaled by humans and animals and is absorbed by plant life and by the sea. It is sometimes referred to as a "greenhouse gas."

Carbon monoxide (CO) A colorless, odorless, highly poisonous gas made up of carbon and oxygen molecules formed during combustion. It is a major air pollutant on the basis of weight.

Carcinogens Potential cancer-causing agents. They include industrial chemicals found in food additives, pesticides, fertilizers, drugs, toys, household cleaners, toiletries, and paints.

Catalyst A material that facilitates or accelerates a chemical reaction while retaining its own properties and without being consumed.

Category 3 (CATIII) A classification of test equipment. Meters classified as CAT III are required for testing electric drive vehicles because of the high voltages, three-phase current, and the potential for high transient voltages.

Cathode The positive electrode. The electrode at which a reduction reaction (gain of electrons) occurs.

CCM Convenience Charge Module. The 110-volt charger provided in the trunk of the lead acid EV1.

Cell A primary galvanic unit, which converts chemical energy directly into electric energy. Normally is made up of a positive and a negative electrode, a separator, and a electrolyte.

Cell, cylindrical Cells whose heights are equal to or greater than their diameters.

Cell, secondary Rechargeable battery cell.

Celsius A temperature scale based on the freezing (0 degrees) and boiling (100 degrees) points of water. Abbreviated as C and formerly known as Centigrade. To convert Celsius to Fahrenheit, multiply the number by 9, divide by 5, and add 32.

Charge inlet The location on an electric vehicle where the recharger is connected.

Charge/Charging Refilling a battery with electrical energy.

Charging process The supplying of electrical energy for conversion to stored chemical energy at a battery or capacitor.

Charging station The device that provides a connection from a power source to a battery for charging.

Chassis ground The use of the vehicle's frame and/or body as a common connection to the negative terminal of a battery.

Chemical energy The energy generated when a chemical combusts, decomposes, or transforms to produce new compounds.

Circuit The path for electrical current. It normally includes a load and control.

Circuit breaker A circuit protection device that opens when excessive current is present in its circuit.

Clean fuel vehicle Refers to vehicles that use low-emission, clean-burning fuels, which include ethanol, hydrogen, liquefied petroleum gas, methanol, natural gas, and reformulated gasoline.

Closed circuit An electrical circuit that has a completed path from the negative of the battery to the positive terminal.

Coin cell A miniature cell. A button- or coin-shaped battery whose diameter is greater than its height.

Cold cranking amps (CCA) rating A common method of rating most automotive starting batteries. This rating is based on the load, in amperes, that a battery is able to deliver for 30 seconds at 0°F (−17.7°C) without its voltage dropping below a predetermined level.

Combustion Rapid oxidation, with the release of energy in the form of heat and light.

Compressed natural gas (CNG) Natural gas that has been condensed by high pressure, typically between 2,000 and 3,600 pounds per square inch.

Compression ignition A form of ignition that initiates combustion in a diesel engine. It is caused by the rapid compression of air within the cylinders that generates the required heat to ignite the fuel as it is injected.

Conductance A battery's ability to conduct current. It is a measurement of the plate surface available in a battery for chemical reaction.

Conductance test A test that measures conductance and provides a reliable indication of a battery's condition and is correlated to battery capacity. Conductance can be used to detect cell defects, shorts, normal aging, and open circuits, which can cause the battery to fail.

Conduction The transfer of heat energy through a material by the motion of adjacent atoms and molecules without changing the position of the particles.

Conductive charging An 110V or 220V recharging method that uses conventional metal-to-metal contact to transfer electricity from a charger to a battery.

Conductor A device that readily allows for current flow.

Continuity A term used to describe the presence of a completed circuit between two points.

Continuously Variable Transmission (CVT) Transmissions that automatically change torque and speed ranges without requiring a change in engine speed. A CVT is a transmission without fixed forward speeds.

Controller A device used to manage the flow of electricity from batteries to motor(s).

Convection Heat transfer by the movement of air.

Convenience charger An integral charger carried on board EVs. These chargers have a standardized connection that allows plugging into any conventional 110V household outlet.

Converted or conversion vehicle A vehicle originally designed to operate on gasoline or diesel that has been modified to run on an alternative fuel or electricity.

Converter Drops high-voltage direct current to DC12V in order to supply electricity to body electrical components, as well as to recharge the 12V battery.

Corrosion inhibitors Additives used to inhibit corrosion in the fuel systems.

CP Charge Port. Located at the nose of EV1, it is where the paddle is inserted to commence inductive charging.

Cranking amps (CA) rating A method of rating automotive starting batteries. This rating is based on the load, in amperes, that a battery is able to deliver for 30 seconds at 32°F (0° C) without its voltage dropping below a predetermined level.

Crude oil Petroleum as found in the earth, before it is refined into various oil products.

Cryogenic Cold, frost. When applied to gases, it refers to low temperatures where the gases are in their liquid phase.

Current The number of electrons flowing past a given point in a given amount of time.

Cycle A cycle is one complete charge/discharge sequence of a battery.

Cycle life Number of cycles a battery will undergo before it no longer can provide its designed electrical power.

Cylinder idling system A system used by Honda that stops combustion in select cylinders in some of its gasoline engines.

Cylindrical battery A battery whose height is greater than its diameter.

Cylindrical cell A type of electrochemical cell in which the electrodes are rolled together and fit into a metal cylinder. A separator soaked in an electrolyte is placed between the plates.

D

Data link connector The connector used to connect equipment into a vehicle's computer system for the purpose of diagnostics.

Dedicated vehicle A vehicle that operates only on one fuel.

Deep cycling A term used to describe the repeated process of a battery discharging and being recharged frequently.

Deep discharge A qualitative term indicating the discharge of a significant percentage of a battery's capacity (50 percent or more).

Delta winding A type of stator winding connection that connects three windings in series and has the appearance of the Greek letter delta. AC generators with delta windings are capable of putting out higher amperages.

Density The mass of a unit volume of a substance.

Depth of discharge The ampere-hours discharged from a cell or battery at a given rate divided by the available capacity under the same specified conditions.

Diagnostic Trouble Codes (DTCs) Numerical codes generated by an electronic control system to indicate a problem in a circuit or subsystem or to indicate a general condition that is out of limits.

Dielectric material A substance that serves as an insulator of electricity.

Diesel fuel Fuel for diesel engines obtained from the distillation of crude oil. It is composed of hydrocarbons and its efficiency is measured by cetane number.

Direct Current (DC) Electricity that flows continuously in one direction.

Direct-Methanol Fuel Cell (DMFC) A type of the PEM fuel cell that uses liquid methanol as the fuel, rather than hydrogen.

Discharge Rate The rate at which a cell or battery is discharged.

Discharge Withdrawal of electrical energy from a cell or battery.

DLC3 Data Link Connector 3.

DMFC Direct Methanol Fuel Cell. A fuel cell that breaks down methanol into its subparts and uses the hydrogen to produce electrical energy.

Drain Withdrawal of current from a cell or battery.

Driveline efficiency The amount of energy produced in an engine or motor that is used for propulsion and not wasted.

Duty cycle The length of time a device is turned on compared to the time it is off. Duty cycle can be expressed as a ratio or as a percentage.

Dynamometer An instrument for measuring mechanical power.

E

E85 E85 is a blend of 85 percent ethanol and 15 percent gasoline. E85 is a domestic renewable energy source.

ECM Engine control module. Is often part of the powertrain control module. The module controls the operation of various engine systems.

ECU Electronic control unit. An electronic unit that monitors and controls a system or subsystems.

Efficiency The ratio of the useful energy delivered by an engine or motor to the energy supplied to it over the same period of time.

Electric generator A device that converts heat, chemical, or mechanical energy into electrical energy.

Electric Motor-assisted Power Steering (EMPS) system The EMPS is a variable ratio, speed sensitive power steering system. The EMPS provides steering assist when the engine is on and when it is off.

Electric propulsion motor An AC or DC electric motor designed for vehicle propulsion.

Electric resistance heater A device that produces heat through electric resistance.

Electric Vehicle (EV) A vehicle that is propelled exclusively by electric power.

Electricity A form of energy that is produced by the controlled movement of electrons.

Electrode A conducting structure within a cell in which electrochemical reactions take place

Electro-chemical reactions Chemical reactions that produce free electrons.

Electrohydraulic brakes An electrically powered braking system with no engine-related vacuum sources for power assist.

Electro-Hydraulic Power Steering (EHPS) An electric motor-driven, variable assist power steering system.

Electrolysis A process that breaks a chemical compound down into its elements by passing a direct current through it.

Electrolyte Normally an aqueous salt solution that permits ionic conduction between the positive and negative electrodes in a battery cell.

Electromotive Force (EMF) The force created by the presence of voltage.

Electrons The particles in an atom that have a negative charge.

Element A substance consisting entirely of atoms with the same atomic number.

Emissions Exhaust emissions are the pollutants emitted by the engine through the tailpipe; high exhaust emissions leads to smog, poor air quality, and global warming.

Energy The capacity for doing work.

Energy density A battery's rated energy per unit of volume. Measured in units of watt-hours per liter (Wh/l).

Energy management Smart onboard systems that optimize driving range and allow the powering of electrical accessories.

Energy Storage Box (ESB) The name for the battery in GM hybrids.

Environmental Protection Agency (EPA) A federal agency charged with protecting the environment. It is responsible for regulating exhaust emissions in automotive and other vehicles.

ETCS-i Electronic throttle control system with intelligence.

Ethanol Also known as ethyl alcohol or grain alcohol. It can be produced from the fermentation of various sugars from carbohydrates found in agricultural crops and cellulosic residues from crops or wood. Used as a gasoline octane enhancer and oxygenate and can also be used in higher concentrations in alternative fuel vehicles.

Ethyl Tertiary Butly Ether (ETBE) A fuel oxygenate that can be added to gasoline and used as an oxygenate in reformulated gasolines.

EV (electric vehicle) A vehicle powered solely by electricity. Energy is normally provided by batteries but may also be provided by photovoltaic (solar) cells or a fuel cell.

F

Fahrenheit A temperature scale in which the boiling point of water is 212 degrees and its freezing point is 32 degrees. To convert Fahrenheit to Celsius, subtract 32, multiply by 5, and divide the product by 9.

Farad (F) The standard measure of capacitance. A one-farad capacitor can store one coulomb of charge at one volt.

Fast charging A battery recharging process that uses high current delivered for a short time.

Fission A release of energy caused by the splitting of an atom's nucleus. This is the energy process used in nuclear power plants to make the heat needed to run steam powered generators.

Fixed value resistors Resistors whose value does not or cannot change.

Flexible Fuel Vehicle (FFV) A vehicle that can operate on either methanol or ethanol and regular gasoline or any combination of the two from the same tank.

Flux density The number of flux lines per square centimeter.

Flux field The magnetic field formed by magnetic lines of force.

Flywheel energy storage A system comprised of a large, heavy flywheel suspended by magnetic bearings and connected to a motor/generator. Kinetic energy is stored as the flywheel spins with the motor. The flywheel's momentum keeps it rotating while it spins the generator to produce electrical energy.

Flywheel power storage See flywheel energy storage.

Fossil fuel Fuel that was formed in the earth from remains of living organisms. These include oil, coal, natural gas, and their by-products.

Frequency The number of cycles that an alternating current moves through in one second.

Fuel A substance that can be used to produce heat.

Fuel cell An electrochemical engine with no moving parts in which hydrogen and oxygen combine in a controlled manner to directly produce an electric current and heat.

Fuel cell stack An assembly of several hundred fuel cells connected in series and layered or stacked next to one another. Each fuel cell produces electricity and the combined output of the cells is used to power a vehicle.

Full hybrid A hybrid vehicle that is able to run on the engine, the batteries, or a combination of the two.

Full-wave rectification The conversion of the total AC voltage signal to a DC voltage signal.

Fuse An electrical device used to protect a circuit against accidental overload or unit malfunction.

Fusible link A type of fuse made of a special wire that melts to open a circuit when current draw is excessive.

G

Gallon A unit of volume. A U.S. gallon has 3.785 liters.

Gas Gaseous fuel (normally natural gas) that is burned to produce heat energy. The word also is used to refer to gasoline.

Gasohol Refers to gasoline that contains 10 percent ethanol by volume.

Gasoline A light petroleum product obtained by refining crude oil and used as a motor vehicle fuel.

Greenhouse effect A warming of the Earth and its atmosphere that results from the trapping of solar radiation by CO_2, water vapor, methane, nitrous oxide, chlorofluorocarbons, and other gases.

Grids The basic framework of the lead-acid battery plates. They are the lead alloy framework that supports the active material of the plate.

Gross Vehicle Weight (GVW) The maximum allowable fully laden weight of the vehicle and its payload.

Ground The negatively charged side of an electrical circuit. A ground can be a wire, the negative side of the battery, or the vehicle chassis.

Ground wire A wire that connects a component to the negative side of the battery through the vehicle's chassis. The use of the ground wire eliminates the need to run a separate wire from the component to the negative terminal of the battery.

H

Half-wave rectification The conversion of half of the total AC voltage signal to a DC voltage signal.

HC Adsorber and Catalyst system (HCAC) A system used on early Prius' to capture hydrocarbons in the exhaust when the three-way catalytic converter was cold and basically ineffective.

Heat engine An engine that converts heat to mechanical energy.

Heat pump An air-conditioning unit that is capable of heating by refrigeration. It transfers from one object to another. It operates in the opposite way as air conditioing.

Heavy-duty vehicle Generally, a vehicle that has a GVWR of more than 26,000 lb.

High Voltage (HV) battery The unit that supplies voltage to the traction motor in an electric drive vehicle. The battery is an assembly of many batteries or battery cells.

Horsepower A unit for measuring the rate of doing work. One horsepower equals about three-fourths of a kilowatt (745.7 watts).

HV High voltage.

HV battery See High-voltage battery.

HVAC Heating ventilation and air conditioning. A system that provides heating, ventilation, and/or cooling.

Hybrid vehicle Vehicles that have two or more sources of energy.

Hybrid Vehicle Control Unit (HV ECU) The electronic control unit that monitors and controls the operation of the hybrid system.

Hydraulic hybrids Similar to a hybrid electric vehicles, except energy for the alternative power source is stored in tanks of hydraulic fluid under pressure. These vehicles have a hydraulic propulsion system.

Hydrocarbons (HC) Any compound that contains hydrogen and carbon. Most often refers to the pollutant HC, which results from incomplete combustion.

Hydrogen H is the chemical symbol for hydrogen. It is the lightest element of the table of elements and the most abundant element in the universe.

Hydrometer Instrument for measuring the density of liquids in relation to water. This tool is commonly used to measure the state of charge in a lead-acid battery by measuring specific gravity of the electrolyte.

I

IGBT See Insulated gate bipolar transistor.

Induction The process of producing electricity through magnetism rather than direct flow through a conductor.

Inductive charging A recharging method that transfers electricity from charger to a battery through a magnetic field.

Infrastructure Refers to the recharging and refueling network necessary to distribute a fuel.

Initial charge The first charging process after the electrolyte has been poured into a dry battery.

Injectors Valves that allow fuel under pressure to enter an engine's intake manifold or combustion chambers.

Insulated Gate Bipolar Transistor (IGBT) A solid state device that converts DC voltage into 3-phase AC voltage to power an electrical motor or other device.

Insulation resistance tester A tester that measures the amount of voltage that can leak through the insulation of a wire or cable.

Insulator A material that does not allow for good current flow.

Integrated Motor Assist (IMA) The motor/generator assembly used in Honda hybrids that fits between the engine and the transmission.

Integrated Starter Alternator Damper (ISAD) Similar to Honda's IMA, this unit replaces the flywheel, generator, and starter motor. It is placed between the engine and transmission in some hybrid vehicles.

Inverter A device that converts AC electricity to DC electricity and DC to AC.

Ion An electrically charged particle or molecule.

J

Joule A unit of work or energy. 1,055 joules equals a British thermal unit.

K

kBtu One thousand (1,000) Btus.

Kerosene A colorless, low-sulfur oil product that burns without producing much smoke.

Kilovolt (kv) One thousand (1,000) volts.

Kilowatt (kW) The international unit to measure power (not just electrical), a kilowatt equals 1,000 watts. One kW equals 1.34 horsepower and 746 watts equals 1 horsepower.

Kilowatt hour (kWh) The standard unit for measuring quantities of energy that are consumed over time. Specifically, it is 1 kilowatt supplied for 1 hour.

Kinetic energy Energy in motion.

L

Lead-acid battery A battery that uses lead oxide and spongy lead electrodes with sulfuric acid as an electrolyte.

LEV See Low emission vehicle.

Linear air/fuel (LAF) sensor An exhaust oxygen sensor that is capable of measuring oxygen content when an engine is running very lean air/fuel mixtures.

Lineman's gloves Electrically insulated gloves. These gloves must be worn whenever working on or around high-voltage circuits.

Liquefied gases Gases that have been changed into liquid form. The most common liquefied gases are natural gas, butane, butylene, ethane, ethylene, propane, and propylene.

Liquefied petroleum gas (LPG) A mixture of hydrocarbons found in natural gas and used primarily as a home heating fuel and motor vehicle fuel.

Liquefied natural gas (LNG) Natural gas that has been condensed to a liquid typically by cooling the gas.

Lithium polymer battery A battery in which lithium is used as the electrochemically active material and the electrolyte is a polymer or polymer-like material that conducts lithium ions.

Lithium ion battery A battery in which lithium is used as the electrochemically active material and the electrolyte is a liquid that conducts lithium ions.

LNG Liquefied natural gas.

Load The amount of electric current drawn from a power source to operate a circuit or device.

Low emission vehicle (LEV) A vehicle certified by the California Air Resources Board to have emissions from 0 to 50,000 miles no higher than 0.075 grams/mile (g/mi) of non-methane organic gases, 3.4 g/mi of carbon monoxide, and 0.2 g/mi of nitrogen oxides.

Low-maintenance battery A battery design that uses the same basic materials as a maintenance-free lead-acid battery, but has vent holes and caps, which allow water to be added to the cells when needed.

LPG See Liquefied petroleum gas.

M

M100 100 percent methanol used as a fuel in dedicated methanol vehicles.

M85 A blend of 85 percent methanol and 15 percent unleaded regular gasoline used as a fuel.

Maintenance-free battery A type of sealed lead-acid battery that does not have a provision for adding water to the cells

Malfunction Indicator Light (MIL) A warning lamp in a vehicle's instrument panel that lets the driver know when the vehicle's electronic control units detected a problem.

MCFC Molten carbonate fuel cell. A fuel cell with a molten alkaline carbonate electrolyte.

Methane A light hydrocarbon that is the main component of natural gas and marsh gas. It is the product of the anaerobic decomposition of organic matter.

Methanol Also known as methyl alcohol, wood alcohol. A liquid fuel formed by catalytically combining CO with hydrogen under high temperature and pressure.

Methyl Tertiary Butyl Ether (MTBE) An ether manufactured by reacting methanol and isobutylene. The resulting ether has high octane and low volatility. MTBE is a fuel oxygenate.

Micro (mild) hybrid A vehicle equipped with stop-start technology combined with regenerative braking. The electric motor/generator never drives the wheels or adds power to the drivetrain.

MIL Malfunction indicator light.

Mild (micro) hybrid A vehicle equipped with stop-start technology combined with regenerative braking. The electric motor/generator never drives the wheels or adds power to the drivetrain.

Miles per kilowatt hour (MPkWh) The unit of measure for fuel efficiency of electric vehicles. The equivalent of miles per gallon for liquid fueled vehicles.

Mobile source emissions Emissions resulting from the operations of any type of motor vehicle.

Molten carbonate fuel-cell (MCFC) A type of fuel cell that uses a liquefied carbonate salt as its electrolyte. This fuel cell operates at very high temperatures.

Motor controller An electronic device that reads accelerator and brake pedal positions and controls the operation and speed of an electric motor.

Motor power inverter (MPI) The unit that converts the direct current from a battery or capacitor into alternating current for the traction motors. The inverter also changes the AC generated by the motors into DC to charge the batteries.

MPa mega Pascals. The metric unit of measure for pressure. One MPa corresponds to a pressure of 10 atmospheres (10 bars).

MTBE See Methyl tertiary butyl ether.

N

Natural Gas Hydrocarbon gas found in the earth, composed of a mixture of gaseous hydrocarbons, primarily methane, occurring naturally in the earth and used principally as a fuel.

Neutron An uncharged particle found in the nucleus of every atom except that of hydrogen.

Newton A metric unit of force. The amount of force it takes to accelerate one kilogram at one meter per second per second.

Nickel Metal Hydride battery (NiMH) A battery made with nickel hydroxide and hydride alloys. The electrolyte is potassium hydroxide.

Nickel-Cadmium battery (NiCad) A battery made with a nickel electrode and a cadmium electrode and use potassium hydroxide as the electrolyte.

Nitrogen oxide (NO$_x$) One of the regulated exhaust emissions of an internal combustion engine. NO$_x$ is produced by the combination of nitrogen and oxygen due to the high temperatures reached in the combustion process.

Nitrogen-oxide-adsorptive catalytic converter This type of catalytic converter is used on some lean-burn engines, in addition to a conventional three-way catalytic converter, to keep NO$_x$ emission levels low.

NO$_x$ Oxides of nitrogen that are a chief component of air pollution that can be produced by the burning of fossil fuels. Also called nitrogen oxides.

O

Octane A rating scale used to grade a gasoline's antiknock properties.

Octane rating A measure of a gasoline's ability to resist ignition by heat.

OEM See Original equipment manufacturer.

Ohm A unit of measure of electrical resistance. One volt can push a current of one ampere through a resistance of one ohm.

OPEC The acronym for the Organization of Petroleum Exporting Countries founded to unify and coordinate petroleum polices of the members.

Open circuit An electrical circuit that has a break in the wire.

Open-circuit voltage (OCV) The no load voltage of a cell or battery measured with a high resistance voltmeter.

Open-loop fuel control A system in which the air/fuel mixture is preset with no feedback correction signal to optimize fuel metering.

Operating voltage The voltage available to be consumed by an electrical component.

Original equipment manufacturer (OEM) A manufacturer that certifies that all of its vehicle components have been installed under its direct supervision by its own assembly processes and are covered by the manufacturer's full warranty protection.

Oxides of nitrogen Various compounds of oxygen and nitrogen that are formed in the cylinders during combustion and are part of the exhaust gas.

Oxygenate A prime ingredient in reformulated gasoline. The increased oxygen content given by oxygenates promotes more complete combustion, thereby reducing tailpipe emissions.

Oxygenated fuels Fuels blended with an additive to increase oxygen content, allowing more thorough combustion for reduced carbon monoxide emissions.

Ozone A type of oxygen that has three atoms per molecule instead of the two. Ozone is a poisonous gas, but the ozone layer in the upper atmosphere shields the Earth from deadly ultraviolet radiation from space.

P

PAFC See Phosphoric acid fuel cell.

Parallel circuit In this type of circuit, there is more than one path for the current to follow.

Parallel drivetrain A drivetrain where both the motor and the engine can apply torque to move the vehicle. The motor can act in reverse as a generator for braking and to charge the batteries.

Parasitic drain The electric drain on a battery while the vehicle is not operating and the "ignition" switch is in the off position.

Particulate matter (PM) Unburned fuel particles that form smoke or soot. A chief component of exhaust emissions from heavy-duty diesel engines.

Particulate trap Diesel vehicle emission control device that traps and incinerates diesel particulate emissions after they are exhausted but before they are expelled into the atmosphere.

PCM Powertrain control module.

PCS Power control system.

PCU Power control unit.

PEB Power electronics bay.

PEMFC See Proton exchange membrane fuel cell.

Petroleum Oil as it is found in its natural state under the ground.

Phosphoric acid fuel cell The most commercially used fuel cell. These fuel cells use liquid phosphoric acid as the electrolyte with electrodes made of carbon paper coated with a platinum catalyst.

Photo biological A process, currently being studied, that produces hydrogen when certain algae and bacteria are exposed to light.

Photo electrolysis The process during which the sun's light is collected at photovoltaic cells and converted to electricity. That electrical energy is then passed through the water to begin electrolysis

Photovoltaic cell A semiconductor that converts light into electricity.

Photovoltaic cells Also called "solar cells." These convert solar energy to electrical energy.

PIM Power Inverter Module.

Planetary carrier Part of a planetary gear set. The carrier has a shaft for each of the planetary pinion gears. The carrier and pinions are considered one unit—the midsize gear member. The planetary pinions surround the sun gear and are surrounded by the ring gear.

Planetary gear unit A combination of three gears in mesh, two of the gears have external teeth, and the third has internal teeth.

Planetary pinion gears Small gears fitted into the planetary carrier of a planetary gear set.

Plug-in Hybrid Electric Vehicles (PHEVs) Full hybrids with larger batteries and the ability to recharge from an electric power grid. They are equipped with a power socket that allows the batteries to be recharged when the engine is not running.

Potentiometer A variable resistor commonly used as a tracking device or position sensor.

Power The rate at which energy is released. Power is measured in kilowatts for an electric vehicle.

Power density A battery rating for the amount of power available per unit of volume. Measured in watts per liter (w/l).

Power inverter Converts the high power of a DC battery pack into the pulsed AC required for powering the electric motor.

Power split device The basic name for the CVT transmission used in Ford and Toyota hybrid vehicles. This device is based on planetary gears and divides the output of the engine and the electric motors to drive the wheels or the generator.

PPM Parts per million. The unit commonly used to represent the degree of pollutant concentration where the concentrations are small.

Primary A cell or battery designed to deliver its rated capacity once and be discarded; not designed to be recharged.

Prismatic cells Electrochemical cells with flat electrodes placed into box with electrolytic separators placed between them.

Propane A gas that is both present in natural gas and refined from crude oil. It is used for heating, lighting, and industrial applications.

Propulsion system The combination of the powertrain and battery system, which converts stored electrical energy into mechanical energy in a vehicle.

Proton Exchange Membrane Fuel Cell (PEMFC) The most commonly used fuel cell in vehicles. This fuel cell easily allows for adjustable outputs, is compact, and is capable of providing high outputs.

Pulse Width Modulation (PWM) The characteristic of a continuous on-and-off cycling of a solenoid for a fixed number of times per second. While the frequency of the cycles remains constant, the ratio of on-time to total cycle time varies, or is modulated.

R

Range The distance that a vehicle can travel on a charge or fuel refill.

Rate of charge Amount of energy in a set period of time that is being added to battery. This is commonly expressed as a ratio of the battery's rated capacity to the charge duration in hours.

Recombinant battery A completely sealed maintenance-free lead-acid battery that uses an electrolyte in a gel form.

Reformulated gasoline (RFG) Gasolines that have had their compositions and/or characteristics altered to reduce vehicular emissions of pollutants, particularly pursuant to the EPA regulations under the CAA.

Regenerative braking A method that captures a vehicle's kinetic energy while it slowing down or being stopped. This captured energy is used to charge batteries and/or an ultra-capacitor.

Regulator A device used to control the input and/or output if something.

Reid Vapor Pressure (RVP) A standard measurement of a liquid's vapor pressure in psi at 100 degrees Fahrenheit. It is an indication of the propensity of the liquid to evaporate.

Relay This is an electromagnetic switch that uses a low current circuit to control a high-current circuit.

Reluctance A force created by the magnetic fields that results from current passing through a conductor. This force opposes the current flow and therefore limits the amount of current that can flow through a coil of wire.

Renewable energy A form of energy that is never exhausted because it is renewed by nature. Typical sources include energy produced by wind, hydroelectric, geothermic, or solar heat power stations, solar cells, and biomass.

Renewable resources Renewable energy resources are naturally replenishable. Renewable energy resources include biomass, hydro, geothermal, solar, and wind.

Resistance The ability of something to resist the flow of current and turning some of it into heat. Resistance is measured in ohms.

Resolver A sensor that monitors the position of the magnetic fields of the rotor.

Reserve Capacity (RC) rating A battery rating system that expresses the length of time, in minutes, that a fully charged starting battery at 80°F (26.7°C) can be discharged at 25 amperes before battery voltage drops below 10.5 volts.

Retrofit To change a vehicle or engine from the way it was built. This usually done by adding equipment such as conversion systems.

Rheostat A variable resistor typically used to control the action of a device.

Ring gear The internally toothed gear in a planetary gear set, it is the largest of the gears in the gear set

Rolling losses The amount of energy lost as a result of tire rolling resistance.

Rolling resistance coefficient A measure of the drag created by friction between the tires of a moving car and the pavement.

Root Mean Square (RMS) A common expression for stating the amount of alternating current measured by a meter.

S

SAE viscosity number A system established by the Society of Automotive Engineers for classifying engine oils and transmission and differential lubricants according to their viscosities.

Sealed lead-acid A type of lead-acid battery in which a special mat material saturated in electrolyte is placed between the battery's plates. There is no free liquid in the battery. Gases produced by the chemical reactions in the battery are "recombined" within the battery.

Secondary A cell or battery designed to be recharged.

Selective Catalytic Reduction (SCR) A process where a reductant is injected into the exhaust stream of a diesel engine and then absorbed onto a catalyst, which breaks down the NO_x to form H_2O and N_2.

Sensor A device used to monitor the activity or condition of a component or system.

Separator Material used as insulation between electrodes of opposite polarity.

Series circuit An electrical circuit that only allows current to flow through one path.

Series drivetrain The series vehicle components include an engine, a generator, batteries, and a motor. The engine does not drive the vehicle directly. Instead, it drives the generator, which charges the battery to supply power to the electric motor that propels the vehicle.

Series-parallel circuit An electrical circuit that has the characteristics of both a series and a parallel circuit.

Service plug A device used to shut off the high-voltage circuit of the HV battery when it is removed for vehicle inspection or maintenance.

Shift position sensor Converts the shift lever position into an electrical signal that is sent to a control module.

Sine wave The resultant wave from AC voltage. The amount the pattern on an oscilloscope moves up and down is equal within one cycle.

Slow charging A battery recharging process that uses low current delivered for a long period of time.

Smart charging The use of computerized charging stations that constantly monitor the battery so charging is done at the best rate and temperature to prolong battery life.

Smog A mixture of smoke and fog. The term is commonly used to refer to air pollutants.

SMR System main relay.

SOC See State of charge.

SOFC See Solid oxide fuel cell.

Solar cell A photovoltaic cell that converts light into electricity.

Solar energy Heat and light radiated from the sun.

Solar power Electricity generated from solar radiation.

Solenoid An electromagnetic device that changes electrical energy into mechanically energy. This is not a motor, as a solenoid does not work continuously; movement only occurs at initial energizing or deenergizing.

Solid Oxide Fuel Cell (SOFC) Fuel cells comprised of a ceramic anode, ceramic cathode, and a solid electrolyte. Their high operating temperatures eliminate the need for expensive catalysts and therefore they have low production costs.

Spark ignition engine An internal combustion engine in which combustion begins by the initiation of an electrical spark.

Specific energy A battery's rated energy per unit weight. Measured in units of watt-hours per kilogram (Wh/kg).

Specific gravity The weight of a volume of any liquid divided by the weight of an equal volume of water at equal temperature and pressure. The ratio of the weight of any liquid to the weight of water, which has a specific gravity of 1.000.

Specific power A battery's rated power per unit weight. Measured in units of watts per kilogram (w/kg).

Starting battery The low-voltage battery used to start the engine and power low-voltage circuits in most hybrid vehicles.

State of Charge (SOC) A rating, expressed in a percentage, of the current capacity of a battery.

Stator The stationary member of an AC motor or generator. It is made up of a number of conductors that serve as an electromagnet or collect the voltage induced by the rotating magnetic field.

Stepped resistors A variable resistor that has fixed resistance values at specific points.

Steam reforming A common procedure for extracting hydrogen from hydrocarbons.

Stirling engine An external combustion engine that converts heat into useable mechanical energy by the heating and cooling of a gas.

SULEV Super ultra low emissions vehicle.

Sulfur dioxide (SO$_2$) An EPA criteria pollutant.

Sun gear A gear located in the center of a planetary gear set and is the smallest of the gears in the set.

Superconductor A synthetic material that has very low or no electrical resistance.

Synfuel Fuel that is artificially made rather than found in nature.

Synchronous motor An electric motor that has a constant speed, regardless of the load placed on it.

T

Tapped resistors Resistors designed to have two or more fixed values, available by connecting wires to the several taps of the resistor.

Thermistor A solid state resistor that changes its resistance according to its heat.

Three-phase voltage A power system comprised of three conductors carrying voltage waveforms that are out of phase with each other. Normally the phase difference is 120 degrees.

Torque The amount of twisting force exerted on a shaft or other item. It is measured in foot-pounds.

Toxic emission Any pollutant that can negatively affect human health or the environment.

Tractive force The amount of power available from the drivetrain. It is calculated by multiplying the engine's torque by the overall gear ratio.

Tractive resistance The forces that work against the movement of a vehicle, such and rolling resistance and aerodynamic drag.

Transmission Control Module (TCM) A control module that directly controls the operation of a transmission/transaxle.

Trickle charge A method of recharging in which a battery is either continuously or intermittently connected to a constant current supply that maintains the battery in a fully or near full charged condition.

Turbocharging A way to increase the quantity of intake air to an internal combustion engine. It uses the pressure of exhaust gases to spin a turbine and impeller that forces air into the engine.

TWC Three-way catalytic converter.

Two-mode hybrid system A hybrid transmission that fits into a standard housing and is basically two planetary gear sets coupled to two electric motors. This results in a continuously variable transmission and motor/generators for hybrid operation. This also allows for two distinct modes of hybrid drive operation: low speed/low load and cruising at highway speeds.

U

ULEV Ultra low emissions vehicle.

Unleaded gasoline Gasoline that does not contain tetraethyl lead and is in compliance with federal and state regulations.

V

Valve-Regulated Lead-Acid (VRLA) batteries A type of recombinant battery in which the oxygen produced on the positive plates of this lead-acid battery is absorbed by the negative plate. That, in turn, decreases the amount of hydrogen produced at the negative plate. The combination of hydrogen and oxygen produces water, which is returned to the electrolyte.

Variable Cylinder Management (VCM) A system with the capability of shutting down various cylinders of the engine when they are not needed.

Variable Fuel Vehicle (VFV) A vehicle that has the capability of running on any combination of gasoline and an alternative fuel. Also called a flexible fuel vehicle.

Variable resistors A resistor that allows for a change in resistance based on the physical movement of a control. The control can be moved by an individual or a component.

Volatile Organic Compound (VOC) An emission regulated by the EPA. These compounds are reactive gases that are released during combustion or the evaporation of fuel. VOCs react with NO_x in the presence of sunlight and form ozone.

Volt A unit of measurement of electromotive force. It is the amount of force required to drive a steady current of one ampere through a resistance of one ohm.

Voltage Electrical pressure resulting from a difference in electrical potential at one point and another.

Voltage drop This term represents the amount of electrical energy that is changed to another form of energy as current passes through an electrical load.

VVT-i Variable valve timing with intelligence.

W

Watt A unit of measure of electric power.

Watt-hour One watt of power expended for one hour.

Wye configuration A type of stator winding connection that connects three windings in parallel and has the appearance of the letter "Y." AC generators with wye windings are capable of putting out higher voltages.

X

Xylene A petroleum-based hydrocarbon used to increase the octane rating of gasoline. It is photochemically reactive and, when present in exhaust emissions, contributes to the formation of smog.

Z

ZEV Zero emissions vehicle.

Zinc-Air battery A battery constructed with a zinc electrode and an air electrode and uses potassium hydroxide as the electrolyte.

INDEX